T0336959

Atlantis Studies in Variational Geometry

Volume 4

More information about this series at http://www.atlantis-press.com

Mike Crampin · David Saunders

Cartan Geometries and their Symmetries

A Lie Algebroid Approach

Mike Crampin
Burnham Market, Norfolk
UK

David Saunders
University of Ostrava
Ostrava
Czech Republic

ISSN 2214-0700 ISSN 2214-0719 (electronic)
Atlantis Studies in Variational Geometry
ISBN 978-94-6239-191-8 ISBN 978-94-6239-192-5 (eBook)
DOI 10.2991/978-94-6239-192-5

Library of Congress Control Number: 2016936967

Printed on acid-free paper

Contents

Introduction

'A Cartan geometry' is R.W. Sharpe's translation of Élie Cartan's term 'un espace généralizé' (see the Preface to his book [36]): the kind of space being generalized is a 'homogeneous space' in the sense of Klein, and an example of the kind of generalization being contemplated is that which takes one from Euclidean to Riemannian space.

A Klein space, or a Klein geometry, is a homogeneous space of a Lie group G, that is, it is a smooth manifold F on which G acts smoothly, effectively and transitively. The definition of a Cartan geometry on a manifold M, modelled on a Klein geometry F of the same dimension, is then supposed to capture the idea that M looks like F 'infinitesimally close to each point'. Thus one needs to associate a copy of F with each point x of M, in such a way that in an appropriate sense F and M are tangent at x. In addition one needs a way of 'connecting' the copies of F at different points; that is to say, one needs a connection of an appropriate kind.

Cartan's own papers on espaces généralizés and 'Cartan connections', such as [11, 12, 13], seem to have acquired a certain reputation for obscurity. Difficulty of comprehension may very well have arisen simply because of the lack, at the time, of a suitable language to express his ideas. The much more recent publication of the book mentioned above did much to bring clarity to the subject. It provides a useful general reference for an account of Cartan geometries which in fact has a considerable history (we might mention Kobayashi's paper [28] of 1956 about the general theory, and slightly later publications such as [32, 39, 40] on particular geometries) and wide current circulation. We feel justified therefore in regarding it as the conventional version. It is couched in the language of forms on principal bundles. S.-S. Chern writes, in his Foreword, of Sharpe's 'beautiful book', a description which few would wish to challenge. It seems to us, however, that one can get even closer to Cartan's ideas by expressing them in another, different, modern language, that of Lie groupoids and algebroids.

In the first place, the notion of associating a copy of a manifold F with each point of a manifold M certainly brings thoughts of fibre bundles to mind; but surely the concept most immediately invoked is that of a bundle $E \to M$ whose standard fibre is F, that is (to put things in a somewhat back-to-front way) of the associated rather than the principal bundle. The group G still has a natural role as group of the bundle. The idea that fibre and base should be tangent at each point of the base may be captured by requiring E to be 'soldered to M' in the technical manner we shall describe in due course.

Suppose given, then, a fibre bundle $E \to M$ with standard fibre F and group G. For any pair of points x, $\tilde{x}$ of M, the fibres E_x, $E_{\tilde{x}}$ of E are copies of F, and we may consider the maps[1] $E_{\tilde{x}} \to E_x$ which, when the fibres are identified with F, belong to G. The collection of all such maps, for all x and $\tilde{x}$, is a Lie groupoid $\mathcal{G}$. The identification of fibres at different points along a path in M, in a way which respects the homogeneous structure of the fibres, can be carried out if we have a path connection Γ on $\mathcal{G}$; that is, a rule for lifting curves c on M to curves c^Γ on $\mathcal{G}$ satisfying certain properties.

These then are the main ingredients of our interpretation of a Cartan geometry as a groupoid: a fibre bundle $E \to M$ with standard fibre F and group G, such that F is a homogeneous space of G of the same dimension as M, and such that E is soldered to M; and a path connection on the groupoid $\mathcal{G}$ of fibre maps of E which belong to G when the fibres are identified with F. (There are in fact certain additional requirements which will be made explicit as required.)

From a given Lie groupoid $\mathcal{G}$ we may construct its Lie algebroid $A\mathcal{G}$, and from a path connection on $\mathcal{G}$ we may construct an infinitesimal connection on $A\mathcal{G}$. So to every Cartan geometry, defined as a Lie groupoid with path connection having certain properties, there corresponds what we shall call an infinitesimal Cartan geometry, which consists of a transitive Lie algebroid with infinitesimal connection having certain properties. In fact an infinitesimal Cartan geometry may be defined independently of a Cartan geometry, by extracting the relevant properties. It is not necessarily the case that from a given infinitesimal Cartan geometry one can reconstruct a Cartan geometry: given what is known about the relation between Lie algebroids and Lie groupoids this should come as no surprise. The fact of this greater generality in the infinitesimal version will explain why we decided to use 'a Lie algebroid approach' as our subtitle rather than 'a Lie groupoid approach'.

We shall show that one may identify the Lie algebroid of an infinitesimal Cartan geometry with a certain Lie algebroid $\mathcal{A}$ of vector fields along fibres of $E \to M$ which are projectable to M; the anchor map is just the projection. Sections of $\mathcal{A}$ are projectable vector fields on E, and the algebroid bracket is just the ordinary bracket of vector fields on E restricted to projectable fields. An infinitesimal connection γ on $\mathcal{A}$ is a section of the anchor map, and therefore associates with each vector field X on M a projectable vector field $\gamma \circ X$ on E, which may be regarded as horizontal. So in discussing infinitesimal Cartan geometries we have the option of using, as well as the language of Lie algebroids, the more familiar language of vector fields and their brackets; and we shall go back and forth between the two as the occasion seems to demand.

---

[1]Our notational choice of $E_{\tilde{x}}$ rather than E_x as the domain fibre is deliberate; we shall comment further on this in due course.

It is easy nowadays to think that the last word about connection theory has been said with the formulation in terms of a connection form on a principal bundle, to associate that formulation with the name of Ehresmann, and to regard Cartan's theory of connections as an oddity that doesn't fit comfortably into the orthodox framework. As Sharpe says in his Preface 'the Ehresmann definition was taken as the definitive one, and Cartan's original notion went into a more or less total eclipse'. There is a certain irony therefore in the fact that Ehresmann was apparently the first to use the expression 'connexion de Cartan', in his paper of 1950 [22] in which he defined connections on fibre bundles. It is significant also that Ehresmann there introduced the notion of a connection in a bundle $E \to M$ as a horizontal distribution, and the associated concept of parallel transport of fibres along curves in M via their horizontal lifts. Where the bundle is equipped with a group G and standard fibre F one may require the connection to have the property that parallel transport between fibres is represented by the action of an element of G when the fibres are identified with F. This version of Ehresmann's theory of connections corresponds exactly to the idea of a path connection in the groupoid $\mathcal{G}$ referred to above.

Sharpe says that 'the simple and compelling geometric origin of a connection on a principal bundle is that it is a generalization of the Maurer-Cartan form' on a Lie group, and the conventional account of Cartan geometry is entirely based on this insight. We, by contrast, take the view that Ehresmann's version of Cartan's theory of connections leads directly to an account in terms of groupoids and algebroids as outlined above. The two approaches are of course essentially equivalent, as we demonstrate below; but the conceptual differences are considerable. C.-M. Marle has recently published a very readable account [31] of Cartan connections, Ehresmann connections, and the relations between them, which approaches matters from a similar direction to ours, while mentioning groupoids only in passing and algebroids not at all. We think of the first part of Chap. 6, where the groupoid and algebroid definitions of Cartan geometries are discussed, as an extension of his ideas.

Cartan published his paper [12] on projective connections in 1924. At around the same time a somewhat different line of research, also described as a theory of projective connections, was being pursued by several other authors, including T.Y. Thomas [41, 42] and J.H.C. Whitehead [45]. This second theory is concerned with the relationship between two linear connections whose geodesics, although having different parametrizations, are geometrically the same; two such connections are said to be projectively related. A brief history of the development of these ideas up to 1930, which names the mathematicians principally involved, can be found in the introductory section of Whitehead's paper. A modern version of the method used by Thomas and Whitehead, captured in the concept of a Thomas–Whitehead projective connection, has been given by Roberts in [35] and developed in [27]. As we shall see, these ideas may be described using the approach to Cartan geometries in terms of groupoids and algebroids. Indeed, a generalization of the projective differential geometry of linear connections was studied by Douglas [21] under the name of the general geometry of paths; it is also known as the projective differential geometry of sprays and incorporates the geometrical theory of systems of second-order ordinary differential equations under point transformations. The Thomas–Whitehead theory may be extended to cover this more general situation, and once again may be expressed in terms of groupoids and algebroids.

Thomas's linear connection theories of projective, and also conformal [43], geometry have been extended in a different way, under the name 'tractor connection' or 'tractor calculus', by Čap and Gover [9, 10], drawing on earlier work of, for example, Bailey et al. [2]. The primary aim of tractor calculus is to develop a theory of linear connections (covariant derivatives) on vector bundles which captures the essential features of a Cartan geometry, conceived of as a species of connection form on a principal bundle: covariant derivative operators are seen as easier to compute with than connection forms. There are specific tractor connections for projective and conformal geometries, to be found in [2] and [9], and a general construction in [10]. Our version of Cartan geometry already contains several vector bundles, starting with the Lie algebroid $\mathcal{A}$ itself; and it turns out to be pretty straightforward to identify a vector bundle with linear connection which plays the role of a general tractor connection and corresponds to what is called the 'adjoint tractor connection' in the references cited. We shall also reproduce the projective and conformal tractor bundles and connections. It should perhaps be mentioned that several other covariant derivatives and similar operators on vector bundles, only indirectly related to tractor connections, find a natural and important place in our theory.

A particular, and fundamental, application of these ideas concerns the symmetries of a geometry. The point of Klein's view of geometry was that it focussed attention on groups. A Cartan geometry can be, and often is, described as a Klein geometry deformed by curvature; conversely, a Klein geometry is a flat Cartan geometry. The group G of the Klein geometry is its group of symmetries. The questions naturally arise of how one defines and determines the symmetries of a general Cartan geometry. For the four standard geometries, namely affine, metric, conformal and projective geometries, these questions have been much studied on a case-by-case basis. Yano's book on Lie derivatives [46] contains a definitive statement of the classical results: these are however expressed in terms of the classical tensor calculus. An account in the spirit of Cartan geometry, taking advantage of modern understanding, and applying uniformly to all geometries rather than to individual examples or classes, seems to be lacking: the most significant recent paper on the subject, by Čap [8], concentrates on the special, if important, class of parabolic geometries. Yet there is a very obvious way of defining symmetry of a Cartan geometry as we interpret it, which is just as a structure-preserving diffeomorphism of E; that is, a fibred diffeomorphism of $E \to M$ which preserves the soldering and the path connection. An infinitesimal symmetry of an infinitesimal Cartan geometry, similarly, is a projectable vector field on E which is tangent to the section of $E \to M$ which defines the soldering, and along which the Lie derivative of the infinitesimal connection vanishes (where the term 'Lie derivative' has to be interpreted suitably of course). It turns out, very conveniently, that this Lie derivative condition may also be interpreted as the vanishing of a certain covariant derivative of the projectable vector field defining the infinitesimal symmetry: the covariant derivative in question has previously been introduced by Blaom in [4]; see also [17].

We hope, in the present volume, to provide a description of this approach to Cartan geometries and their symmetries in a structured manner. The volume falls essentially into four parts, where as we progress from one part to the next we reduce the generality of our discussion.

The first part, Chaps. 1 and 2, reviews the ideas of Lie groupoid and Lie algebroid, and the associated concepts of connection. Much of it is based on the text of Mackenzie [30], with various extensions as required for the subsequent exposition. Then in the second part, Chaps. 3–5, we consider what might be called 'pre-Cartan geometries', where we consider in particular the Lie groupoids of fibre morphisms of a given fibre bundle, and the connections on such groupoids together with their symmetries. In this second part we also see how the infinitesimal approach, using Lie algebroids rather than Lie groupoids, and in particular using Lie algebroids of projectable vector fields along the fibres of the bundle, may be of benefit.

In the third part, Chaps. 6–10, we introduce Cartan geometries proper, together with a number of tools we shall use to study them. We take, as particular examples, the four classical types of geometry: affine, projective, Riemannian and conformal geometry. We also see how our approach can start to fit into a more general theory. Finally, in Chaps. 11–12, we specialize to the geometries (affine and projective) associated with path spaces and geodesics, and consider their symmetries and other properties.

Throughout this work we shall use a variety of techniques; in particular, we shall frequently use the representation of a geometric object in a given system of local coordinates in order to establish local properties of that object. We shall endeavour, however, to avoid defining such objects using coordinates, in order to obviate the need to confirm that the object 'transforms correctly' when using a different system of coordinates. Unless stated otherwise, we shall use the summation convention for repeated indices in each formula.

We shall always require differentiable manifolds to be of class C^∞, Hausdorff and second countable (and so paracompact); we do not in general require manifolds to be connected, although some results do require connectedness. The algebra of smooth functions (that is, functions of class C^∞) on a manifold M will be denoted by $C^\infty(M)$.

With one class of exceptions, all maps between manifolds will be smooth. The exception is that a map $c : J \to M$, where $J \subset \mathbb{R}$ is an interval of strictly positive length, will need to be only continuous and piecewise-smooth: there might be finitely many points $t \in J$ where c is not smooth, although we shall still require all left and right derivatives to exist at those points. Such a map c will be called a *curve*; a curve which happens to be smooth will be called, explicitly, a *smooth curve*.

While this book was being written the first author was a Guest Professor at Ghent University; he wishes to thank the Department of Mathematics for its hospitality. The second author wishes to acknowledge support from grant no. 14-02476S 'Variations, Geometry and Physics' of the Czech Science Foundation.

Chapter 1
Lie Groupoids and Lie Algebroids

The idea of a groupoid is a generalization of that of a group, where not every pair of elements can be combined: see [6] for a history of the concept. The axioms for a groupoid are the same as those for a category where every morphism is an isomorphism, with the additional condition that the collection of 'objects' must be a set, so that a groupoid is in fact a 'small category'. A fruitful application arises in algebraic topology, where the fundamental group $\pi_1(X, x)$ of a topological space depends upon the choice of a basepoint $x \in X$, whereas the fundamental groupoid $\Pi(X)$, containing homotopy classes of paths (rather than loops) in X, does not. Lie groupoids, where the groupoid is a differentiable manifold and the associated maps are smooth, together with some additional conditions described below, were used by Ehresmann [23] in the context of principal bundles.

To each Lie groupoid there is associated a Lie algebroid, in much the same way as to each Lie group one associates a Lie algebra, capturing the infinitesimal aspects of the construction. Lie algebroids are vector bundles with some additional structure, and were first used by Pradines [34]. In the first two sections of this chapter we give the basic definitions of Lie groupoids and Lie algebroids, stating some properties without proof; a comprehensive account of the general theory may be found in the work of Mackenzie [30]. Some of the proofs in this chapter and the next are versions of those given in the latter reference, and are included here for completeness.

1.1 Groupoids and Lie Groupoids

A *groupoid* is a set $\mathcal{G}$ with a partial multiplication (often denoted by a dot, as in $\varphi_1 \cdot \varphi_2$), another set M of *identities* with an injection $1 : M \to \mathcal{G}$, two surjective maps $\alpha, \beta : \mathcal{G} \to M$ (the *source* and *target* maps respectively) and a bijection $\iota : \mathcal{G} \to \mathcal{G}$ (the *inverse* map). We say that the pair of elements (φ_1, φ_2) is *admissible* if the product $\varphi_1 \cdot \varphi_2$ is defined. We frequently write 1_M instead of $1(M) \subset \mathcal{G}$ and 1_x instead of $1(x)$ for $x \in M$; we also usually write φ^{-1} instead of $\iota(\varphi)$ for $\varphi \in \mathcal{G}$. This collection of entities must satisfy the following conditions in order to be a groupoid:

M. Crampin and D. Saunders, *Cartan Geometries and their Symmetries*,
Atlantis Studies in Variational Geometry 4, DOI 10.2991/978-94-6239-192-5_1

- the pair (φ_1, φ_2) is admissible if, and only if, the condition $\alpha(\varphi_1) = \beta(\varphi_2)$ holds, and then

$$\alpha(\varphi_1 \cdot \varphi_2) = \alpha(\varphi_2), \qquad \beta(\varphi_1 \cdot \varphi_2) = \beta(\varphi_1);$$

- if either triple product $(\varphi_1 \cdot \varphi_2) \cdot \varphi_3$ or $\varphi_1 \cdot (\varphi_2 \cdot \varphi_3)$ exists then both do, and they are equal;
- $\alpha(1_x) = \beta(1_x) = x$ for all $x \in M$;
- $\alpha(\varphi)$, $\beta(\varphi)$ satisfy $\varphi \cdot 1_{\alpha(\varphi)} = \varphi$ and $1_{\beta(\varphi)} \cdot \varphi = \varphi$;
- both (φ, φ^{-1}) and (φ^{-1}, φ) are admissible for all $\varphi \in \mathcal{G}$, and $\varphi^{-1} \cdot \varphi = 1_{\alpha(\varphi)}$ and $\varphi \cdot \varphi^{-1} = 1_{\beta(\varphi)}$.

We should caution the reader that the requirement $\alpha(\varphi_1) = \beta(\varphi_2)$ for the existence of $\varphi_1 \cdot \varphi_2$ is a convention, and an alternative convention would be to require $\alpha(\varphi_2) = \beta(\varphi_1)$. For our choice we say that 'the source is on the right, and the target on the left'. We make this choice because the elements of our groupoids will be maps (in fact fibre morphisms, as described in Chap. 3), and the arguments of our maps are on the right so that the source of a groupoid element should be on the right.

It will sometimes be convenient to use the left and right *translation maps*

$$l_\varphi : \mathcal{G}^{(\varphi *)} \to \mathcal{G}, \quad l_\varphi(\varphi_2) = \varphi \cdot \varphi_2 \quad \text{where } \mathcal{G}^{(\varphi *)} = \{\varphi_2 \in \mathcal{G} \,:\, \beta(\varphi_2) = \alpha(\varphi)\}$$
$$r_\varphi : \mathcal{G}^{(* \varphi)} \to \mathcal{G}, \quad r_\varphi(\varphi_1) = \varphi_1 \cdot \varphi \quad \text{where } \mathcal{G}^{(* \varphi)} = \{\varphi_1 \in \mathcal{G} \,:\, \alpha(\varphi_1) = \beta(\varphi)\}.$$

We may also define morphisms. If $\mathcal{G}$, $\mathcal{H}$ are groupoids with sets M, N of identities then a pair of maps $\mathcal{P} : \mathcal{G} \to \mathcal{H}$, $p : M \to N$ specify a *groupoid morphism* if

- both $\alpha_{\mathcal{H}} \circ \mathcal{P} = p \circ \alpha_{\mathcal{G}}$ and $\beta_{\mathcal{H}} \circ \mathcal{P} = p \circ \beta_{\mathcal{G}}$; and
- if (φ_1, φ_2) is an admissible pair, so that $\alpha_{\mathcal{H}} \mathcal{P}(\varphi_1) = p\,\alpha_{\mathcal{G}}(\varphi_1) = p\,\beta_{\mathcal{G}}(\varphi_2) = \beta_{\mathcal{H}} \mathcal{P}(\varphi_2)$ and that therefore $\big(\mathcal{P}(\varphi_1), \mathcal{P}(\varphi_2)\big)$ is also an admissible pair, we have in addition

$$\mathcal{P}(\varphi_1 \cdot \varphi_2) = \mathcal{P}(\varphi_1) \cdot \mathcal{P}(\varphi_2).$$

If $M = N$ and $p = \mathrm{id}_M$ we say that $\mathcal{P}$ is a *basepoint-preserving* morphism. It is evident that the composition of two morphisms, obtained by composing the pairs of individual maps, is again a morphism, and that the identity $(\mathrm{id}_{\mathcal{G}}, \mathrm{id}_M)$ is a morphism. A groupoid isomorphism is, of course, a morphism with a two-sided inverse morphism under this composition.

A *Lie groupoid* is a groupoid where

- both $\mathcal{G}$ and M are C^∞ differentiable manifolds;
- the subset $\mathcal{G}^{(2)} \subset \mathcal{G} \times \mathcal{G}$ of admissible pairs is a submanifold;
- the map $\mu : \mathcal{G}^{(2)} \to \mathcal{G}$ given by $\mu(\varphi_1, \varphi_2) = \varphi_1 \cdot \varphi_2$ and the maps 1, α, β and ι are all smooth;
- the map 1 is an immersion, the maps α, β are surjective submersions, and the map ι is a diffeomorphism.

A consequence of the last property is that the fibres $\alpha^{-1}(1_\varphi)$ and $\beta^{-1}(1_\varphi)$ passing through φ are closed submanifolds of $\mathcal{G}$. For any φ the right translation map r_{φ_2} (where $\alpha(\varphi) = \beta(\varphi_2)$) maps the α-fibre $\alpha^{-1}(1_\varphi)$ onto the α-fibre $\alpha^{-1}(1_{\varphi_2})$, and similarly l_{φ_1} maps β-fibres to β-fibres. For any $x \in M$ the *vertex group* at x is the subset

$$\mathcal{G}_x = \{\varphi \in \mathcal{G} : \alpha(\varphi) = \beta(\varphi) = x\};$$

it is obvious that $\mathcal{G}_x$ is a Lie group under the restriction of the groupoid multiplication. If $\mathcal{G}$, $\mathcal{H}$ are Lie groupoids then the groupoid morphism $(\mathcal{P}, p)$ from $\mathcal{G}$ to $\mathcal{H}$ is a *Lie groupoid morphism* if $\mathcal{P}$ is smooth. As the source map $\alpha_\mathcal{G}$ is a smooth surjective submersion, given any $x \in M$ there will be a local section s of $\alpha_\mathcal{G}$ defined on a neighbourhood U of x; it will follow that the restriction of p to U satisfies

$$p|_U = p \circ \alpha_\mathcal{G} \circ s = \alpha_\mathcal{H} \circ \mathcal{P} \circ s,$$

showing that $p|_U$ is smooth and hence that p is smooth. A Lie groupoid morphism is an isomorphism if it is a groupoid isomorphism, and if its inverse is also a Lie groupoid morphism.

There are several standard examples of Lie groupoids.

- Any Lie group G with identity element e_G, where $M = \{e_G\}$ and $1 : M \to G$ is the inclusion.
- Any Cartesian product $M \times M$ with partial multiplication[1] $(x, \tilde{x}) \cdot (y, \tilde{y}) = (x, \tilde{y})$ defined when $\tilde{x} = y$, identity map $1 : M \to M \times M$ given by $1_x = (x, x)$, and inverse map given by $(x, \tilde{x})^{-1} = (\tilde{x}, x)$. This is called the *pair groupoid*.
- A *trivial Lie groupoid* on M with group G is the product $M \times G \times M$ with partial multiplication $(x, g, \tilde{x}) \cdot (y, h, \tilde{y}) = (x, gh, \tilde{y})$ defined when $\tilde{x} = y$, identity map $1 : M \to M \times G \times M$ given by $1_x = (x, 1_G, x)$, and inverse map given by $(x, g, \tilde{x})^{-1} = (\tilde{x}, g^{-1}, x)$.
- The *gauge groupoid* of a principal G-bundle $\Pi : P \to M$, comprising equivalence classes of pairs $[(p, \tilde{p})] \in P \times P$, where $(p_1, \tilde{p}_1) \sim (p_2, \tilde{p}_2)$ when there is an element $g \in G$ with $(p_1 g, \tilde{p}_1 g) = (p_2, \tilde{p}_2)$ (so that p_1 and p_2 are necessarily in the same fibre of Π, as are $\tilde{p}_1$ and $\tilde{p}_2$); the partial multiplication is given by $[(p_1, \tilde{p}_1)] \cdot [(p_2, \tilde{p}_2)] = [(p_1, \tilde{p}_2 g)]$ where $\tilde{p}_1, p_2$ are in the same fibre and $g \in G$ is the unique element satisfying $\tilde{p}_1 g = p_2$; the identity map is $1_x = [(p, p)]$ for any $p \in P_x$; and the inverse map is $[(p, \tilde{p})]^{-1} = [(\tilde{p}, p)]$.
- If G is a Lie group and $l_g : M \to M$ denotes a left action of G on a manifold M then the corresponding *action groupoid* is the groupoid structure on $G \times M$ given by taking $\alpha(g, x) = x$ and $\beta(g, x) = l_g(x)$, so that the product $(g, x) \cdot (h, y)$ is defined when $x = l_h(y)$; we then take $(g, x) \cdot (h, y) = (gh, y)$. The identity map is $1 : M \to G \times M$ with $1_x = (1_G, x)$, and $(g, x)^{-1} = (g^{-1}, l_g(x))$. In the special

[1] We shall often write the source of a typical groupoid element as $\tilde{x}$ and the target as x. This choice is deliberate, and is related to our use, when considering the Lie algebroid of a Lie groupoid, of vector fields vertical over the source projection α.

case where $M = G$ and the action is left translation then the action groupoid is the same as the pair groupoid $G \times G$. More generally, if M is a left homogeneous space G/H, so that $G \to G/H$ is a principal H-bundle, then the action groupoid is canonically isomorphic to the gauge groupoid; the isomorphism is given by $(g_1, g_2H) \leftrightarrow [(g_1, g_2)]$.

A *Lie subgroupoid* of $\mathcal{G}$ is a pair of submanifolds $\mathcal{H} \subset \mathcal{G}$ and $N \subset M$ with the properties that

- $\alpha(\varphi), \beta(\varphi) \in N$ for each $\varphi \in \mathcal{H}$, and the restricted maps $\alpha|_{\mathcal{H}}, \beta|_{\mathcal{H}} : \mathcal{H} \to N$ are surjective submersions;
- the subset $\mathcal{H}^{(2)} \subset \mathcal{H} \times \mathcal{H}$ of admissible pairs is a submanifold;
- if $\varphi \in \mathcal{H}$ and $(\varphi_1, \varphi_2) \in \mathcal{H}^{(2)}$ then $\varphi_1 \cdot \varphi_2, \varphi^{-1} \in \mathcal{H}$, and if $x \in N$ then $1_x \in \mathcal{H}$.

It is clear that a Lie subgroupoid is itself a Lie groupoid using the restrictions of the source, target, identity and inverse maps. For example, if $U \subset M$ is a nonempty open subset then

$$\mathcal{G}_U = \{\varphi \in \mathcal{G} : \alpha(\varphi), \beta(\varphi) \in U\}$$

is a Lie subgroupoid of $\mathcal{G}$. In the special case where $N = M$ we say that $\mathcal{H}$ is a *wide Lie subgroupoid* of $\mathcal{G}$; for example, the action groupoid $G \times M$ is canonically isomorphic to a wide Lie subgroupoid of the trivial groupoid $M \times G \times M$, where the isomorphism is given by $(g, x) \leftrightarrow (l_g(x), g, x)$ with l_g being the action of G on M.

We shall say that a Lie groupoid $\mathcal{G}$ is *locally trival* if the map $(\beta, \alpha) : \mathcal{G} \to M \times M$ is a surjective submersion. This may seem different from the concept of local triviality for fibre bundles, where the meaning is that the bundle is locally isomorphic to a trivial bundle, but in fact our condition implies a similar property for Lie groupoids, and if the manifold M is connected then the two properties are equivalent.

Lemma 1.1.1 *If $\mathcal{G}$ is a locally trivial Lie groupoid then, fixing an arbitrary $x \in M$, for each $y \in M$ there is an open neighbourhood U of y and a basepoint-preserving isomorphism between the Lie subgroupoid $\mathcal{G}_U$ and the trivial Lie groupoid $U \times \mathcal{G}_x \times U$.*

Conversely, if $\mathcal{G}$ is a Lie groupoid with a connected manifold M of identities having the property that, for some vertex group $G = \mathcal{G}_x$, each $y \in M$ has a neighbourhood U such that $\mathcal{G}_U$ is basepoint-preserving isomorphic to $U \times \mathcal{G}_x \times U$, then $\mathcal{G}$ is locally trivial.

Proof Suppose first that $\mathcal{G}$ is locally trivial, and fix $x \in M$. Let $\alpha^{-1}(x)$ be the α-fibre at x, and let $\beta_x : \alpha^{-1}(x) \to M$ be the restriction of β to this fibre. As (β, α) is a surjective submersion, it follows that β_x is a surjective submersion. Thus, given $y \in M$, there is a neighbourhood U of y and a local section $\sigma : U \to \alpha^{-1}(x)$ of β_x; we do not, of course, require $x \in U$. We now define $\mathcal{P} : U \times \mathcal{G}_x \times U \to \mathcal{G}_U$ by

$$\mathcal{P}(z, g, \tilde{z}) = \sigma(z) \cdot g \cdot \left(\sigma(\tilde{z})\right)^{-1}. \tag{1}$$

It is immediate from the properties of σ that this groupoid product is defined, and that the result is an element of $\mathcal{G}_U$. Furthermore, given $\varphi \in \mathcal{G}_U$, the product

$$g_\varphi = \left(\sigma\beta(\varphi)\right)^{-1} \cdot \varphi \cdot \left(\sigma\alpha(\varphi)\right)$$

is defined, and clearly $\alpha(g_\varphi) = \beta(g_\varphi) = x$ so that $g_\varphi \in \mathcal{G}_x$; by construction the map $\varphi \mapsto (\beta(\varphi), g_\varphi, \alpha(\varphi))$ is the inverse of $\mathcal{P}$. As the operations used in the construction of $\mathcal{P}$ and its inverse are smooth, the map is a diffeomorphism. It is evident from the defining formula (1) that $\mathcal{P}$ preserves the groupoid operations, and as

$$\mathcal{P}(z, 1_x, z) = \sigma(z) \cdot 1_x \cdot \left(\sigma(z)\right)^{-1} = 1_z$$

we see that $\mathcal{P}$ is a basepoint-preserving isomorphism of Lie groupoids.

Now suppose, conversely, that $\mathcal{G}$ has this property of being locally isomorphic to trivial groupoids with a fixed group G. Given an isomorphism $\mathcal{G}_U \cong U \times G \times U$, it is immediate that the restriction $(\beta, \alpha)|_{\mathcal{G}_U} : \mathcal{G}_U \to U \times U$ is a submersion, so that $(\beta, \alpha) : \mathcal{G} \to M \times M$ is a submersion. To see that (β, α) is surjective when M is connected, consider $(x, \tilde{x}) \in M \times M$ and let $c : I = [0, 1] \to M$ be a curve with $c(0) = \tilde{x}$ and $c(1) = x$. For each $y \in c(I)$ let U_y be the corresponding open neighbourhood; by compactness we may choose finitely many of these, $U_1, U_2, \ldots, U_m$, covering $c(I)$. We may assume that each $c(I) \cap U_i$ is connected, and that the U_i are indexed so that $\tilde{x} \in U_1$ and $x \in U_m$, and so that we may find $u_i \in c(I) \cap U_i \cap U_{i+1}$ for $1 \leq i \leq m - 1$. Let $\mathcal{P}_i$ be the isomorphisms $\mathcal{G}_{U_i} \to U_i \times G \times U_i$, and let $\varphi_i \in \mathcal{G}_{U_i}$ be given by

$$\varphi_1 = \mathcal{P}_1(u_1, 1_G, \tilde{x}), \qquad \varphi_i = \mathcal{P}_i(u_i, 1_G, u_{i-1}) \quad (2 \leq i \leq m - 1), \qquad \varphi_m = \mathcal{P}_m(x, 1_G, u_{m-1})$$

where $1_G \in G$ is the identity. The product

$$\varphi = \varphi_m \cdot \varphi_{m-1} \cdots \varphi_2 \cdot \varphi_1$$

is therefore defined, and $(\beta, \alpha)(\varphi) = (x, \tilde{x})$. □

The condition in the lemma, that M should be connected, is necessary. Let $\mathbb{R}_0$ be the punctured real line, excluding zero, and let $\mathcal{G} \subset \mathbb{R}_0 \times \mathbb{R}_0$ be given by $(x, y) \in \mathcal{G}$ when $xy > 0$. It is evident that $\mathcal{G}$ is a wide Lie subgroupoid of the pair groupoid, and indeed $\mathcal{G} = (\mathbb{R}_+ \times \mathbb{R}_+) \cup (\mathbb{R}_- \times \mathbb{R}_-)$ is locally isomorphic to trivial groupoids. However $(1, -1)$ does not lie in the image of $(\beta, \alpha) : \mathcal{G} \to \mathbb{R}_0 \times \mathbb{R}_0$, so that $\mathcal{G}$ is not transitive, and is therefore not locally trivial according to our definition.

We remark finally that any Lie groupoid $\mathcal{G}$ is by definition a fibred manifold over M in two different ways, so that we should like to be able to consider submanifolds of $\mathcal{G}$ that are the images of sections of both projections $\alpha, \beta : \mathcal{G} \to M$. We may do this by considering sections Φ of α with the property that the composition $\phi = \beta \circ \Phi$ is a diffeomorphism of M. Such a map is called a *bisection* of $\mathcal{G}$. The roles of source and target maps can be interchanged, and the map $\Phi \circ \phi^{-1} : M \to \mathcal{G}$ is a section of β whose composition $\alpha \circ (\Phi \circ \phi^{-1}) = \phi^{-1}$ is a diffeomorphism.

We say that a bisection Φ is *vertical at* x if x is a fixed point of ϕ, so that $\Phi(x) \in \mathcal{G}_x$, the vertex group at x, and that Φ is a *vertical bisection* if it is vertical at each $x \in M$.

Proposition 1.1.2 *The set of global bisections of $\mathcal{G}$ forms a group $\mathsf{B}(\mathcal{G})$, with product defined by $(\Phi_1 * \Phi_2)(x) = \Phi_1\phi_2(x) \cdot \Phi_2(x)$, identity $1 : x \mapsto 1_x$ and inverse $\Phi^{-1}(x) = \big(\Phi\phi^{-1}(x)\big)^{-1}$. The correspondence $\mathsf{B}(\mathcal{G}) \to \mathsf{Diff}(M)$ given by $\Phi \mapsto \phi = \beta \circ \Phi$ is a group homomorphism. The set $\mathsf{B}^{\mathsf{v}}(\mathcal{G})$ of vertical bisections is a normal subgroup of $\mathsf{B}(\mathcal{G})$.*

Proof We see first that $\alpha\big((\Phi_1 * \Phi_2)(x)\big) = \alpha\Phi_2(x) = x$, so that $\Phi_1 * \Phi_2$ is a section of α, and that $\beta\big((\Phi_1 * \Phi_2)(x)\big) = \beta\Phi_1\phi_2(x) = \phi_1\phi_2(x)$, so that $\beta \circ (\Phi_1 * \Phi_2) = \phi_1 \circ \phi_2$ is a diffeomorphism; thus $\Phi_1 * \Phi_2$ is a bisection.

We next check associativity, and see that

$$
\begin{aligned}
\big((\Phi_1 * \Phi_2) * \Phi_3\big)(x) &= (\Phi_1 * \Phi_2)(\phi_3(x)) \cdot \Phi_3(x) = \Big(\Phi_1\phi_2\phi_3(x) \cdot \Phi_2\phi_3(x)\Big) \cdot \Phi_3(x) \\
&= \Phi_1\phi_2\phi_3(x) \cdot \Big(\Phi_2\phi_3(x) \cdot \Phi_3(x)\Big) = \Phi_1\big((\phi_2 \circ \phi_3)(x)\big) \cdot (\Phi_2 * \Phi_3)(x) \\
&= \big(\Phi_1 * (\Phi_2 * \Phi_3)\big)(x)
\end{aligned}
$$

so that $(\Phi_1 * \Phi_2) * \Phi_3 = \Phi_1 * (\Phi_2 * \Phi_3)$.

It is immediate that 1 is a bisection with $\beta \circ 1 = \mathrm{id}_M$, and that $1 * \Phi = \Phi * 1 = \Phi$.

We also see that $\alpha\Phi^{-1}(x) = \beta\Phi\phi^{-1}(x) = \phi\phi^{-1}(x) = x$, so that Φ^{-1} is a section of α, and $\beta\Phi^{-1}(x) = \alpha\Phi\phi^{-1}(x) = \phi^{-1}(x)$, so that $\beta \circ \Phi^{-1} = \phi^{-1}$ is a diffeomorphism; thus Φ^{-1} is a bisection. It is clear that $(\Phi^{-1} * \Phi)(x) = \Phi^{-1}\phi(x) \cdot \Phi(x) = \big(\Phi\phi^{-1}(x)\big)^{-1} \cdot \Phi(x) = 1_{\alpha(\Phi(x))} = 1_x$, so that $\Phi^{-1} * \Phi = 1$, and similarly $\Phi * \Phi^{-1} = 1$.

Thus $\mathsf{B}(\mathcal{G})$ is a group, and the relation $\beta \circ (\Phi_1 * \Phi_2) = \phi_1 \circ \phi_2$ defines a group homomorphism $\mathsf{B}(\mathcal{G}) \to \mathsf{Diff}(M)$.

Finally if $\Phi_1, \Phi_2 \in \mathsf{B}^{\mathsf{v}}(\mathcal{G})$ then $\phi_1 = \phi_2 = \mathrm{id}_M$, so that $\phi_1 \circ \phi_2 = \mathrm{id}_M$ and $\phi_1^{-1} = \mathrm{id}_M$ and hence $\Phi_1 * \Phi_2, \Phi_1^{-1} \in \mathsf{B}^{\mathsf{v}}(\mathcal{G})$. If instead $\Phi_1 \in \mathsf{B}^{\mathsf{v}}(\mathcal{G})$ and $\Phi_2 \in \mathsf{B}(\mathcal{G})$ then $\phi_2 \circ \phi_1 \circ \phi_2^{-1} = \mathrm{id}_M$ so that $\Phi_2 * \Phi_1 * \Phi_2^{-1} \in \mathsf{B}^{\mathsf{v}}(\mathcal{G})$. Thus $\mathsf{B}^{\mathsf{v}}(\mathcal{G})$ is a normal subgroup of $\mathsf{B}(\mathcal{G})$. □

We also consider local bisections Φ defined on any nonempty open subset $U \subset M$. If Φ_1, Φ_2 are local bisections defined on the nonempty open sets U_1, U_2 then the product $\Phi_1 * \Phi_2$ is defined on $\phi_2^{-1}(U_1) \cap U_2$; if this intersection is empty then the product is not defined. The set of local bisections of $\mathcal{G}$ is therefore like a pseudogroup, where the group properties hold whenever they are defined.

We shall see in Chap. 5 that bisections can be considered as 'generalized elements' of Lie groupoids.

1.2 Lie Algebroids

A *Lie algebroid* is a vector bundle $\tau : \mathcal{A} \to M$ together with a morphism $\rho : \mathcal{A} \to TM$ of vector bundles over the identity on M, called the *anchor map* of the Lie algebroid, and a bracket $[\![\xi, \eta]\!]$ defined on sections of τ. These entities must satisfy the following conditions:

- the bracket is skew-symmetric and bilinear over the real numbers, so that

$$[\![\eta, \xi]\!] = -[\![\xi, \eta]\!], \qquad [\![\lambda\xi + \zeta, \eta]\!] = \lambda[\![\xi, \eta]\!] + [\![\zeta, \eta]\!]$$

  for $\lambda \in \mathbb{R}$ and sections ξ, η, ζ ;
- the bracket satisfies the Jacobi identity, so that

$$[\![\xi, [\![\eta, \zeta]\!]]\!] + [\![\eta, [\![\zeta, \xi]\!]]\!] + [\![\zeta, [\![\xi, \eta]\!]]\!] = 0;$$

- the bracket of sections is related to the Lie bracket of vector fields on M by the anchor, so that

$$\rho \circ [\![\xi, \eta]\!] = [\rho \circ \xi, \rho \circ \eta], \qquad [\![\xi, f\eta]\!] = f[\![\xi, \eta]\!] + \big((\rho \circ \xi)f\big)\eta$$

  for any function $f \in C^\infty(M)$, where for example $\rho \circ \xi : M \to TM$ is the vector field on M obtained from the section $\xi : M \to \mathcal{A}$ of the Lie algebroid.

The vector space of global sections of a Lie algebroid must therefore be a (generally infinite-dimensional) Lie algebra, although of course this condition on its own is not sufficient as we also require each section to act as a derivation on $C^\infty(M)$ through the anchor map.

Given vector bundle coordinates (x^i, z^A) on $\mathcal{A}$ we obtain functions ρ^i_A, Θ^A_{BC} on M from the anchor and the bracket, given by

$$\rho^i_A = \dot{x}^i \circ \rho \circ e_A, \qquad \Theta^A_{BC} = z^A \circ [\![e_B, e_C]\!]$$

where e_A are the local sections of the algebroid dual to the linear fibre coordinates z^A. The functions Θ^A_{BC} are called the *structure functions* of the Lie algebroid, by analogy with the structure constants of a Lie algebra.

There are several standard examples of Lie algebroids.

- Any Lie algebra $\mathfrak{g}$ with zero element $0_\mathfrak{g}$ may be regarded as a Lie algebroid $\mathfrak{g} \to \{0_\mathfrak{g}\}$.
- The tangent bundle $TM \to M$ of any manifold is a Lie algebroid, with the usual bracket of vector fields and with the identity as the anchor map.
- Let $\mathfrak{g}$ be a Lie algebra. The Whitney sum $TM \oplus_M (M \times \mathfrak{g}) \to M$ is a vector bundle over M, and it becomes a Lie algebroid with anchor map given by $\rho(v, a) = v$ and bracket of sections given by

$$[\![(X,\xi),(Y,\eta)]\!] = \big([X,Y],(\mathrm{id}, X(\bar{\xi}) - Y(\bar{\eta}) - \{\bar{\xi},\bar{\eta}\}\big)$$

where

- X, Y are vector fields on M,
- ξ and η are sections of $(M \times \mathfrak{g}) \to M$,
- $\bar{\xi}$ and $\bar{\eta}$ are the corresponding maps $M \to \mathfrak{g}$,
- $\{\bar{\xi},\bar{\eta}\}$ denotes the pointwise Lie algebra bracket, and
- $X(\bar{\eta})$ denotes the Lie derivative of the algebra-valued function $\bar{\eta}$, and similarly for $Y(\bar{\xi})$.

This is called the *trivial Lie algebroid*.

- Let $\mathfrak{g}$ be a Lie algebra and let the Lie algebra homomorphism $\mathfrak{g} \to \mathfrak{X}(M)$, $a \mapsto a^\sharp$ be an infinitesimal action on a manifold M, so that $a^\sharp$ is a vector field on M and, for any $x \in M$, $a^\sharp_x \in T_xM$ is the corresponding tangent vector. For any map $\varepsilon : M \to \mathfrak{g}$ define a vector field $\varepsilon^\sharp \in \mathfrak{X}(M)$ by $\varepsilon^\sharp_x = (\varepsilon(x))^\sharp_x$. The *action algebroid* corresponding to this infinitesimal action is the trivial vector bundle $M \times \mathfrak{g} \to M$, with anchor map given by $\rho(x,a) = a^\sharp_x$ and bracket of sections given by

$$[\![\xi,\eta]\!](x) = \big(x,\ \xi^\sharp_x(\bar{\eta}) - \eta^\sharp_x(\bar{\xi}) + \{\bar{\xi}(x),\bar{\eta}(x)\}\big)$$

where

- ξ and η are sections of the trivial vector bundle,
- $\bar{\xi}$ and $\bar{\eta}$ are the corresponding maps $M \to \mathfrak{g}$,
- $\{\bar{\xi}(x),\bar{\eta}(x)\} \in \mathfrak{g}$ denotes the Lie algebra bracket, and
- $\xi^\sharp_x(\bar{\eta}) \in \mathfrak{g}$ is the derivative by the tangent vector $\xi^\sharp_x$ of the algebra-valued function $\bar{\eta}$ and similarly for $\eta^\sharp_x(\bar{\xi})$.

- A particular type of Lie algebroid arises when we have a vector bundle $\tau : \mathcal{K} \to M$ where each fibre $\mathcal{K}_x$ is itself a Lie algebra, and where there is a fixed Lie algebra $\mathfrak{g}$ and a family of local trivializations[2] $\mathrm{T}_U : U \times \mathfrak{g} \to \tau^{-1}(U)$ where each $\mathrm{T}_U\big|_x$ is a Lie algebra isomorphism. The Lie bracket of sections is defined pointwise, so that $[\![\kappa,\lambda]\!](x) = \{\kappa(x),\lambda(x)\}_x$ (the Lie algebra bracket of the fibre $\mathcal{K}_x$), and the anchor $\rho : \mathcal{K} \to TM$ is the zero map. Such a Lie algebroid is called a *Lie algebra bundle*; we sometimes abbreviate this to LAB, particularly when referring to LAB-morphisms. We often denote the bracket on such a bundle by $\{\kappa,\lambda\}$ rather than $[\![\kappa,\lambda]\!]$ to emphasize its pointwise nature.

The definition of a morphism for Lie algebroids is more tricky than for more general vector bundles (or, indeed, for Lie groupoids or Lie algebra bundles). The difficulty arises because we would like morphisms to act also on sections, and then to preserve the bracket of sections. A vector bundle morphism projecting to the identity on M gives an action on sections by composition, and so in this case we can say that a *Lie*

[2]There are different conventions in the literature regarding the direction of trivialization maps. We shall consistently regard the trivial (product) manifold as the domain of the trivialization map rather than the codomain, as this simplifies many formulæ.

algebroid morphism over the identity is a vector bundle morphism (P, id_M) satisfying the condition that $[\![P \circ \xi, P \circ \eta]\!] = P \circ [\![\xi, \eta]\!]$. In particular, the anchor map (ρ, id_M) is always a morphism over the identity between the Lie algebroids $\mathcal{A}$ and TM. In general, though, a vector bundle morphism need not have an induced action on sections at all. We will, however, be able to avoid this issue, and our definition of a 'full morphism' for Lie algebroids will be given in Sect. 1.4.

Given a general Lie algebroid $\tau : \mathcal{A} \to M$, there is no reason why its anchor map ρ should be surjective; but if it is then we say that the Lie algebroid is *transitive*.

We shall in fact confine our attention almost exclusively to transitive algebroids in what follows. Similarly there is no reason why ρ should be injective: indeed, its kernel, which we shall usually denote by $\mathcal{K}$, will play an important role.

Proposition 1.2.1 *If $\mathcal{A}$ is a transitive Lie algebroid then $\mathcal{K}$ is a Lie subalgebroid of $\mathcal{A}$.*

Proof Since ρ is a morphism of vector bundles over the identity and has constant rank when $\mathcal{A}$ is transitive, its kernel is a vector sub-bundle of $\mathcal{A}$, and in particular $\mathcal{K}$ is a vector bundle. If κ and λ are sections of $\mathcal{K}$, so that $\rho \circ \kappa = \rho \circ \lambda$, then $\rho \circ [\![\kappa, \lambda]\!] = [\rho \circ \kappa, \rho \circ \lambda] = 0$, so $[\![\cdot, \cdot]\!]$ restricts to sections of $\mathcal{K}$ and defines a bracket which clearly makes $\mathcal{K}$ a Lie algebroid. □

We call $\mathcal{K}$ with the Lie algebroid structure obtained by restriction of that of $\mathcal{A}$ the *kernel of the transitive Lie algebroid* $\mathcal{A}$, so that we have a short exact sequence of vector bundles

$$0 \to \mathcal{K} \to \mathcal{A} \xrightarrow{\rho} TM \to 0,$$

each of which is a Lie algebroid. The anchor of $\mathcal{K}$ is, of course, the zero map, so it follows that the bracket of $\mathcal{K}$ is linear over $C^\infty(M)$:

$$[\![\kappa, f\lambda]\!] = f[\![\kappa, \lambda]\!].$$

This entails that for $x \in M$, $[\![\kappa, \lambda]\!](x)$ depends only on $\kappa(x)$ and $\lambda(x)$. Thus the restriction of $[\![\cdot, \cdot]\!]$ to $\mathcal{K}$ actually equips it with a fibrewise bracket, not just a bracket of sections. We denote the bracket on $\mathcal{K}$ by $\{\cdot, \cdot\}$, to draw attention to this property.

It is clear that for each $x \in M$, $\{\cdot, \cdot\}_x$ is a Lie algebra bracket, so that $\mathcal{K}$ is a bundle of Lie algebras, but to show that that $\mathcal{K}$ is actually a Lie algebra bundle needs further work, because we need to see that different fibres are isomorphic as Lie algebras and that there are local trivializations which respect the Lie algebra structures.

It is in fact the case that the kernel $\mathcal{K}$ of a transitive Lie algebroid $\mathcal{A}$ is a Lie algebra bundle under the restriction of the bracket on sections of $\mathcal{A}$ [30, Theorem 6.5.1] and we shall use this in the sequel; the general proof is quite intricate, and we don't give it here because this property will actually be assumed as part of the definition of an infinitesimal Cartan geometry—which is the case of most interest; but we do derive the result for another important special case later in this chapter, in Corollary 1.3.3.

1.3 The Lie Algebroid of a Lie Groupoid

There is a fundamental relationship between Lie groupoids and Lie algebroids which is similar to the relationship between Lie groups and Lie algebras. There are, however, significant differences in the conventions involved, and it important to be aware of these.

The Lie algebra $\mathfrak{g}$ of a Lie group G may be regarded as $T_{1_G}G$, the tangent space to G at the identity $1_G \in G$. This is, however, simply a vector space, and the Lie bracket of two elements $a, b \in T_{1_G}G \cong \mathfrak{g}$ is normally obtained by using the left-invariant vector fields a^{L}, b^{L} on G defined by $a^{\mathsf{L}}_g = l_{g*}(a)$ and similarly for b^{L}. Writing $\{\cdot, \cdot\}$ for the Lie bracket on $\mathfrak{g}$, the definition is then given as $\{a, b\} = [a^{\mathsf{L}}, b^{\mathsf{L}}]_{1_G}$.

It is also possible to use right-invariant vector fields a^{R}, b^{R} to define a bracket $\{\cdot, \cdot\}_{\mathsf{R}}$ on $T_{1_G}G$, and we shall call the Lie algebra $\mathfrak{g}_{\mathsf{R}}$ with this alternative bracket the *opposite* Lie algebra of G; indeed $\{a, b\}_{\mathsf{R}} = \{b, a\} = -\{a, b\}$. Often, though, we are concerned only with the vector space structure of $\mathfrak{g}$, and in those circumstances we omit the subscript.

For Lie algebroids there is again a choice of using left-invariant or right-invariant vector fields to define a bracket. There is also, though, a choice to be made of whether a Lie algebroid element, thought of as a tangent vector to the Lie groupoid, should be regarded as tangent to the source fibre or the target fibre. Together with the choice (from the remark on p. 2) of taking the source of a groupoid element on the right or on the left, there are now three binary choices to be made in adopting a convention. The choices are not, however, all independent. When 'spreading out' a tangent vector, regarded as the jet of a curve in a source or target fibre of the groupoid, we should act on the fixed end of the curve. For the groupoid of fibre morphisms which we introduce in Chap. 3, the fixed end is the source; and with the source on the right, it follows that we must use right translation on Lie groupoids to generate the Lie algebroid bracket. This, indeed, is the convention used in [30].

It is important, therefore, to be aware that our convention for Lie groupoids and Lie algebroids differs from our convention (the standard one) for Lie groups and Lie algebras, though it is worth mentioning that [30] uses the same convention (right invariance) for both. We shall draw attention to the consequences of this difference where appropriate.

We now explain how to obtain a Lie algebroid from a Lie groupoid.

Each Lie groupoid $\mathcal{G}$ defines a Lie algebroid $A\mathcal{G}$ by taking the normal bundle of 1_M in $\mathcal{G}$. If $x \in M$, the fibre of $A\mathcal{G}$ over x is the quotient space $T_{1_x}\mathcal{G}/1_*(T_xM)$, so that an element of $A\mathcal{G}$ is a coset $1_*(T_xM)+\boldsymbol{a}$ where $\boldsymbol{a} \in T_{1_x}\mathcal{G}$. We may choose a representative $\boldsymbol{a}$ of this coset satisfying $\alpha_*(\boldsymbol{a}) = 0$ because α is a surjective submersion; such a representative is unique because if $\boldsymbol{a}_0 \in T_xM$ then $\alpha_* 1_*(\boldsymbol{a}_0) = \boldsymbol{a}_0$, so that if $\alpha_*(\boldsymbol{a}) = 0$ where $\boldsymbol{a} \in 1_*(T_xM)$ then $\boldsymbol{a} = 0$. With this point of view, $A\mathcal{G} = \ker \alpha_*|_{1_M}$.

We note that each right-invariant vector field $\mathbf{X}$ on $\mathcal{G}$ tangent to the fibres of α determines a section $\mathbf{X}|_{1_M}$ of $A\mathcal{G} \to M$ by restriction to the diffeomorphic image $1_M \subset \mathcal{G}$. Conversely, each section ξ of $A\mathcal{G} \to M$ defines a right-invariant vector field ξ^{R} on $\mathcal{G}$ by $\xi^{\mathsf{R}}_\varphi = r_{\varphi*}(\xi^{\mathsf{R}}_{1_{\beta(\varphi)}}) = r_{\varphi*}\xi\beta(\varphi)$; as $\alpha_*(\xi^{\mathsf{R}}_{1_{\beta(\varphi)}}) = 0$ we may choose a curve c

in the fibre $\alpha^{-1}(1_{\beta(\varphi)})$ such that $\dot{c}(0) = \xi^{\mathsf{R}}_{1_{\beta(\varphi)}}$, so that $\alpha \circ r_\varphi \circ c$ is a constant curve at $\alpha(\varphi) \in M$, showing that

$$\alpha_*(\xi^{\mathsf{R}}_\varphi) = \alpha_* r_{\varphi *}(\xi^{\mathsf{R}}_{1_{\beta(\varphi)}}) = (\alpha \circ r_\varphi)_*(\xi^{\mathsf{R}}_{1_{\beta(\varphi)}}) = (\alpha \circ r_\varphi)_* \dot{c}(t) = 0.$$

If the right-invariant vector field $\mathbf{X}$ is defined only on some open subset $U \subset \mathcal{G}$ then we have to use a slightly more complicated approach which reduces to the previous case when $U = \mathcal{G}$. For any $\varphi \in U$ we define the value of the corresponding section at $\alpha(\varphi)$ by translating the vector $\mathbf{X}_\varphi$ to the identity element $1_{\alpha(\varphi)}$, in other words by taking $r_{\varphi^{-1}*}(\mathbf{X}_\varphi)$; if $\varphi_1, \varphi_2 \in U$ satisfy $\alpha(\varphi_1) = \alpha(\varphi_2)$ then $r_{\varphi_1^{-1}*}(\mathbf{X}_{\varphi_1}) = r_{\varphi_2^{-1}*}(\mathbf{X}_{\varphi_2})$ by right invariance.

We now use the Lie bracket of right-invariant vector fields (which is itself right-invariant) to define a bracket of sections by setting

$$[\![\xi, \eta]\!]^{\mathsf{R}} = [\xi^{\mathsf{R}}, \eta^{\mathsf{R}}];$$

as both ξ^{R} and η^{R} are tangent to the fibres of α it follows that $[\xi^{\mathsf{R}}, \eta^{\mathsf{R}}]$ is also tangent to the fibres of α. We also define the anchor $\rho : A\mathcal{G} \to TM$ by setting $\rho = \beta_*$. The required properties of both the bracket and the anchor follow from the properties of vector fields on $\mathcal{G}$. We call the Lie algebroid constructed from $\mathcal{G}$ in this way the *Lie algebroid of the Lie groupoid* $\mathcal{G}$.

There are several standard examples of Lie algebroids arising from Lie groupoids.

- When a Lie group G is regarded as a Lie groupoid, the Lie algebroid is the vector bundle over a point $\mathfrak{g}_{\mathsf{R}} \to \{0_{\mathfrak{g}_{\mathsf{R}}}\}$, where $\mathfrak{g}_{\mathsf{R}}$ is the opposite Lie algebra of G. A section of this Lie algebroid is just an element of $\mathfrak{g}_{\mathsf{R}}$, and the bracket of sections comes from *right*-invariant vector fields on G. This is the prime example of a phenomenon we shall meet on several occasions, where a construction relating a Lie groupoid $\mathcal{G}$ to a Lie group G gives rise to an infinitesimal construction relating the Lie algebroid $A\mathcal{G}$ to the opposite Lie algebra $\mathfrak{g}_{\mathsf{R}}$.
- The pair groupoid $M \times M$ has a Lie algebroid which may be identified with the tangent bundle algebroid $TM \to M$.
- The *Atiyah algebroid* of a principal G-bundle $P \to M$ is the quotient of TP by the derivative action of G with bracket induced by the bracket of vector fields on P, and is the Lie algebroid of the gauge groupoid.
- Given a left action of the Lie group G on the manifold M, the Lie algebroid of the action groupoid $G \times M$ is the action algebroid $M \times \mathfrak{g}_{\mathsf{R}} \to M$ of the corresponding infinitesimal action of the opposite Lie algebra $\mathfrak{g}_{\mathsf{R}}$ on M.

Proposition 1.3.1 *The Lie algebroid of a locally trivial Lie groupoid is transitive.*

Proof If $\mathcal{G}$ is locally trivial, so that $(\alpha, \beta) : \mathcal{G} \to M \times M$ is a surjective submersion, then $(\alpha_*, \beta_*) : T\mathcal{G} \to TM \times TM$ will be surjective. Restricting to the submanifold $\ker \alpha_* \subset T\mathcal{G}$ gives a surjective map $(0, \beta_*) : \ker \alpha_* \to 0_M \times TM$, and then further restricting to $A\mathcal{G} = \ker \alpha_*|_{1_M}$ gives a surjective map $\rho = \beta_* : A\mathcal{G} \to 0_M \times_M TM \cong TM$. □

The Lie algebroid of the trivial Lie groupoid $M \times G \times M$ will be of particular interest, and may be identified with a trivial Lie algebroid.

Proposition 1.3.2 *The Lie algebroid of the trivial Lie groupoid $\mathcal{G} = M \times G \times M$ may be identified with the vector bundle $TM \oplus_M (M \times \mathfrak{g}_{\mathrm{R}})$ over M, with anchor map given by $\rho(X_x, \xi(x)) = X_x$ and bracket*

$$[\![(X, \xi), (Y, \eta)]\!] = \big([X, Y], (\mathrm{id}, X(\bar{\eta}) - Y(\bar{\xi}) + \{\bar{\xi}, \bar{\eta}\}_{\mathrm{R}})\big)$$

where

- *$\mathfrak{g}$ is the Lie algebra of G and $\mathfrak{g}_{\mathrm{R}}$ is its opposite,*
- *X, Y are vector fields on M,*
- *ξ, η are sections of the trivial Lie algebra bundle $M \times \mathfrak{g}_{\mathrm{R}} \to M$,*
- *$\bar{\xi}$, $\bar{\eta}$ are the corresponding maps $M \to \mathfrak{g}_{\mathrm{R}}$,*
- *$\{\bar{\xi}, \bar{\eta}\}_{\mathrm{R}}$ denotes the pointwise Lie bracket on $\mathfrak{g}_{\mathrm{R}}$, and*
- *$X(\bar{\eta})$ denotes the Lie derivative of $\bar{\eta}$ as an algebra-valued function, and similarly for $Y(\bar{\xi})$.*

Proof Let $e \in G$ be the identity of the group. For each $x \in M$ the tangent space $T_{1_x}\mathcal{G}$ at the identity element $1_x = (x, e, x)$ may be written as $T_xM \oplus \mathfrak{g}_{\mathrm{R}} \oplus T_xM$ (noting that $\mathfrak{g}_{\mathrm{R}} = \mathfrak{g}$ as vector spaces), and if $\boldsymbol{a} = (v, a, \tilde{v}) \in T_{1_x}\mathcal{G}$ satisfies $\alpha_*(\boldsymbol{a}) = 0$ then $\tilde{v} = 0$. Thus $A_x\mathcal{G}$ may be identified with $T_xM \oplus \mathfrak{g}_{\mathrm{R}}$, and $A\mathcal{G}$ may be identified with the vector bundle $TM \oplus_M (M \times \mathfrak{g}_{\mathrm{R}}) \to M$.

To establish the bracket formula we consider vector fields on $\mathcal{G}$. If $\boldsymbol{X} \in \mathfrak{X}(\mathcal{G})$ we may put $\boldsymbol{X} = (X, \Xi, \tilde{X})$ where $\tilde{X}$ is a vector field along α, X is a vector field along β and Ξ is a vector field along the projection $\mathcal{G} \to G$; requiring $\alpha_* \circ \boldsymbol{X} = 0$ implies that $\tilde{X} = 0$.

Now suppose that $\boldsymbol{X}$ is right invariant (thus corresponding to a section of $A\mathcal{G}$), so that

$$\boldsymbol{X}_{(x,g,\tilde{x})} = \boldsymbol{r}_* \boldsymbol{X}_{(x,e,x)} \in T_xM \oplus T_gG \oplus T_{\tilde{x}}M,$$

where $\boldsymbol{r} : \alpha^{-1}(x) \to \alpha^{-1}(\tilde{x})$ is the groupoid right translation map $\boldsymbol{r}_{(x,g,\tilde{x})}$ in $\mathcal{G}$. Let c be a curve in $\alpha^{-1}(x) = M \times G \times \{x\}$ satisfying $\dot{c}(0) = \boldsymbol{X}_{(x,e,x)}$ so that, writing $c(t) = (c_1(t), c_2(t), x)$, we have $\dot{c}_1(0) = X_{(x,e,x)} \in T_xM$ and $\dot{c}_2(0) = \Xi_{(x,e,x)} \in \mathfrak{g}_{\mathrm{R}}$. Then

$$\boldsymbol{r}\big(c(t)\big) = c(t) \cdot (x, g, \tilde{x}) = \big(c_1(t), c_2(t)g, \tilde{x}\big) = \big(c_1(t), (r_g \circ c_2)(t), \tilde{x}\big)$$

where r_g is the group right translation map in G, so that

$$\begin{aligned}(X_{(x,g,\tilde{x})}, \Xi_{(x,g,\tilde{x})}, 0_{\tilde{x}}) &= \boldsymbol{r}_*(X_{(x,e,x)}, \Xi_{(x,e,x)}, 0_x) \\ &= \boldsymbol{r}_*\big(\dot{c}(0)\big) \\ &= (\boldsymbol{r} \circ c)^{\cdot}(0)\end{aligned}$$

$$= \big(\dot{c}_1(0), r_{g*}\dot{c}_2(0), 0_{\tilde{x}}\big)$$
$$= (X_{(x,e,x)}, r_{g*}\Xi_{(x,e,x)}, 0_{\tilde{x}}).$$

Consequently $X_{(x,g,\tilde{x})} = X_{(x,e,x)} \in T_xM$, so that we may define a vector field $\bar{X}$ on M satisfying $X = \bar{X} \circ \beta$ by putting $\bar{X}_x = X_{(x,e,x)}$. Similarly $\Xi_{(x,g,\tilde{x})} = r_{g*}\Xi_{(x,e,x)} \in T_gG$, so that Ξ may be considered as a right-invariant vector field on G depending on a parameter in M; we may therefore define a section ξ of $M \times \mathfrak{g}_{\mathsf{R}} \to M$ by $\xi(x) = \Xi_{(x,e,x)}$, so that the vector field $\boldsymbol{X} = (X, \Xi, 0)$ on $\mathcal{G}$ corresponds to the section $(\bar{X}, \xi)$ of $A\mathcal{G}$.

We now describe the vector field bracket $[\boldsymbol{X}, \mathbf{Y}] = [(X, \Xi, 0), (Y, H, 0)]$ using flows, and hence obtain the Lie algebroid bracket $[\![(X, \xi), (Y, \eta)]\!]$.

- For each $x \in M$ define the right-invariant vector field Ξ_x on G by $\Xi_x|_g = r_{g*}\Xi_{(x,e,x)}$, so that the vector field $(0, \Xi, 0)$ on $\mathcal{G}$ has a global flow $\big(x, \exp(t\,\Xi_x)g, \tilde{x}\big)$. Similarly $(0, H, 0)$ has global flow $(x, \exp(tH_x)g, \tilde{x})$, so that $[(0, \Xi, 0), (0, H, 0)] = (0, [\Xi, H], 0)$. It follows that

$$[\![(0, \xi), (0, \eta)]\!] = \big(0, (\mathrm{id}, \{\bar{\xi}, \bar{\eta}\}_{\mathsf{R}})\big) = \big(0, (\mathrm{id}, \{\bar{\xi}, \bar{\eta}\}_{\mathsf{R}})\big).$$

  Note that, as the vector fields Ξ_x and H_x on G are right invariant, their bracket defines the opposite bracket $\{\cdot, \cdot\}_{\mathsf{R}}$ on the $\mathfrak{g}_{\mathsf{R}}$-valued maps $\bar{\xi}, \bar{\eta}$.
- If ϕ_t is the local flow of $\bar{X}$ then $(\phi_t(x), g, \tilde{x})$ is the local flow of $(X, 0, 0)$. Similarly if ψ_t is the local flow of $\bar{Y}$ then $(\psi_t(x), g, \tilde{x})$ is the local flow of $(Y, 0, 0)$, so that

$$[\![(X, 0), (Y, 0)]\!] = \big([X, Y], (\mathrm{id}, 0)\big).$$

- Finally we see that $[(X, 0, 0), (0, H, 0)] = \big(0, (X(\bar{\eta}))^{\mathsf{R}}, 0\big)$ where $(X(\bar{\eta}))^{\mathsf{R}}$ is the right-invariant vector field on G (parametrized by M) generated by the $\mathfrak{g}_{\mathsf{R}}$-valued function $X(\bar{\eta})$. Similarly $[(0, \Xi, 0), (Y, 0, 0)] = \big(0, -(Y(\bar{\xi}))^{\mathsf{R}}, 0\big)$, so that

$$[\![(X, 0), (0, \eta)]\!] = \big(0, (\mathrm{id}, X(\bar{\eta}))\big), \qquad [\![(0, \xi), (Y, 0)]\!] = \big(0, (\mathrm{id}, -Y(\bar{\xi}))\big).$$

The bracket formula now follows immediately. □

Corollary 1.3.3 *If $\mathcal{G}$ is a trivial Lie groupoid then the kernel $K\mathcal{G}$ of its Lie algebroid is a Lie algebra bundle. More generally, if $\mathcal{G}$ is a locally trivial Lie groupoid then the kernel of its Lie algebroid is a Lie algebra bundle.*

Proof The first assertion follows immediately from the restricted bracket formula

$$[\![(0, \xi), (0, \eta)]\!] = \big(0, (\mathrm{id}, \{\bar{\xi}, \bar{\eta}\}_{\mathsf{R}})\big).$$

The second follows by adapting the proof of the proposition to the restriction $\boldsymbol{X}|_{(U\times G\times U)}$ to a local trivialization using Lemma 1.1.1. □

Corollary 1.3.4 *If there is a left action l_g of G on M, and if the action groupoid $\mathcal{H} = G \times M$ is regarded as a Lie subgroupoid of the trivial Lie groupoid $\mathcal{G} = M \times G \times M$, then the Lie algebroid $A\mathcal{H}$ may be identified with a Lie subalgebroid of $A\mathcal{G} = TM \oplus_M (M \times \mathfrak{g}_\mathsf{R})$ isomorphic to the action algebroid $M \times \mathfrak{g}_\mathsf{R} \to M$.*

Proof We first need to be clear about the definition of the Lie algebra action arising from the left Lie group action $l_g : M \to M$.

Given an element $a \in \mathfrak{g}$, we define the tangent vector $a^\sharp_x \in T_x M$ by

$$a^\sharp_x = \frac{d}{dt}(l_{\exp ta}(x))\Big|_{t=0}$$

and thus obtain a vector field $a^\sharp \in \mathfrak{X}(M)$, the *fundamental vector field* corresponding to a. It is well known that the map $a \mapsto a^\sharp$ is an antihomomorphism $\mathfrak{g} \to \mathfrak{X}(M)$, and so does not define a Lie algebra action of $\mathfrak{g}$ on M. The same formula may, however, be used to define a map $\mathfrak{g}_\mathsf{R} \to \mathfrak{X}(M)$ using the opposite of the bracket on $\mathfrak{g}$, and this second map is indeed a Lie algebra homomorphism. We therefore regard the map $a \mapsto a^\sharp$ as defining a Lie algebra action of the opposite algebra $\mathfrak{g}_\mathsf{R}$ on M.

Now let $i : \mathcal{H} \to \mathcal{G}$ be the inclusion map given by $i(g, x) = (l_g(x), g, x)$; we find that a tangent vector $(w, a, 0) \in A_x\mathcal{G} \subset T_{1_x}\mathcal{G}$ is contained in $i_*(A_x\mathcal{H})$ exactly when $w = a^\sharp_x$. It follows that we may identify $A_x\mathcal{H}$ with $\mathfrak{g}_\mathsf{R}$, and that the bracket

$$[\![(X, \xi), (Y, \eta)]\!] = \big([X, Y], (\mathrm{id}, X(\bar{\eta}) - Y(\bar{\xi}) + \{\bar{\xi}, \bar{\eta}\}_\mathsf{R})\big)$$

on $A\mathcal{G}$ restricts to the action algebroid bracket

$$[\![\xi, \eta]\!](x) = \big(x, \bar{\xi}^\sharp_x(\bar{\eta}) - \bar{\eta}^\sharp_x(\bar{\xi}) + \{\bar{\xi}(x), \bar{\eta}(x)\}_\mathsf{R}\big)$$

on sections ξ, η of $M \times \mathfrak{g}_\mathsf{R} \to M$. □

The bracket formula given in Proposition 1.3.2 may be familiar; it is the bracket in the semidirect sum of two Lie algebras, in this case the Lie algebras of sections of the Lie algebroid TM and of the trivial Lie algebra bundle $M \times \mathfrak{g}_\mathsf{R}$. We shall have more to say about semidirect sums in Sect. 2.6.

1.4 Projections on Lie Groupoids and Lie Algebroids

We mentioned in Sect. 1.2 that the general definition of a morphism for Lie algebroids involved some difficulties. We shall avoid these difficulties by restricting attention where necessary to a particular class of Lie algebroid morphisms, namely those satisfying a property which may be described more generally in terms of fibred manifolds (and so applies also to morphisms of Lie groupoids). Given this property, we can then say when objects defined on Lie groupoids or Lie algebroids are 'projectable'.

Let $\pi_E : E \to M$ and $\pi_F : F \to N$ be two fibred manifolds (so that both π_E and π_F are surjective submersions) and let (P, p) be a fibred map between them, so that $P : E \to F$ and $p : M \to N$ with $\pi_F \circ P = p \circ \pi_E$. We shall say that the fibred map is *full* if p is a surjective submersion and P is a *fibrewise surjective submersion*: that is to say, that for each $x \in M$ the restriction of P to a map $\pi_E^{-1}(x) \to \pi_F^{-1}(p(x))$ is a surjective submersion.

For Lie groupoids we shall say that a morphism is a *full Lie groupoid morphism* if it is full as a fibred map between the fibred manifolds $\alpha_{\mathcal{G}} : \mathcal{G} \to M$ and $\alpha_{\mathcal{H}} : \mathcal{H} \to N$ given by their source projections.

Given two Lie groupoids and a full morphism, we may consider when a bisection is projectable. If $(\mathcal{P}, p)$ is a full morphism, where $\mathcal{P} : \mathcal{G} \to \mathcal{H}$ and $p : M \to N$, then $\Phi \in \mathsf{B}(\mathcal{G})$ is said to be *projectable* if, whenever $x_1, x_2 \in M$ satisfy $p(x_1) = p(x_2)$, then

$$\mathcal{P}\Phi(x_1) = \mathcal{P}\Phi(x_2), \qquad p\phi^{-1}(x_1) = p\phi^{-1}(x_2).$$

Note that if Φ is projectable then $p\phi(x_1) = p\phi(x_2)$ automatically, because

$$p\phi(x_1) = p\beta_{\mathcal{G}}\Phi(x_1) = \beta_{\mathcal{H}}\mathcal{P}\Phi(x_1) = \beta_{\mathcal{H}}\mathcal{P}\Phi(x_2) = p\beta_{\mathcal{G}}\Phi(x_2) = p\phi(x_2).$$

We now show that a projectable bisection Φ of $\mathcal{G}$ does indeed project to a bijection of $\mathcal{H}$. Define the map $\Phi^{\mathcal{P}} : N \to \mathcal{H}$ by, for $y \in N$, choosing $x \in G$ with $p(x) = y$ and setting $\Phi^{\mathcal{P}}(y) = \mathcal{P}\Phi(x)$; by projectability this does not depend on the choice of x.

Proposition 1.4.1 *If Φ is a projectable bisection of $\mathcal{G}$ then $\Phi^{\mathcal{P}}$ is a bisection of $\mathcal{H}$. If $\widehat{\mathcal{P}}$ denotes the map $\mathsf{B}(\mathcal{G}) \to \mathsf{B}(\mathcal{H})$ given by $\widehat{\mathcal{P}}(\Phi) = \Phi^{\mathcal{P}}$ then $\widehat{\mathcal{P}}$ is a group homomorphism.*

Proof Let Φ be projectable. Smoothness of $\Phi^{\mathcal{P}}$ at any $y \in N$ may be shown by choosing a neighbourhood U of y and a local section $\varsigma : U \to M$, so that $\Phi^{\mathcal{P}}\big|_U = \mathcal{P} \circ \Phi \circ \varsigma$. Then, on U,

$$\alpha_{\mathcal{H}} \circ \Phi^{\mathcal{P}} = \alpha_{\mathcal{H}} \circ \mathcal{P} \circ \Phi \circ \varsigma = p \circ \alpha_{\mathcal{G}} \circ \Phi \circ \varsigma = p \circ \varsigma = \mathrm{id}_U$$

so that, on N,

$$\alpha_{\mathcal{H}} \circ \Phi^{\mathcal{P}} = \mathrm{id}_N .$$

Given $\Phi^{\mathcal{P}}$, define the smooth map $\phi^{\mathcal{P}} : N \to N$ by $\phi^{\mathcal{P}} = \beta_{\mathcal{H}} \circ \Phi^{\mathcal{P}}$. In addition, define a map $\xi : N \to N$ by, for $y \in N$, choosing $x \in M$ with $p(x) = y$ and setting $\xi(y) = p\phi^{-1}(x)$; by projectability this does not depend on the choice of x. Smoothness of ξ at any $y \in N$ may be shown by choosing a neighbourhood U of y and a local section $\varsigma : U \to M$, so that $\xi|_U = p \circ \phi^{-1} \circ \varsigma$. Observe that, for any $y \in N$ and any $x \in M$ with $p(x) = y$,

$$\xi\phi^{\mathcal{P}}(y) = \xi\beta_{\mathcal{H}}\Phi^{\mathcal{P}}(y) = \xi\beta_{\mathcal{H}}\mathcal{P}\Phi(x) = \xi p\beta_{\mathcal{G}}\Phi(x) = \xi p\phi(x) = p\phi^{-1}\phi(x) = p(x) = y$$

and that

$$\begin{aligned}\phi^{\mathcal{P}}\xi(y) &= \phi^{\mathcal{P}}p\phi^{-1}(x) = \beta_{\mathcal{H}}\Phi^{\mathcal{P}}p\phi^{-1}(x) = \beta_{\mathcal{H}}\mathcal{P}\Phi\phi^{-1}(x)\\ &= p\beta_{\mathcal{G}}\Phi\phi^{-1}(x) = p\phi\phi^{-1}(x) = p(x) = y\end{aligned}$$

so that $\phi^{\mathcal{P}}$ is a diffeomorphism of N; thus $\Phi^{\mathcal{P}}$ is a bisection of $\mathcal{H}$.

If Φ_1, Φ_2 are both projectable then, taking $x_1, x_2 \in M$ with $p(x_1) = p(x_2)$, we have $p\phi_2(x_1)) = p\phi_2(x_2))$ and so

$$\begin{aligned}\mathcal{P}\big(\Phi_1 * \Phi_2(x_1)\big) &= \mathcal{P}\big(\Phi_1\phi_2(x_1)\cdot\Phi_2(x_1)\big) = \mathcal{P}\Phi_1\phi_2(x_1)\cdot\mathcal{P}\Phi_2(x_1)\\ &= \mathcal{P}\Phi_1\phi_2(x_2)\cdot\mathcal{P}\Phi_2(x_2) = \mathcal{P}\big(\Phi_1\phi_2(x_2)\cdot\Phi_2(x_2)\big) = \mathcal{P}\big(\Phi_1 * \Phi_2(x_2)\big);\end{aligned}$$

in addition we have $p(\phi_1^{-1}(x_1)) = p(\phi_1^{-1}(x_2))$ and so

$$p\big(\phi_1\circ\phi_2)^{-1}(x_1)\big) = p\phi_2^{-1}\phi_1^{-1}(x_1) = p\phi_2^{-1}\phi_1^{-1}(x_2) = p\big(\phi_1\circ\phi_2)^{-1}(x_2)\big);$$

thus the product bisection $\Phi_1 * \Phi_2$ is again projectable.

Finally let $y \in N$ and take $x \in M$ with $p(x) = y$; then

$$\begin{aligned}\big(\widehat{\mathcal{P}}(\Phi_1) * \widehat{\mathcal{P}}(\Phi_2)\big)(y) &= \widehat{\mathcal{P}}(\Phi_1)\big(\beta_{\mathcal{H}}\widehat{\mathcal{P}}(\Phi_2)(y)\big)\cdot\widehat{\mathcal{P}}(\Phi_2)(y)\\ &= \widehat{\mathcal{P}}(\Phi_1)\big(\beta_{\mathcal{H}}\mathcal{P}\Phi_2(x)\big)\cdot\mathcal{P}\Phi_2(x)\\ &= \widehat{\mathcal{P}}(\Phi_1)\big(p\beta_{\mathcal{G}}\Phi_2(x)\big)\cdot\mathcal{P}\Phi_2(x)\\ &= \widehat{\mathcal{P}}(\Phi_1)\big(p\phi_2(x)\big)\cdot\mathcal{P}\Phi_2(x)\\ &= \mathcal{P}\Phi_1\phi_2(x)\cdot\mathcal{P}\Phi_2(x)\\ &= \mathcal{P}\big(\Phi_1\phi_2(x)\cdot\Phi_2(x)\big)\\ &= \mathcal{P}\big((\Phi_1 * \Phi_2)(x)\big)\\ &= \big(\widehat{\mathcal{P}}(\Phi_1 * \Phi_2)\big)(y)\end{aligned}$$

so that $\widehat{\mathcal{P}}(\Phi_1) * \widehat{\mathcal{P}}(\Phi_2) = \widehat{\mathcal{P}}(\Phi_1 * \Phi_2)$, showing that the map $\widehat{\mathcal{P}} : \mathsf{B}(\mathcal{G}) \to \mathsf{B}(\mathcal{H})$ is a group homomorphism. □

A similar result holds for local bisections where the product operations are defined.

We now turn to Lie algebroids. In order to define a full morphism between $\tau_{\mathcal{A}} : \mathcal{A} \to M$ and $\tau_{\mathcal{B}} : \mathcal{B} \to N$, we start with a vector bundle morphism (P, p) which is full as a map of fibred manifolds. We observe that, as P is linear on each fibre of its domain, it is sufficient to check that p is a surjective submersion and that P is fibrewise surjective, as P will then automatically be a fibrewise submersion.

Given a full morphism, we say that a section ξ of $\mathcal{A}$ is *projectable* if, whenever $x_1, x_2 \in M$ with $p(x_1) = p(x_2)$, we have $P\big(\xi(x_1)\big) = P\big(\xi(x_2)\big)$. If ξ is projectable, we then define its projection $\xi^P : N \to \mathcal{B}$ by, for $y \in N$, choosing $x \in M$ with $p(x) = y$

and setting $\xi^P(y) = P(\xi(x))$; as ξ is projectable this does not depend on the choice of x. As

$$\tau_{\mathcal{B}}(\xi^P(y)) = \tau_{\mathcal{B}}(P(\xi(x))) = p(\tau_{\mathcal{A}}(\xi(x))) = p(x) = y$$

we see that ξ^P is indeed a section of $\mathcal{B}$. Smoothness at any $y \in N$ may be shown by choosing a neighbourhood U of y and a local section $\varsigma : U \to M$ of p, so that $\xi^P|_U = P \circ \xi \circ \varsigma$. As an example, a vector field X on M (regarded as a fibred manifold over N) is projectable when regarded as a section of the tangent bundle Lie algebroid, exactly when it is projectable in the usual sense; for simplicity we write its projection as X^p rather than X^{p_*}. As another example, if $M = N$ and $p = \mathrm{id}_M$ then any section ξ of $\mathcal{A}$ is projectable with projection $\xi^P = P \circ \xi$.

We now say that the full vector bundle morphism (P, p) is a *full Lie algebroid morphism* if $\rho_{\mathcal{B}} \circ P = p_* \circ \rho_{\mathcal{A}}$ and if, whenever ξ and η are projectable sections of $\mathcal{A}$, the section $[\![\xi, \eta]\!]$ is also projectable and satisfies $[\![\xi, \eta]\!]^P = [\![\xi^P, \eta^P]\!]$.

(It is, in fact, possible to give a definition of Lie algebroid morphism which reduces to ours if the underlying vector bundle morphism is full, but which can be used without any surjectivity conditions. In its most general form this definition is quite complex [30, Definition 4.3.1] and we shall not need it.)

Lemma 1.4.2 *If ξ is a projectable section of $\mathcal{A}$ then $\rho_{\mathcal{A}} \circ \xi$ is a projectable vector field on M; in addition $(\rho_{\mathcal{A}} \circ \xi)^p = \rho_{\mathcal{B}} \circ \xi^P$.*

Proof If $x_1, x_2 \in M$ satisfy $p(x_1) = p(x_2)$ then

$$p_*\big((\rho_{\mathcal{A}} \circ \xi)_{x_1}\big) = p_*\rho_{\mathcal{A}}\xi(x_1) = \rho_{\mathcal{B}}P\xi(x_1) = \rho_{\mathcal{B}}P\xi(x_2) = p_*\rho_{\mathcal{A}}\xi(x_2) = p_*\big((\rho_{\mathcal{A}} \circ \xi)_{x_2}\big)$$

so that $\rho_{\mathcal{A}} \circ \xi$ is projectable. If $y \in N$, and if $x \in M$ satisfies $p(x) = y$, then

$$(\rho_{\mathcal{A}} \circ \xi)^p_y = p_*(\rho_{\mathcal{A}} \circ \xi)_x = p_*\rho_{\mathcal{A}}\xi(x) = \rho_{\mathcal{B}}P\xi(x) = \rho_{\mathcal{B}}\xi^P(y)$$

so that $(\rho_{\mathcal{A}} \circ \xi)^p = \rho_{\mathcal{B}} \circ \xi^P$. □

As one might expect, transitive Lie algebroids behave well with respect to full morphisms.

Proposition 1.4.3 *Let $\mathcal{A} \to M$ and $\mathcal{B} \to N$ be Lie algebroids, and let (P, p) be a full Lie algebroid morphism from $\mathcal{A}$ to $\mathcal{B}$. If $\mathcal{A}$ is transitive then so is $\mathcal{B}$.*

Proof If $\mathcal{A}$ is transitive then $\rho_{\mathcal{A}}$ is surjective, so it is fibrewise surjective as it projects to the identity on M. But p is a submersion as (P, p) is full, so that p_* is fibrewise surjective, and therefore the composite $p_* \circ \rho_{\mathcal{A}} = \rho_{\mathcal{B}} \circ P$ is also fibrewise surjective. But P is fibrewise surjective as (P, p) is full, so that $\rho_{\mathcal{B}}$ is fibrewise surjective, showing that $\mathcal{B}$ is transitive. □

Full morphisms of Lie groupoids give rise to full morphisms of their Lie algebroids.

Lemma 1.4.4 *Let $\mathcal{G}$ and $\mathcal{H}$ be Lie groupoids, and let $(\mathcal{P}, p)$ be a full morphism with $\mathcal{P} : \mathcal{G} \to \mathcal{H}$ and $p : M \to N$. Then the restriction of $\mathcal{P}_*$ to $A\mathcal{G} \subset T\mathcal{G}$ takes its values in $A\mathcal{H} \subset T\mathcal{H}$.*

Proof Take $x \in M$; then

$$\mathcal{P}(1_x) = \mathcal{P}(1_x \cdot 1_x) = \mathcal{P}(1_x) \cdot \mathcal{P}(1_x)$$

so that $\mathcal{P}(1_x) = 1_{p(x)}$. Also, if $v \in A_x\mathcal{G}$ then

$$\alpha_{\mathcal{H}*}\mathcal{P}_*(v) = \mathcal{P}_*\alpha_{\mathcal{G}*}(v) = 0$$

so that $\mathcal{P}_*(v) \in A_{p(x)}\mathcal{H}$. □

Proposition 1.4.5 *The pair $(\mathcal{P}_*, p)$ is a full Lie algebroid morphism from $A\mathcal{G}$ to $A\mathcal{H}$.*

Proof We are given that p is a surjective submersion, so we have to show that $\mathcal{P}_*$ restricted to $A\mathcal{G}$ is fibrewise surjective; we shall see that this follows from the fact that the Lie groupoid morphism $(\mathcal{P}, p)$ is full, so that $\mathcal{P}$ is a fibrewise surjective submersion with respect to $\alpha_{\mathcal{G}}$ and $\alpha_{\mathcal{H}}$.

Take $y \in N$ and $\boldsymbol{b} \in A_y\mathcal{H}$, and let $x \in M$ satisfy $p(x) = y$. Then $\boldsymbol{b} \in T_{1_y}\mathcal{H}$ with $\alpha_{\mathcal{H}*}(\boldsymbol{b}) = 0$, so that there is an element $\boldsymbol{a} \in T_{1_x}\mathcal{G}$ with $\mathcal{P}_*(\boldsymbol{a}) = \boldsymbol{b}$ and $\alpha_{\mathcal{G}*}(\boldsymbol{a}) = 0$; thus $\boldsymbol{a} \in A_x\mathcal{G}$. □

1.5 Pullbacks of Lie Groupoids and Lie Algebroids

If $\pi : E \to N$ is a fibre bundle and if $p : M \to N$ is a surjective submersion then the pullback of π by p is a fibre bundle $p^*\pi : p^*E \to M$ where p^*E is the fibre product manifold $M \times_N E$ and $p^*\pi$ is projection on the first factor. We may perform similar constructions with locally trivial Lie groupoids and transitive Lie algebroids, and again the appropriate transversality conditions hold so that the total spaces are submanifolds of Cartesian products. We shall see that, in each case, the canonical map from the pullback object to the original object is a full morphism as we have defined it earlier.

So let $\mathcal{H}$ be a locally trivial Lie groupoid with a manifold of identities N and source and target maps $\alpha, \beta : \mathcal{H} \to N$, and let $p : M \to N$ be a surjective submersion. The *pullback groupoid* is the double fibre product manifold

$$p^{**}\mathcal{H} = \{(x, \varphi, \tilde{x}) \in M \times \mathcal{H} \times M : p(\tilde{x}) = \alpha(\varphi),\ p(x) = \beta(\varphi)\}$$

with manifold of identities M. We may define a generalization of the trivial groupoid structure on $p^{**}\mathcal{H}$ where

- the source and target projections are $\alpha^p(x, \varphi, \tilde{x}) = \tilde{x}$ and $\beta^p(x, \varphi, \tilde{x}) = x$;

- if $\alpha^p(x_1, \varphi_1, \tilde{x}_1) = \beta^p(x_2, \varphi_2, \tilde{x}_2)$ then $\tilde{x}_1 = x_2$, so that

$$\alpha(\varphi_1) = p(\tilde{x}_1) = p(x_2) = \beta(\varphi_2)$$

  and the product $(x_1, \varphi_1, \tilde{x}_1) \cdot (x_2, \varphi_2, \tilde{x}_2) = (x_1, \varphi_1 \cdot \varphi_2, \tilde{x}_2)$ is defined;
- the identities are $1_x = (x, 1_{p(x)}, x)$; and
- the inverses are $(x, \varphi, \tilde{x})^{-1} = (\tilde{x}, \varphi^{-1}, x)$.

(Indeed if $\mathcal{H}$ is just a Lie group H, and $N = \{e\}$ where $e \in H$ is the identity, then there is only one map $p : M \to N$ and $p^{**}H$ is just the trivial groupoid $M \times H \times M$.)

Proposition 1.5.1 *The groupoid $p^{**}\mathcal{H}$ is a locally trivial Lie groupoid.*

Proof By transversality $p^{**}\mathcal{H}$ is a manifold. The source and target maps α^p, β^p are surjective because α, β are surjective, and they are submersions as restrictions of the Cartesian product projections; thus $p^{**}\mathcal{H}^{(2)} \subset p^{**}\mathcal{H} \times p^{**}\mathcal{H}$ is a submanifold by transversality again. It is also clear that the identity map is an immersion and that the inverse map is a diffeomorphism. Thus $p^{**}\mathcal{H}$ is a Lie groupoid. The map $(\alpha^p, \beta^p) : p^{**}\mathcal{H} \to M \times M$ is surjective because (α, β) is surjective, and is a submersion as the restriction of a Cartesian product projection, so that $p^{**}\mathcal{H}$ is locally trivial. □

Proposition 1.5.2 *If the map $\mathcal{P} : p^{**}\mathcal{H} \to \mathcal{H}$ is defined by $\mathcal{P}(x, \varphi, \tilde{x}) = \varphi$ then $(\mathcal{P}, p)$ is a full Lie groupoid morphism.*

Proof We are given that p is a surjective submersion.

For each $\tilde{x} \in M$ the α^p-fibre of $p^{**}\mathcal{H}$ at $\tilde{x}$ is

$$\begin{aligned}(\alpha^p)^{-1}(\tilde{x}) &= \{(x, \varphi, \tilde{x}) \in M \times \mathcal{H} \times M : \alpha(\varphi) = p(\tilde{x}),\ p(x) = \beta(\varphi)\} \\ &= \{(x, \varphi) \in M \times \alpha^{-1}(\tilde{x}) : p(x) = \beta(\varphi)\};\end{aligned}$$

but for each $\varphi \in \alpha^{-1}\big(p(\tilde{x})\big)$ there is an element $x \in M$ such that $p(y) = \beta(\varphi)$ because p is surjective, so that $(x, \varphi, \tilde{x}) \in (\alpha^p)^{-1}(\tilde{x})$ with $\mathcal{P}(x, \varphi, \tilde{x}) = \varphi$, showing that $\mathcal{P}$ is fibrewise surjective. Also, if $w \in T_\varphi \alpha^{-1}\big(p(\tilde{x})\big)$ there is an element $v \in T_x M$ such that $p_*(v) = \beta_*(w)$ because p is a submersion, so that $(v, w, 0) \in T_{(x,\varphi,\tilde{x})}(\alpha^p)^{-1}(\tilde{x})$ with $\mathcal{P}_*(v, w, 0) = w$, showing that $\mathcal{P}$ is a fibrewise submersion. □

Now suppose that $\tau : \mathcal{B} \to N$ is a transitive Lie algebroid with anchor map $\rho : \mathcal{B} \to TN$, and as before let $p : M \to N$ be a surjective submersion, so that $p_* : TM \to TN$ is a vector bundle morphism. Retaining $p^*\mathcal{B}$ for the usual fibre product $M \times_N \mathcal{B}$, we shall write $p^{**}\mathcal{B}$ for the fibre product

$$p^{**}\mathcal{B} = TM \times_{TN} \mathcal{B} = \{(v, \boldsymbol{b}) \in TM \times \mathcal{B} : p_*(v) = \rho(\boldsymbol{b})\}$$

and let $\tau^p : p^{**}\mathcal{B} \to M$ be given by $\tau^p(v, \boldsymbol{b}) = \tau_M(v)$ where $\tau_M : TM \to M$ is the tangent bundle projection. If $\tau^p(v_1, \boldsymbol{b}_1) = \tau^p(v_2, \boldsymbol{b}_2)$ then $\tau_M(v_1) = \tau_M(v_2)$ and in addition

$$\tau(\boldsymbol{b}_1) = \tau_N \rho(\boldsymbol{b}_1) = \tau_N p_*(v_1) = p\tau_M(v_1) = p\tau_M(v_2) = \tau_N p_*(v_2) = \tau_N \rho(\boldsymbol{b}_2) = \tau(\boldsymbol{b}_2)$$

so that we may define a vector space structure on each fibre $p^{**}\mathcal{B}_x$ giving, in view of the conditions on p and ρ, a vector bundle structure on τ^p; the anchor map $\rho^p : p^{**}\mathcal{B} \to TM$ is given by $\rho^p(v, \boldsymbol{b}) = v$.

In order to define a pullback bracket, it is necessary to be clear about the meaning of a section of $p^{**}\mathcal{B} \to M$. Such a section will be a pair (X, Ξ) where X is a vector field on M and Ξ is a section of $\mathcal{B}$ along p, such that $p_* \circ X = \rho \circ \Xi$. As Ξ is a map $M \to \mathcal{B}$ satisfying $\tau \circ \Xi = p$, we may write it as a sum of terms $f(\xi \circ p)$ where ξ is a section of $\mathcal{B} \to N$ and where f is a function on M. We shall therefore define a bracket

$$[\![(X, f(\xi \circ p)), (Y, g(\eta \circ p))]\!] = \Big([X, Y], fg\big([\![\xi, \eta]\!] \circ p\big) + X(g)(\eta \circ p) - Y(f)(\xi \circ p)\Big).$$

Proposition 1.5.3 *The vector bundle $\tau^p : p^{**}\mathcal{B} \to M$ is a transitive Lie algebroid.*

Proof The anchor map ρ^p is clearly a vector bundle morphism; it is fibrewise surjective because ρ is fibrewise surjective. The bracket is clearly $\mathbb{R}$-bilinear and skew-symmetric; a straightforward calculation shows that it satisfies the Jacobi identity. If h is a function on M then

$$\begin{aligned}
[\![(X, f(\xi \circ p)), h(Y, g(\eta \circ p))]\!] &= [\![(X, f(\xi \circ p)), (hY, hg(\eta \circ p))]\!] \\
&= \Big([X, hY], fhg\big([\![\xi, \eta]\!] \circ p\big) + X(hg)(\eta \circ p) \\
&\quad - (hY)(q)(\xi \circ p)\Big) \\
&= \Big(h[X, Y] + X(h)Y, fhg\big([\![\xi, \eta]\!] \circ p\big) \\
&\quad + \big(hX(g) + gX(h)\big)(\eta \circ p) - (hY)(f)(\xi \circ p)\Big) \\
&= h\Big([X, Y], fg\big([\![\xi, \eta]\!] \circ p\big) + X(g)(\eta \circ p) - Y(f)(\xi \circ p)\Big) \\
&\quad + X(h)\Big(Y, g(\eta \circ p)\Big)
\end{aligned}$$

where $X = \rho^p \circ (X, f(\xi \circ p))$. □

Proposition 1.5.4 *If the map $P : p^{**}\mathcal{B} \to \mathcal{B}$ is defined by $P(v, \boldsymbol{b}) = \boldsymbol{b}$ then (P, p) is a full Lie algebroid morphism.*

Proof We are given that p is a surjective submersion. Also, for each $x \in M$ the fibre of $p^{**}\mathcal{B}$ at x is

$$(\tau^p)^{-1}(x) = \{(v, \boldsymbol{b}) \in TM \times \mathcal{B} : p_*(v) = \rho(\boldsymbol{b}), \tau_M(v) = x\},$$

so for each $\boldsymbol{b} \in \tau^{-1}\big(p(x)\big)$ there is an element $v \in T_xM$ such that $p_*(v) = \rho(\boldsymbol{b})$ because p is a surjective submersion, so that $(v, \boldsymbol{b}) \in (\tau^p)^{-1}(x)$ and $P(v, \boldsymbol{b}) = \boldsymbol{b}$; thus P is fibrewise surjective. □

The pullbacks of Lie groupoids and their Lie algebroids are, of course, related.

Proposition 1.5.5 *If $\tau : A\mathcal{H} \to N$ is the Lie algebroid of a locally trivial Lie groupoid $\mathcal{H}$ and if $p : M \to N$ is a surjective submersion then the Lie algebroid of the pullback Lie groupoid $p^{**}\mathcal{H}$ may be identified with the pullback Lie algebroid $p^{**}(A\mathcal{H})$.*

Proof An element of $A_x(p^{**}\mathcal{H})$ is a tangent vector

$$(v, \boldsymbol{b}, w) \in T_{(x,1_{p(x)},x)}(p^{**}\mathcal{H}) \subset T_{(x,1_{p(x)},x)}(M \times \mathcal{H} \times M) = T_xM \oplus T_{1_{p(x)}}\mathcal{H} \oplus T_xM$$

satisfying $\alpha^p_*(v, \boldsymbol{b}, w) = 0$. The condition of being tangent to $p^{**}\mathcal{H}$ at $(x, 1_{p(x)}, x)$ gives

$$p_*(w) = \alpha_*(\boldsymbol{b}), \qquad p_*(v) = \beta_*(\boldsymbol{b})$$

and then the condition $\alpha^p_*(v, \boldsymbol{b}, w) = 0$ gives $w = 0$ and hence $\alpha_*(\boldsymbol{b}) = 0$. Thus $\boldsymbol{b} \in A_{p(x)}\mathcal{H}$, and we may identify $(v, \boldsymbol{b}, 0)$ with $(v, \boldsymbol{b}) \in T_xM \oplus A_{p(x)}\mathcal{H}$. As $\rho(\boldsymbol{b}) = \beta_*(\boldsymbol{b}) = p_*(v)$ we see that $(v, \boldsymbol{b}) \in p^{**}(A\mathcal{H})_x$. This correspondence is evidently a bijection.

If $\big(X, f(\xi \circ p)\big)$ and $\big(Y, g(\eta \circ p)\big)$ are sections of $p^{**}(A\mathcal{H})$, where ξ, η are sections of $A\mathcal{H}$, then their bracket is given by

$$[\![(X, f(\xi \circ p)), (Y, g(\eta \circ p))]\!] = \Big([X, Y], fg\big([\![\xi, \eta]\!] \circ p\big) + X(g)(\eta \circ p) - Y(f)(\xi \circ p)\Big);$$

the proof that this matches the bracket of the corresponding sections of $A(p^{**}\mathcal{H})$ is a straightforward generalization of that given in the proof of Proposition 1.3.2 where the central term is now a Lie groupoid rather than a Lie group and we take a fibre product instead of an ordinary product. □

1.6 Lie Derivatives on Lie Algebroids

It is an important property of vector fields on any differentiable manifold that one can use them to define Lie derivatives of tensor fields over the manifold, and in particular that for vector fieids X, Y the Lie derivative $\mathcal{L}_X Y$ is equal to the Lie bracket $[X, Y]$ defined as the commutator of derivative operators. We use this last observation to motivate the definition of the 'Lie derivative' by a section of a Lie algebroid of various tensorial objects defined over the Lie algbroid.

We have defined the algebroid bracket on the Lie algebroid $A\mathcal{G}$ of a Lie groupoid $\mathcal{G}$ by

$$[\![\xi, \eta]\!]^{\mathsf{R}} = [\xi^{\mathsf{R}}, \eta^{\mathsf{R}}],$$

so it is reasonable to think of $[\![\xi, \cdot]\!]$ as a Lie derivative operator $\mathcal{L}_\xi$ acting on sections of the Lie algebroid. We now show how to extend such a Lie derivative operator to certain tensorial objects, beginning with a particular case which will be of fundamental importance later, namely vector bundle morphisms $TM \to A\mathcal{G}$ projecting to the identity on M. That is, we use the Lie algebroid bracket to construct, for each such vector bundle morphism $Q : TM \to A\mathcal{G}$, another vector bundle morphism $\mathcal{L}_\xi Q : TM \to A\mathcal{G}$ which it is again reasonable to regard as its Lie derivative, by setting

$$\mathcal{L}_\xi Q \circ Y = [\![\xi, Q \circ Y]\!] - Q \circ [\rho \circ \xi, Y]$$

for any vector field Y on M.

To justify the description of the latter operator as a Lie derivative, we shall define a $(1, 1)$ tensor field Q^{R} on $\mathcal{G}$ using the formula

$$Q^{\mathsf{R}}_\varphi(v) = r_{\varphi *} Q_{\beta(\varphi)} \beta_*(v)$$

for each $\varphi \in \mathcal{G}$ and each $v \in T_\varphi \mathcal{G}$. This formula makes sense because $\beta_*(v) \in T_{\beta(\varphi)} M$ so that

$$Q_{\beta(\varphi)} \beta_*(v) \in A_{\beta(\varphi)} \mathcal{G} \subset T_{1_{\beta(\varphi)}} \mathcal{G},$$

and therefore $Q^{\mathsf{R}}_\varphi(v) \in T_\varphi \mathcal{G}$ as required. In the particular case where $\boldsymbol{a} \in A_x \mathcal{G} \subset T_{1_x} \mathcal{G}$ we have

$$Q^{\mathsf{R}}_{1_x}(\boldsymbol{a}) = r_{1_x *} Q_{\beta(1_x)} \beta_*(\boldsymbol{a}) = Q_x \beta_*(\boldsymbol{a}) = Q_x \rho(\boldsymbol{a}).$$

Now let Y be a vector field on M, and let η be any section of $A\mathcal{G} \to M$ satisfying $\rho \circ \eta = Y$ (using the transitivity of $A\mathcal{G}$). From

$$\beta_*(\eta^{\mathsf{R}}_\varphi) = \beta_* r_{\varphi *}(\eta^{\mathsf{R}}_{1_{\beta(\varphi)}}) = \beta_*(\eta^{\mathsf{R}}_{1_{\beta(\varphi)}}) = \beta_*(\eta_{\beta(\varphi)}) = \rho(\eta_{\beta(\varphi)}) = Y_{\beta(\varphi)}$$

we see that

$$(Q^{\mathsf{R}} \eta^{\mathsf{R}})_\varphi = Q^{\mathsf{R}}_\varphi(\eta^{\mathsf{R}}_\varphi) = r_{\varphi *} Q_{\beta(\varphi)} \beta_*(\eta^{\mathsf{R}}_\varphi) = r_{\varphi *} Q_{\beta(\varphi)}(Y_{\beta(\varphi)}) = r_{\varphi *}((QY)_{\beta(\varphi)}),$$

so that $Q^{\mathsf{R}} \eta^{\mathsf{R}}$ is the right-invariant vector field on $\mathcal{G}$ corresponding to the algebroid section $Q \circ Y$. We may therefore write $Q^{\mathsf{R}} \eta^{\mathsf{R}} = (Q \circ Y)^{\mathsf{R}}$, and it follows that

$$[\xi^{\mathsf{R}}, Q^{\mathsf{R}} \eta^{\mathsf{R}}]_{1_x} = [\![\xi, Q \circ Y]\!]_x.$$

In addition

$$\begin{aligned}
\left(Q^{\mathsf{R}}[\xi^{\mathsf{R}}, \eta^{\mathsf{R}}]\right)_{1_x} &= Q^{\mathsf{R}}_{1_x}\left([\xi^{\mathsf{R}}, \eta^{\mathsf{R}}]_{1_x}\right) = Q_x \rho\left([\xi^{\mathsf{R}}, \eta^{\mathsf{R}}]_{1_x}\right) \\
&= Q_x \rho\left([\![\xi, \eta]\!]_x\right) = Q_x\left([\rho \circ \xi, \rho \circ \eta]_x\right) = Q_x\left([\rho \circ \xi, Y]_x\right) \\
&= \left(Q[\rho \circ \xi, Y]\right)_x
\end{aligned}$$

so that the tensor field $\mathcal{L}_{\xi^{\mathsf{R}}} Q^{\mathsf{R}}$ on $\mathcal{G}$ satisfies

$$\begin{aligned}(\mathcal{L}_{\xi^{\mathsf{R}}} Q^{\mathsf{R}})_{1_x}\eta_x &= (\mathcal{L}_{\xi^{\mathsf{R}}} Q^{\mathsf{R}})_{1_x}\eta^{\mathsf{R}}_{1_x} \\ &= \big((\mathcal{L}_{\xi^{\mathsf{R}}} Q^{\mathsf{R}})\eta^{\mathsf{R}}\big)_{1_x} \\ &= [\xi^{\mathsf{R}}, Q^{\mathsf{R}}\eta^{\mathsf{R}}]_{1_x} - \big(Q^{\mathsf{R}}[\xi^{\mathsf{R}}, \eta^{\mathsf{R}}]\big)_{1_x} \\ &= [\![\xi, Q\circ Y]\!]_x - \big(Q\circ[\rho\circ\xi, Y]\big)_x.\end{aligned}$$

It therefore makes sense to define the Lie derivative $\mathcal{L}_\xi Q$ by

$$(\mathcal{L}_\xi Q)_x Y_x = (\mathcal{L}_{\xi^{\mathsf{R}}} Q^{\mathsf{R}})_{1_x}\eta_x = [\![\xi, Q\circ Y]\!]_x - \big(Q\circ[\rho\circ\xi, Y]\big)_x.$$

Now suppose that $\mathcal{A}$ is a general Lie algebroid, not necessarily arising from a Lie groupoid $\mathcal{G}$. We may again interpret the operator $\mathcal{L}_\xi = [\![\xi, \cdot]\!]$ as a Lie derivative operator on sections of $\mathcal{A}$, and if Q is a vector bundle morphism $TM \to \mathcal{A}$ projecting to the identity on M then we can still define a new vector bundle morphism $\mathcal{L}_\xi Q : TM \to \mathcal{A}$ by setting

$$\mathcal{L}_\xi Q \circ Y = [\![\xi, Q\circ Y]\!] - Q\circ[\rho\circ\xi, Y]$$

for any vector field Y on M.

Lemma 1.6.1 *If Q is a vector bundle morphism $TM \to A\mathcal{G}$ projecting to the identity on M then the Lie derivative $\mathcal{L}_\xi Q$ is also a vector bundle morphism. The composite $\rho\circ Q$ is a type $(1,1)$ tensor field on M (regarded as an endomorphism of TM) satisfying*

$$\mathcal{L}_{\rho\circ\xi}(\rho\circ Q) = \rho\circ\mathcal{L}_\xi Q.$$

Proof To prove the first assertion it is sufficient to show that, for any vector field Y on M and any function f on M,

$$\mathcal{L}_\xi Q\circ(fY) = (f\mathcal{L}_\xi Q)\circ Y;$$

but

$$\begin{aligned}[\![\xi, Q\circ(fY)]\!] - Q\circ[\rho\circ\xi, fY] &= [\![\xi, f(Q\circ Y)]\!] - Q\circ\Big(f[\rho\circ\xi, Y] + \big((\rho\circ\xi)f\big)Y\Big) \\ &= f[\![\xi, Q\circ Y]\!] - fQ\circ[\rho\circ\xi, Y].\end{aligned}$$

For the second assertion we see that, for any vector field Y on M,

$$\mathcal{L}_{\rho\circ\xi}(\rho\circ Q)\circ Y = \mathcal{L}_{\rho\circ\xi}(\rho\circ Q\circ Y) - \rho\circ Q\circ\mathcal{L}_{\rho\circ\xi}Y$$

whereas

$$\rho \circ \mathcal{L}_\xi Q \circ Y = \rho \circ \big([\![\xi, Q \circ Y]\!] - Q \circ [\rho \circ \xi, Y] \big)$$
$$= [\rho \circ \xi, \rho \circ Q \circ Y] - \rho \circ Q \circ [\rho \circ \xi, Y]. \qquad \square$$

We may use a similar formula to define the Lie derivative of a vector bundle morphism $T : \mathcal{A} \to \mathcal{A}$, projecting to the identity (or, equivalently, a section of $\mathcal{A} \otimes \mathcal{A}^*$), by setting

$$\mathcal{L}_\xi T \circ \eta = [\![\xi, T \circ \eta]\!] - T \circ [\![\xi, \eta]\!]$$

for any section η of $\mathcal{A}$; indeed if

$$T : \mathcal{A} \times_M \mathcal{A} \times_M \cdots \times_M \mathcal{A} \to \mathcal{A}$$

is any tensorial map then we may define its Lie derivative $\mathcal{L}_\xi T$ by setting

$$(\mathcal{L}_\xi T) \circ (\eta_1, \eta_2, \ldots, \eta_r) = [\![\xi, T \circ (\eta_1, \eta_2, \ldots, \eta_r)]\!] - \sum_{k=1}^{r} T \circ (\eta_1, \eta_2, \ldots, [\![\xi, \eta_k]\!], \ldots, \eta_r).$$

The usual formulæ for Lie derivatives of tensor products hold in both cases, with the proofs being analogues of those for ordinary tensor fields on manifolds. We also have the following result on the commutator of Lie derivatives.

Proposition 1.6.2 *If ξ, η are sections of $\mathcal{A}$ and T is any tensorial map $\mathcal{A} \times_M \mathcal{A} \times_M \cdots \times_M \mathcal{A} \to \mathcal{A}$ then*

$$\mathcal{L}_\xi \mathcal{L}_\eta T - \mathcal{L}_\eta \mathcal{L}_\xi T = \mathcal{L}_{[\![\xi, \eta]\!]} T.$$

If Q is any tensorial map $TM \times_M TM \times_M \cdots \times_M TM \to \mathcal{A}$ then

$$\mathcal{L}_\xi \mathcal{L}_\eta Q - \mathcal{L}_\eta \mathcal{L}_\xi Q = \mathcal{L}_{[\![\xi, \eta]\!]} Q.$$

Proof For any sections $\zeta_1, \zeta_2, \ldots, \zeta_r$ of $\mathcal{A}$ we see first that

$$(\mathcal{L}_\xi \mathcal{L}_\eta T) \circ (\zeta_1, \zeta_2, \ldots, \zeta_r)$$
$$= [\![\xi, (\mathcal{L}_\eta T) \circ (\zeta_1, \zeta_2, \ldots, \zeta_r)]\!] - \sum_{k=1}^{r} (\mathcal{L}_\eta T) \circ (\zeta_1, \zeta_2, \ldots, [\![\xi, \zeta_k]\!], \ldots, \zeta_r)$$
$$= [\![\xi, [\![\eta, T \circ (\zeta_1, \zeta_2, \ldots, \zeta_r)]\!]]\!]$$
$$- \sum_{k=1}^{r} [\![\xi, T \circ (\zeta_1, \zeta_2, \ldots, [\![\eta, \zeta_k]\!], \ldots, \zeta_r)]\!] \qquad (1)$$
$$- \sum_{k=1}^{r} [\![\eta, T \circ (\zeta_1, \zeta_2, \ldots, [\![\xi, \zeta_k]\!], \ldots, \zeta_r)]\!] \qquad (2)$$

$$+\sum_{\substack{k,l=1\\k\neq l}}^{r} T\circ(\zeta_1,\zeta_2,\dots,[\![\xi,\zeta_k]\!],\dots,[\![\eta,\zeta_l]\!],\dots,\zeta_r) \qquad (3)$$

$$+\sum_{k=1}^{r} T\circ(\zeta_1,\zeta_2,\dots,[\![\eta,[\![\xi,\zeta_k]\!]]\!],\dots,\zeta_r)$$

so that in the expansion of $(\mathcal{L}_\xi\mathcal{L}_\eta T-\mathcal{L}_\eta\mathcal{L}_\xi T)\circ(\zeta_1,\zeta_2,\dots,\zeta_r)$ terms (1), (2) and (3) cancel and we are left with

$$\begin{aligned}(\mathcal{L}_\xi\mathcal{L}_\eta T-\mathcal{L}_\eta\mathcal{L}_\xi T)\circ(\zeta_1,\zeta_2,\dots,\zeta_r) &= [\![\xi,[\![\eta,T\circ(\zeta_1,\zeta_2,\dots,\zeta_r)]\!]]\!]\\ &\quad+\sum_{k=1}^{r} T\circ(\zeta_1,\zeta_2,\dots,[\![\eta,[\![\xi,\zeta_k]\!]]\!],\dots,\zeta_r)\\ &\quad-[\![\eta,[\![\xi,T\circ(\zeta_1,\zeta_2,\dots,\zeta_r)]\!]]\!]\\ &\quad-\sum_{k=1}^{r} T\circ(\zeta_1,\zeta_2,\dots,[\![\xi,[\![\eta,\zeta_k]\!]]\!],\dots,\zeta_r).\end{aligned}$$

On the other hand, we also see that

$$\begin{aligned}(\mathcal{L}_{[\![\xi,\eta]\!]}T)\circ(\zeta_1,\zeta_2,\dots,\zeta_r) &= [\![[\![\xi,\eta]\!],T\circ(\zeta_1,\zeta_2,\dots,\zeta_r)]\!]\\ &\quad-\sum_{k=1}^{r} T\circ(\zeta_1,\zeta_2,\dots,[\![[\![\xi,\eta]\!],\zeta_k]\!],\dots,\zeta_r)\\ &= [\![[\![\xi,\eta]\!],T\circ(\zeta_1,\zeta_2,\dots,\zeta_r)]\!]\\ &\quad-\sum_{k=1}^{r} T\circ(\zeta_1,\zeta_2,\dots,[\![[\![\xi,\eta]\!],\zeta_k]\!],\dots,\zeta_r)\\ &= [\![\xi,[\![\eta,T\circ(\zeta_1,\zeta_2,\dots,\zeta_r)]\!]]\!]\\ &\quad+\sum_{k=1}^{r} T\circ(\zeta_1,\zeta_2,\dots,[\![\eta,[\![\xi,\zeta_k]\!]]\!],\dots,\zeta_r)\\ &\quad-[\![\eta,[\![\xi,T\circ(\zeta_1,\zeta_2,\dots,\zeta_r)]\!]]\!]\\ &\quad-\sum_{k=1}^{r} T\circ(\zeta_1,\zeta_2,\dots,[\![\xi,[\![\eta,\zeta_k]\!]]\!],\dots,\zeta_r),\end{aligned}$$

establishing the first result.

The proof of the second result follows a similar pattern, using vector fields $Z_1,Z_2,\dots,Z_n$ instead of sections $\zeta_1,\zeta_2,\dots,\zeta_n$ and using the anchor map ρ as appropriate. □

Chapter 2
Connections on Lie Groupoids and Lie Algebroids

In this chapter we shall describe several different notions of connection. As well as introducing connections on Lie groupoids (path connections) and on Lie algebroids (infinitesimal connections) we shall see how these ideas are related to the classical concepts of covariant derivatives and, more generally, connections on vector bundles.

2.1 Path Connections on Lie Groupoids

There are several related concepts of connection in the general theory of fibre bundles. One such concept involves the lifting of a curve in the base manifold M to 'horizontal' curves in E, in a way that is consistent with the group action on the fibres of the bundle. The various lifts can then be used to determine diffeomorphisms between the fibres at different points on the base curve. It is possible to describe a similar notion for a locally trivial Lie groupoid, and we shall see in the next chapter, when we consider groupoids of fibre morphisms, how the two concepts are related.

The lifting operations we are going to consider will have the important property that the lift must be invariant under reparametrization. This property allows us to start by considering the lift of a curve defined on the specific interval $[0, 1]$, rather than on an arbitrary interval. We shall then show how this is sufficient for us to specify the lift of a curve defined on the whole of $\mathbb{R}$, or indeed, on an arbitrary nonempty open interval. We will also be able to specify the lift of a vector field by lifting its flow.

We recall our convention that a curve need be only continuous and piecewise smooth, rather than smooth on the whole of its domain.

Let $\mathcal{G}$ be a locally trivial Lie groupoid with manifold M of identities. A *path connection* on $\mathcal{G}$ is a map Γ taking each curve $c : [0, 1] \to M$ to a hboxcurve $c^\Gamma : [0, 1] \to \mathcal{G}$, satisfying the following conditions (these are slightly different from those given in [30], but are more directly suited to our purposes):

- $c^\Gamma(0) = 1_{c(0)}$;
- $\alpha c^\Gamma(t) = c(0)$ and $\beta c^\Gamma(t) = c(t)$ for all $t \in [0, 1]$;

M. Crampin and D. Saunders, *Cartan Geometries and their Symmetries*,
Atlantis Studies in Variational Geometry 4, DOI 10.2991/978-94-6239-192-5_2

- if $[a, b] \subset [0, 1]$ and $\chi : [0, 1] \to [a, b]$ is a (smooth) diffeomorphism then $c^\Gamma \circ \chi = r_\varphi \circ (c \circ \chi)^\Gamma$ where $\varphi = c^\Gamma \chi(0)$;
- if c is smooth at $t \in [0, 1]$ with tangent vector $\dot{c}(t) \in T_{c(t)}M$ then c^Γ is smooth at t with tangent vector $\dot{c}^\Gamma(t) \in T_{c^\Gamma(t)}\mathcal{G}$;
- if $c_1(0) = c_2(0) = x$, say, and if $\dot{c}_1(0) = \dot{c}_2(0) \in T_x M$, then $\dot{c}_1^\Gamma(0) = \dot{c}_2^\Gamma(0) \in T_{1_x}\mathcal{G}$;
- the correspondence $T_x M \to T_{1_x}\mathcal{G}$ given by $\dot{c}(0) \mapsto \dot{c}^\Gamma(0)$ is linear, and determines a smooth map $\gamma : TM \to T\mathcal{G}$.

We may now use the path connection to lift curves defined on $\mathbb{R}$. The idea is to imagine the complete curve as a sequence of short curves defined on intervals $[n, n+1]$,[1] and to translate the curves in the sequence so that they are defined on $[0, 1]$. We shall, in stages, prove an appropriate reparametrization property for the lifted curve on $\mathbb{R}$. So, given a curve $c : \mathbb{R} \to M$, we define for each $n \in \mathbb{Z}$ the curve $c_n : [0, 1] \to M$ by $c_n(t) = c(n + t)$, and then define the map $c^\Gamma : \mathbb{R} \to \mathcal{G}$ by

$$c^\Gamma(t) = \left(c_n^\Gamma(t - n)\right) \cdot \varphi_n$$

where $t \in [n, n + 1)$ and where

$$\varphi_n = \begin{cases} 1_{c(0)} & (n = 0) \\ c_{n-1}^\Gamma(1) \cdot \varphi_{n-1} & (n > 0) \\ \left(c_n^\Gamma(1)\right)^{-1} \cdot \varphi_{n+1} & (n < 0). \end{cases}$$

We observe that all the groupoid products are defined, and that $\varphi_n = c_{n-1}^\Gamma(1) \cdot \varphi_{n-1}$ for any $n \in \mathbb{Z}$.

Lemma 2.1.1 *The map $c^\Gamma : \mathbb{R} \to \mathcal{G}$ is continuous and satisfies the conditions that $c^\Gamma(0) = 1_{c(0)}$, and that $\alpha c^\Gamma(t) = c(0)$ and $\beta c^\Gamma(t) = c(t)$ for all $t \in \mathbb{R}$.*

Proof It is immediate that c^Γ is continuous at all points except, possibly, integer points $n \in \mathbb{R}$. It is also immediate that $\lim_{t \to n+} c^\Gamma(t) = c^\Gamma(n)$; to demonstrate continuity from the left we note that

$$\lim_{t \to n-} c^\Gamma(t) = \lim_{t \to n-} c_{n-1}^\Gamma(t - (n - 1)) \cdot \varphi_{n-1} = c_{n-1}^\Gamma(1) \cdot \varphi_{n-1}$$

using the continuity of c_{n-1}^Γ on I, whereas

$$c^\Gamma(n) = c_n^\Gamma(0) \cdot \varphi_n = 1_{c_n(0)} \cdot \varphi_n = \varphi_n = c_{n-1}^\Gamma(1) \cdot \varphi_{n-1}.$$

We also see that

[1] We have temporarily suspended our convention that $n = \dim M$, and for the rest of this section we shall use n as an arbitrary integer.

$$c^\Gamma(0) = c_0^\Gamma(0) \cdot \varphi_0 = 1_{c_0(0)} \cdot 1_{c(0)} = 1_{c(0)},$$

that

$$\alpha c^\Gamma(t) = \alpha(\varphi_n) = \alpha(1_{c(0)}) = c(0)$$

and that

$$\beta c^\Gamma(t) = \beta c_n^\Gamma(t-n) = c_n(t-n) = c(t). \qquad \square$$

We have not yet shown that c^Γ is piecewise smooth, as in principle it could fail to be smooth at infinitely many integral values of t, although we shall see in due course that this cannot happen. We can, though, use the continuity of c^Γ to see that $\varphi_n = c^\Gamma(n)$, because

$$c^\Gamma(n) = \lim_{\varepsilon\to 0} c^\Gamma(n-\varepsilon) = \lim_{\varepsilon\to 0} c_{n-1}^\Gamma(1-\varepsilon)\cdot\varphi_{n-1} = c_{n-1}^\Gamma(1)\cdot\varphi_{n-1} = \varphi_n.$$

We now establish the reparametrization property in detail.

Lemma 2.1.2 *If $b > 0$ and $\chi : [0,1] \to [0,b]$ is an increasing diffeomorphism then $c^\Gamma \circ \chi = (c\circ\chi)^\Gamma$ as maps $I \to \mathcal{G}$.*

Proof Let n be the largest integer not greater than b. If $n = 0$ then $[0,b] \subset [0,1]$ so that the result is an immediate consequence of the reparametrization property of c^Γ; we may therefore assume that $n \geq 1$.

Put $a_k = \chi^{-1}(k)$ for $k = 0, 1, \ldots, n$. Let $\theta_k : [0,1] \to [a_k, a_k+1] \subset I$ be given by $\theta_k(s) = \chi^{-1}(s+k)$, so that θ_k is a diffeomorphism; as $\chi(\theta_k(s)) = s+k$, we see that $c\big(\chi(\theta_k(s))\big) = c(s+k) = c_k(s)$ for $s \in I$, so that

$$c\circ\chi\circ\theta_k = c_k.$$

We also know, using the reparametrization property of the curve $(c\circ\chi)^\Gamma : [0,1] \to \mathcal{G}$, that

$$(c\circ\chi)^\Gamma\theta_k(s) = (c\circ\chi\circ\theta_k)^\Gamma(s)\cdot(c\circ\chi)^\Gamma\theta_k(0) = c_k^\Gamma(s)\cdot(c\circ\chi)^\Gamma(a_k).$$

We now argue recursively on k. Suppose that $(c\circ\chi)^\Gamma(a_k) = c^\Gamma(k)$, so that $(c\circ\chi)^\Gamma(\theta_k(s)) = c_k^\Gamma(s)\cdot c^\Gamma(k)$ for $s \in [0,1]$; then, putting $t = \theta_k(s) \in [a_k, a_{k+1}]$ so that $s = \chi(t) - k$, we see that

$$(c\circ\chi)^\Gamma(t) = c_k^\Gamma(\chi(t)-k)\cdot c^\Gamma(k) = c^\Gamma\chi(t)$$

using the definition of c^Γ for values $\chi(t) \in [n, n+1)$ and the continuity of c^Γ at $n+1$. To justify the recursive step we put $t = a_{k+1}$ to see that $(c\circ\chi)^\Gamma(a_{k+1}) =$

$c^\Gamma\chi(a_{k+1}) = c^\Gamma(k+1)$, and we note that $a_0 = 0$ so that $(c\circ\chi)^\Gamma(a_0) = 1_{c(0)} = c^\Gamma(0)$ to start the recursion.

We have shown, therefore, that $c^\Gamma\chi(t) = (c\circ\chi)^\Gamma(t)$ for all $t \in [0, a_n]$; a slightly modified version of the same argument shows that the result also holds where $t \in [a_n, 1]$ and $\chi(t) \in [n, b]$, so we conclude that $c^\Gamma \circ \chi = (c\circ\chi)^\Gamma$. □

Corollary 2.1.3 *If $b > 0$ and $\chi : [0, b] \to [0, 1]$ is an increasing diffeomorphism then $c^\Gamma \circ \chi = (c\circ\chi)^\Gamma$ as maps $[0, b] \to \mathcal{G}$.*

Proof As $\chi^{-1} : [0, 1] \to [0, b]$ we have

$$(c\circ\chi)^\Gamma \circ \chi^{-1} = (c\circ\chi\circ\chi^{-1})^\Gamma = c^\Gamma$$

so that

$$(c\circ\chi)^\Gamma = c^\Gamma \circ \chi.$$ □

Now for any $b \in \mathbb{R}$ let $\mathrm{t}_b : \mathbb{R} \to \mathbb{R}$ denote the translation map $\mathrm{t}_b(t) = b + t$, for any $b > 0$ let $\mathbf{s}_b : \mathbb{R} \to \mathbb{R}$ denote the scaling map $\mathbf{s}_b(t) = bt$, and let $\mathbf{r} : \mathbb{R} \to \mathbb{R}$ denote the reflection map $\mathbf{r}(t) = -t$.

Lemma 2.1.4 *If $k \in \mathbb{Z}$ then*

$$c^\Gamma \circ \mathrm{t}_k = r_{c^\Gamma(k)} \circ (c\circ \mathrm{t}_k)^\Gamma.$$

Proof If $k = 0$ there is nothing to prove; we consider the case $k > 0$. Take $t \in [n, n+1)$ so that $\mathrm{t}_k(t) \in [k+n, k+n+1)$, and note that $c\circ \mathrm{t}_k = c_k$. Now for any $j \in \mathbb{Z}$ we have

$$(c\circ \mathrm{t}_k)_j(t) = c_j(t+k) = c(t+j+k) = c_{j+k}(t)$$

so that $(c\circ \mathrm{t}_k)_j = c_{j+k}$. Thus from the definition,

$$\begin{aligned}(c\circ \mathrm{t}_k)^\Gamma(t) &= (c\circ \mathrm{t}_k)_n^\Gamma(t-n)\cdot(c\circ \mathrm{t}_k)_{n-1}^\Gamma(1)\cdot\ldots\cdot(c\circ \mathrm{t}_k)^\Gamma(1)\\ &= c_{k+n}^\Gamma(t-1)\cdot c_{k+n-1}^\Gamma(1)\cdot\ldots\cdot c_k^\Gamma(1)\end{aligned}$$

if $n \geq 0$, with a similar (but ascending rather than descending) formula when $n < 0$. Similarly, when $k+n \geq 0$ we have

$$\begin{aligned}c^\Gamma(\mathrm{t}_k(t)) &= c_{k+n}^\Gamma(\mathrm{t}_k(t)-k-n)\cdot c_{k+n-1}^\Gamma(1)\cdot\ldots\cdot c^\Gamma(1)\\ &= c_{k+n}^\Gamma(t-n)\cdot c_{k+n-1}^\Gamma(1)\cdot\ldots\cdot c^\Gamma(1)\end{aligned}$$

so that in this case

$$c^\Gamma(\mathrm{t}_k(t)) = (c\circ \mathrm{t}_k)^\Gamma(t)\cdot c_{k-1}^\Gamma(1)\cdot\ldots\cdot c^\Gamma(1) = (c\circ \mathrm{t}_k)^\Gamma(t)\cdot c^\Gamma(k).$$

The same result holds, with appropriate modifications to the details of the calculations, for all $k \in \mathbb{Z}$ and all $n \in Z$ because the recurrence formula $\varphi_j = c^{\Gamma}_{j-1}(1) \cdot \varphi_{j-1}$ (or $\varphi_j = (c \circ \mathsf{t}_k)^{\Gamma}_{j-1}(1) \cdot \varphi_{j-1}$) holds for any $j \in \mathbb{Z}$. □

Lemma 2.1.5 *For the reflection map* $\mathbf{r} : [-1, 1] \to [-1, 1]$ *we have*

$$(c \circ \mathbf{r})^{\Gamma} = c^{\Gamma} \circ \mathbf{r}.$$

Proof Let $\chi : [0, 1] \to [0, 1]$ be the diffeomorphism $\chi(t) = 1 - t$, so that $\mathbf{r} : [0, 1] \to [-1, 0]$ satisfies $\mathbf{r} = \mathsf{t}_{-1} \circ \chi$. Then

$$\begin{aligned} c^{\Gamma} \circ \mathbf{r} &= c^{\Gamma} \circ \mathsf{t}_{-1} \circ \chi \\ &= r_{c^{\Gamma}(-1)} \circ (c \circ \mathsf{t}_{-1})^{\Gamma} \circ \chi \\ &= r_{c^{\Gamma}(-1)} \circ r_{(c \circ \mathsf{t}_{-1})^{\Gamma}(1)} \circ (c \circ \mathsf{t}_{-1} \circ \chi)^{\Gamma} \\ &= r_{c^{\Gamma}(-1)} \circ r_{(c \circ \mathsf{t}_{-1})^{\Gamma}(1)} \circ (c \circ \mathbf{r})^{\Gamma}; \end{aligned}$$

but

$$(c \circ \mathsf{t}_{-1})^{\Gamma}(1) \cdot c^{\Gamma}(-1) = c^{\Gamma}(\mathsf{t}_{-1}(1)) = c^{\Gamma}(0) = 1_{c(0)}.$$

We then see that $\mathbf{r}^{-1} : [-1, 0] \to [0, 1]$ also satisfies $(c \circ \mathbf{r})^{\Gamma} = c^{\Gamma} \circ \mathbf{r}$, showing that the relationship holds for all $t \in [-1, 1]$. □

Lemma 2.1.6 *If* $b \in (0, 1)$ *then*

$$c^{\Gamma} \circ \mathsf{t}_b = r_{c^{\Gamma}(b)} \circ (c \circ \mathsf{t}_b)^{\Gamma}.$$

Proof We first consider the case where $t \in [n, n + 1 - b]$ for some $n \in \mathbb{Z}$, so that $\mathsf{t}_b(t) \in [n + b, n + 1]$. We note that

$$\mathsf{t}_n \mathsf{s}_b \mathsf{t}_1 \mathsf{s}_{b^{-1}} \mathsf{t}_{-n}(t) = b(b^{-1}(t - n) + 1) + n = t + b = \mathsf{t}_b(t)$$

where

$$[n, n+1-b] \xrightarrow{\mathsf{t}_{-n}} [0, 1-b] \xrightarrow{\mathsf{s}_{b^{-1}}} [0, b^{-1} - 1] \xrightarrow{\mathsf{t}_1} [1, b^{-1}] \xrightarrow{\mathsf{s}_b} [b, 1] \xrightarrow{\mathsf{t}_n} [n+b, n+1],$$

so that

$$\begin{aligned} c^{\Gamma} \circ \mathsf{t}_b &= c^{\Gamma} \circ \mathsf{t}_n \circ \mathsf{s}_b \circ \mathsf{t}_1 \circ \mathsf{s}_{b^{-1}} \circ \mathsf{t}_{-n} \qquad (1) \\ &= r_{c^{\Gamma}(n)} \circ (c \circ \mathsf{t}_n)^{\Gamma} \circ \mathsf{s}_b \circ \mathsf{t}_1 \circ \mathsf{s}_{b^{-1}} \circ \mathsf{t}_{-n} \\ &= r_{c^{\Gamma}(n)} \circ (c \circ \mathsf{t}_n \circ \mathsf{s}_b)^{\Gamma} \circ \mathsf{t}_1 \circ \mathsf{s}_{b^{-1}} \circ \mathsf{t}_{-n} \\ &= r_{c^{\Gamma}(n)} \circ r_{(c \circ \mathsf{t}_n \circ \mathsf{s}_b)^{\Gamma}(1)} \circ (c \circ \mathsf{t}_n \circ \mathsf{s}_b \circ \mathsf{t}_1)^{\Gamma} \circ \mathsf{s}_{b^{-1}} \circ \mathsf{t}_{-n} \end{aligned}$$

$$= r_{c^\Gamma(n)} \circ r_{(c\circ t_n \circ s_b)^\Gamma(1)} \circ (c \circ t_n \circ s_b \circ t_1 \circ s_{b^{-1}})^\Gamma \circ t_{-n} \quad (2)$$
$$= r_{c^\Gamma(n)} \circ r_{(c\circ t_n \circ s_b)^\Gamma(1)} \circ r_{(c\circ t_n \circ s_b \circ t_1 \circ s_{b^{-1}})^\Gamma(-n)} \circ (c \circ t_n \circ s_b \circ t_1 \circ s_{b^{-1}} \circ t_{-n})^\Gamma$$
$$= r_{c^\Gamma(n)} \circ r_{(c\circ t_n \circ s_b)^\Gamma(1)} \circ r_{(c\circ t_n \circ s_b \circ t_1 \circ s_{b^{-1}})^\Gamma(-n)} \circ (c \circ t_b)^\Gamma$$

where step (1) is justified by Corollary 2.1.3 with $\chi = s_b$ because $t_1 s_{b^{-1}} t_{-n}(t) \in [1, b^{-1}] \subset [0, b^{-1}]$, and step (2) is justified by Lemma 2.1.2 with $\chi = s_{b^{-1}}$ because $t_{-n}(t) \in [0, 1-b] \subset [0, 1]$. We now note that

$$\begin{aligned}(c \circ t_n \circ s_b)^\Gamma(1) \cdot c^\Gamma(n) &= (c \circ t_n)^\Gamma s_b(1) \cdot c^\Gamma(n)\\ &= (c \circ t_n)^\Gamma(b) \cdot c^\Gamma(n)\\ &= c^\Gamma t_n(b)\\ &= c^\Gamma(n+b)\end{aligned}$$

and $t_n \circ s_b \circ t_1 \circ s_{b^{-1}} = t_{n+b}$ so that

$$\begin{aligned}(c \circ t_n \circ s_b \circ t_1 \circ s_{b^{-1}})^\Gamma(-n) \cdot (c \circ t_n \circ s_b)^\Gamma(1) \cdot c^\Gamma(n) &= (c \circ t_{n+b})^\Gamma(-n) \cdot c^\Gamma(n+b)\\ &= c^\Gamma t_{n+b}(-n)\\ &= c^\Gamma(b)\end{aligned}$$

from which we see in this case that $c^\Gamma \circ t_b = r_{c^\Gamma(b)} \circ (c \circ t_b)^\Gamma$.

Suppose, instead, that $t \in [n+1-b, n+1]$ for some $n \in \mathbb{Z}$, so that $t_b(t) \in [n+1, n+1+b]$. The argument used above fails at step (1) because the image of s_b is not contained in I so that we cannot appeal to Corollary 2.1.3. Instead we note that

$$t_{n+1} s_b \mathbf{r} t_{-1} s_{b^{-1}} t_1 \mathbf{r} t_{-n}(t) = n+1-b(b^{-1}(1-(t-n))-1) = t+b = t_b(t),$$

where

$$[n+1-b, n+1] \xrightarrow{t_{-n}} [1-b, 1] \xrightarrow{\mathbf{r}} [-1, b-1] \xrightarrow{t_1} [0, b] \xrightarrow{s_{b^{-1}}}$$
$$\xrightarrow{s_{b^{-1}}} [0, 1] \xrightarrow{t_{-1}} [-1, 0] \xrightarrow{\mathbf{r}} [0, 1] \xrightarrow{s_b} [0, b] \xrightarrow{t_{n+1}} [n+1, n+1+b],$$

and a similar but slightly longer argument, now in addition using Lemma 2.1.5, shows that once again $c^\Gamma \circ t_b = r_{c^\Gamma(b)} \circ (c \circ t_b)^\Gamma$. □

Corollary 2.1.7 *If $a \in \mathbb{R}$ then*

$$c^\Gamma \circ t_a = r_{c^\Gamma(a)} \circ (c \circ t_a)^\Gamma.$$

Proof If $a \in \mathbb{Z}$ then the result has already been established; so let $a = k + b$ where $b \in (0, 1)$ and $k \in \mathbb{Z}$. Then

$$c^\Gamma \circ \mathrm{t}_a = c^\Gamma \circ \mathrm{t}_b \circ \mathrm{t}_k = r_{c^\Gamma(b)} \circ (c \circ \mathrm{t}_b)^\Gamma \circ \mathrm{t}_k = r_{c^\Gamma(b)} \circ r_{(c \circ \mathrm{t}_b)^\Gamma(k)} \circ (c \circ \mathrm{t}_b \circ \mathrm{t}_k)^\Gamma$$

and

$$c \circ \mathrm{t}_b \circ \mathrm{t}_k = c \circ \mathrm{t}_{k+b} = c \circ \mathrm{t}_a,$$
$$(c \circ \mathrm{t}_b)^\Gamma(k) \cdot c^\Gamma(b) = c^\Gamma \circ \mathrm{t}_b(k) = c^\Gamma(k+b) = c^\Gamma(a). \qquad \square$$

We therefore obtain the general reparametrization result.

Proposition 2.1.8 *If $[a, b] \subset \mathbb{R}$ and if $\chi : [0, 1] \to [a, b]$ is a diffeomorphism then*

$$c^\Gamma \circ \chi = r_\varphi \circ (c \circ \chi)^\Gamma$$

where $\varphi = c^\Gamma(\chi(0))$.

Proof Suppose first that χ is an increasing diffeomorphism, so that $\chi(0) = a$. Put $\chi_0 = \mathrm{t}_{-a} \circ \chi$, so that $\chi_0 : [0, 1] \to [0, b-a]$ is again an increasing diffeomorphism. Considering the curve $c_a = c \circ \mathrm{t}_a : \mathbb{R} \to \mathcal{G}$ we have

$$c^\Gamma \circ \mathrm{t}_a = r_{c^\Gamma(a)} \circ c_a^\Gamma, \qquad c_a^\Gamma \circ \chi_0 = (c_a \circ \chi_0)^\Gamma$$

and therefore

$$c^\Gamma \circ \chi = c^\Gamma \circ \mathrm{t}_a \circ \chi_0 = r_{c^\Gamma(a)} \circ c_a^\Gamma \circ \chi_0 = r_{c^\Gamma(a)} \circ (c_a \circ \chi_0)^\Gamma = r_{c^\Gamma(a)} \circ (c \circ \chi)^\Gamma.$$

If instead χ is a decreasing diffeomorphism then $\chi \circ \mathbf{r}$ is an increasing diffeomorphism, so that

$$c^\Gamma \circ \chi \circ \mathbf{r} = r_{c^\Gamma(\chi(\mathbf{r}(0)))} \circ (c \circ \chi \circ \mathbf{r})^\Gamma = r_{c^\Gamma\chi(0)} \circ (c \circ \chi)^\Gamma \circ \mathbf{r}$$

and the result follows. $\square$

Corollary 2.1.9 *If c is smooth at $t \in \mathbb{R}$ then so is c^Γ; thus c^Γ is piecewise smooth.*

Proof If $t \notin \mathbb{Z}$ and c is smooth at t then c^Γ is smooth at t by construction, because the curve c_n is smooth at $t - n$, its lift c_n^Γ is smooth at t_n, and groupoid multiplication is smooth.

Consider $n \in \mathbb{Z}$ and put $a = n - 1$, $b = n + 1$. Define the diffeomorphism $\chi : [0, 1] \to [a, b]$ by $\chi(t) = 2t + n - 1$, so that

$$c^\Gamma = r_{c^\Gamma(n-1)} \circ (c \circ \chi)^\Gamma \circ \chi^{-1}.$$

Suppose that c is smooth at n, so that $c \circ \chi$ is smooth at $\frac{1}{2}$ and therefore $(c \circ \chi)^\Gamma$ is smooth at $\frac{1}{2}$. As χ^{-1} is smooth at n and groupoid multiplication is smooth, it follows that c^Γ is smooth at n. It follows that there are only finitely many values of t for which

c^Γ is not smooth. At the points where c^Γ is not smooth, all left and right derivatives must nevertheless exist because they do for the curves c_n and c_n^Γ; thus c^Γ is piecewise smooth. □

It is clear that a similar approach can be used to give the lift of a curve defined on an arbitrary open interval containing zero. We may then define the lift of a vector field by lifting the curves in its flow. This is not completely straightforward, as the integral curve through $x \in M$ will lift to give a curve through $1_x \in \mathcal{G}$, resulting in a vector field along 1_M; we may, however, extend this to a vector field on the whole of $\mathcal{G}$ by right translation.

Proposition 2.1.10 *If a vector field X on M has a global flow $\psi : M \times \mathbb{R} \to M$ then the map $\psi^\Gamma : \mathcal{G} \times \mathbb{R} \to \mathcal{G}$ defined by*

$$\psi^\Gamma(\varphi, t) = \psi^\Gamma_\varphi(t) = r_\varphi(\psi_{\beta(\varphi)})^\Gamma(t),$$

where $\psi_{\beta(\varphi)}$ is the curve in M given by $\psi_{\beta(\varphi)}(t) = \psi(\beta(\varphi), t)$, is the flow of a vector field X^Γ on $\mathcal{G}$. The same formula, where the map ψ^Γ is defined on a proper subset of $\mathcal{G} \times \mathbb{R}$, may be used where X does not have a global flow.

Proof Fix $s \in \mathbb{R}$. If $t \in \mathbb{R}$ then

$$\psi_{\beta(\varphi)}\mathsf{t}_s(t) = \psi_{\beta(\varphi)}(s+t) = \psi_{s+t}\beta(\varphi) = \psi_t\psi_s\beta(\varphi) = \psi_t\psi_{\beta(\varphi)}(s) = \psi_{\psi_{\beta(\varphi)}(s)}(t)$$

so that $\psi_{\beta(\varphi)} \circ \mathsf{t}_s = \psi_{\psi_{\beta(\varphi)}(s)}$. Thus

$$\left(\psi_{\beta(\varphi)}\right)^\Gamma \circ \mathsf{t}_s = r_{(\psi_{\beta(\varphi)})^\Gamma(s)} \circ \left(\psi_{\beta(\varphi)} \circ \mathsf{t}_s\right)^\Gamma = r_{(\psi_{\beta(\varphi)})^\Gamma(s)} \circ \left(\psi_{\psi_{\beta(\varphi)}(s)}\right)^\Gamma$$

using Lemma 2.1.7, so that

$$\begin{aligned}
\psi^\Gamma_{\psi^\Gamma_\varphi(s)} &= \psi^\Gamma_{(r_\varphi \circ (\psi_{\beta(\varphi)})^\Gamma)(s)} \\
&= \psi^\Gamma_{(\psi_{\beta(\varphi)})^\Gamma(s) \cdot \varphi} \\
&= r_{(\psi_{\beta(\varphi)})^\Gamma(s) \cdot \varphi} \circ \left(\psi_{\beta((\psi_{\beta(\varphi)})^\Gamma(s) \cdot \varphi)}\right)^\Gamma \\
&= r_\varphi \circ r_{(\psi_{\beta(\varphi)})^\Gamma(s)} \circ \left(\psi_{\psi_{\beta(\varphi)}(s)}\right)^\Gamma \\
&= r_\varphi \circ \left(\psi_{\beta(\varphi)}\right)^\Gamma \circ \mathsf{t}_s \\
&= \psi^\Gamma_\varphi \circ \mathsf{t}_s
\end{aligned}$$

and we see, putting $\psi^\Gamma_t(\varphi) = \psi^\Gamma(\varphi, t)$, that

$$\psi^\Gamma_t\psi^\Gamma_s(\varphi) = \psi^\Gamma_t\psi^\Gamma_\varphi(s) = \psi^\Gamma_{\psi^\Gamma_\varphi(s)}(t) = \psi^\Gamma_\varphi \circ \mathsf{t}_s(t) = \psi^\Gamma_\varphi(s+t) = \psi^\Gamma_{s+t}(\varphi).$$

As in addition

$$\psi_0^\Gamma(\varphi) = \psi_\varphi^\Gamma(0) = r_\varphi\big((\psi_{\beta(\varphi)})^\Gamma(0)\big) = r_\varphi(1_{\psi_{\beta(\varphi)}(0)}) = r_\varphi(1_{\beta(\varphi)}) = \varphi,$$

we see that ψ_t^Γ satisfies the pseudogroup property for a flow. As each curve ψ_x is smooth, it follows that each curve $\psi_\varphi^\Gamma = r_\varphi \circ (\psi_{\beta(\varphi)})^\Gamma$ is smooth and hence defines a tangent vector

$$X_\varphi^\Gamma = \dot{\psi}_\varphi^\Gamma(0) = r_{\varphi *}(\dot{\psi}_{\beta(\varphi)})^\Gamma(0) \in T_\varphi \mathcal{G}.$$

We must now show that the resulting vector field X^Γ on $\mathcal{G}$ given by $\varphi \mapsto X_\varphi^\Gamma$ is smooth. If $\varphi = 1_x$ for $x \in M$ then

$$X_{1_x}^\Gamma = (\dot{\psi}_x)^\Gamma(0) = \gamma\dot{\psi}_x(0) = \gamma(X_x)$$

where $\gamma : TM \to T\mathcal{G}$ is the smooth map associated with Γ. Thus in general

$$X_\varphi^\Gamma = r_{\varphi *}\gamma(X_{\beta(\varphi)}),$$

showing that $X^\Gamma : \mathcal{G} \to T\mathcal{G}$ is a smooth map.

The argument where X does not have a global flow is similar, taking account of the domains of the maps involved. □

2.2 Infinitesimal Connections on Lie Algebroids

A path connection on $\mathcal{G}$ is the groupoid version of a fibre bundle connection on $E \to M$ given by horizontal lifts of curves. The latter may also, though, be given in infinitesimal terms as a type $(1, 1)$ tensor field on E with the property that it maps lifts of tangent vectors in TM in a well defined way to horizontal tangent vectors in TE (that is, to vectors tangent to horizontal curves in E). This infinitesimal approach permits a definition of the curvature of the connection, as the extent to which the induced mapping from vector fields on M to vector fields on E fails to preserve the Lie bracket. We apply a similar approach to path connections on Lie groupoids, although here it is sufficient to define the infinitesimal version on the associated Lie algebroid.

We have, in fact, already seen the infinitesimal version of a path connection on a locally trivial Lie groupoid: it is the map $\gamma : TM \to T\mathcal{G}$ associated with a path connection Γ.

Lemma 2.2.1 *The map γ associated with the path connection Γ on $\mathcal{G}$ satisfies $\alpha_* \circ \gamma = 0$ and $\beta_* \circ \gamma = \mathrm{id}_{TM}$.*

Proof Take $v \in T_xM$ with $c(0) = x, \dot{c}(0) = v$. From $\alpha c^\Gamma(t) = c(0)$ and $\beta c^\Gamma(t) = c(t)$ we see that $\alpha_*\dot{c}^\Gamma(0) = 0$ and $\beta_*\dot{c}^\Gamma(0) = \dot{c}(0)$. □

Corollary 2.2.2 *The map* γ *takes its values in the Lie algebroid* $A\mathcal{G} = \ker \alpha_*|_{1_M}$. □

We shall therefore say that an *infinitesimal connection* on a general transitive Lie algebroid $\mathcal{A}$, not necessarily arising from a Lie groupoid, is a vector bundle morphism $\gamma : TM \to \mathcal{A}$ that is also a section of the anchor map $\rho : \mathcal{A} \to TM$. Thus an infinitesimal connection is a particular type of algebroid-valued 1-form on M.

We have mentioned that a transitive Lie algebroid may be written as a term in a short exact sequence

$$0 \to \mathcal{K} \xrightarrow{j} \mathcal{A} \xrightarrow{\rho} TM \to 0$$

where j is the inclusion map. We may therefore define, for any infinitesimal connection γ, an associated map $\omega : \mathcal{A} \to \mathcal{K}$, its *kernel projection*, which is the unique vector bundle morphism satisfying $\omega \circ j = \mathrm{id}_{\mathcal{K}}$ and $j \circ \omega + \gamma \circ \rho = \mathrm{id}_{\mathcal{A}}$. (In [30] the kernel projection is called the *reform* of the infinitesimal connection.) We also note the following result.

Lemma 2.2.3 *If* $\gamma : TM \to \mathcal{A}$ *is an infinitesimal connection and* ξ *is any section of* $\mathcal{A}$ *then the Lie derivative* $\mathcal{L}_\xi \gamma$ *takes its values in the kernel* $\mathcal{K}$.

Proof This follows from Lemma 1.6.1 because

$$\rho \circ \mathcal{L}_\xi \gamma = \mathcal{L}_{\rho\circ\xi}(\rho \circ \gamma) = \mathcal{L}_{\rho\circ\xi}\, \mathrm{id}_{TM} = 0.$$ □

We may now, for any infinitesimal connection $\gamma : TM \to \mathcal{A}$, define its *curvature* $R^\gamma : \mathfrak{X}(M) \times \mathfrak{X}(M) \to \sec(\mathcal{A})$ by

$$R^\gamma(X, Y) = \gamma \circ [X, Y] - [\![\gamma \circ X, \gamma \circ Y]\!],$$

so that R^γ measures the failure of γ to be a morphism of Lie algebroids. If $R^\gamma = 0$ the infinitesimal connection γ is said to be *flat*.

Lemma 2.2.4 *The curvature* R^γ *is tensorial, so that it may be represented by a map* $TM \times_M TM \to \mathcal{A}$ *(and therefore may be regarded as an* $\mathcal{A}$*-valued 2-form on* M*); in fact it takes its values in the kernel bundle* $\mathcal{K}$.

Proof The tensorial property follows from

$$\begin{aligned}
R^\gamma(X, fY) &= \gamma \circ [X, fY] - [\![\gamma \circ X, \gamma \circ fY]\!] \\
&= \gamma \circ \big(f[X, Y] + (Xf)Y\big) - [\![\gamma \circ X, f(\gamma \circ Y)]\!] \\
&= \gamma \circ f[X, Y] + (Xf)(\gamma \circ Y) - f[\![\gamma \circ X, (\gamma \circ Y)]\!] - \big((\rho \circ \gamma \circ X)f\big)(\gamma \circ Y) \\
&= \gamma \circ f[X, Y] - f[\![\gamma \circ X, (\gamma \circ Y)]\!] \\
&= fR^\gamma(X, Y).
\end{aligned}$$

As R^γ is obviously skew-symmetric, it is an $\mathcal{A}$-valued 2-form.

From the properties of the anchor map we also see that

$$\rho \circ [\![\gamma \circ X, \gamma \circ Y]\!] = [\rho \circ \gamma \circ X, \rho \circ \gamma \circ Y] = [X, Y]$$

and $\rho \circ \gamma \circ [X, Y] = [X, Y]$, so that $\rho \circ R^\gamma(X, Y) = 0$, showing that R^γ is in fact a $\mathcal{K}$-valued 2-form. □

Lemma 2.2.5 *The curvature R^γ is given in terms of the kernel projection ω of γ by*

$$R^\gamma(\rho \circ \xi, \rho \circ \eta) = -\omega \circ [\![\xi, \eta]\!] + [\![\xi, \omega \circ \eta]\!] + [\![\omega \circ \xi, \eta]\!] - [\![\omega \circ \xi, \omega \circ \eta]\!];$$

in addition the relationship

$$\omega \circ [\![\xi, \eta]\!] - [\![\omega \circ \xi, \omega \circ \eta]\!] = R^\gamma(\rho \circ \xi, \rho \circ \eta) + \mathcal{L}_\xi \gamma \circ (\rho \circ \eta) - \mathcal{L}_\eta \gamma \circ (\rho \circ \xi)$$

holds.

We appeal here to the definition of the Lie derivative by a section ξ of $\mathcal{A}$ given in Sect. 1.6: note that γ is a vector bundle morphism $TM \to \mathcal{A}$, that is, it is an example—indeed the prime example—of the first case discussed in that section.

Proof As $[\rho \circ \xi, \rho \circ \eta] = \rho \circ [\![\xi, \eta]\!]$, the first formula follows from

$$\begin{aligned} R^\gamma(\rho \circ \xi, \rho \circ \eta) &= \gamma \circ \rho \circ [\![\xi, \eta]\!] - [\![\gamma \circ \rho \circ \xi, \gamma \circ \rho \circ \eta]\!] \\ &= [\![\xi, \eta]\!] - \omega \circ [\![\xi, \eta]\!] - [\![\xi - \omega \circ \xi, \eta - \omega \circ \eta]\!] \\ &= -\omega \circ [\![\xi, \eta]\!] + [\![\xi, \omega \circ \eta]\!] + [\![\omega \circ \xi, \eta]\!] - [\![\omega \circ \xi, \omega \circ \eta]\!]. \end{aligned}$$

Rearranging, we obtain

$$[\![\omega \circ \xi, \omega \circ \eta]\!] + \omega \circ [\![\xi, \eta]\!] = [\![\xi, \omega \circ \eta]\!] + [\![\omega \circ \xi, \eta]\!] - R^\gamma(\rho \circ \xi, \rho \circ \eta)$$

so that the second formula follows from

$$\begin{aligned} -[\![\omega \circ \xi, \omega \circ \eta]\!] + \omega \circ [\![\xi, \eta]\!] &= -[\![\xi, \omega \circ \eta]\!] - [\![\omega \circ \xi, \eta]\!] + 2\omega \circ [\![\xi, \eta]\!] + R^\gamma(\rho \circ \xi, \rho \circ \eta) \\ &= [\![\xi, \gamma \circ \rho \circ \eta]\!] + [\![\gamma \circ \rho \circ \xi, \eta]\!] \\ &\quad - 2\gamma \circ \rho \circ [\![\xi, \eta]\!] + R^\gamma(\rho \circ \xi, \rho \circ \eta) \\ &= \mathcal{L}_\xi \gamma \circ (\rho \circ \eta) - \mathcal{L}_\eta \gamma \circ (\rho \circ \xi) + R^\gamma(\rho \circ \xi, \rho \circ \eta). \end{aligned}$$ □

As a tensorial map $TM \times_M TM \to \mathcal{K} \subset \mathcal{A}$ the curvature R^γ has a Lie derivative by any section ξ of $\mathcal{A}$ given by

$$(\mathcal{L}_\xi R^\gamma)(X, Y) = [\![\xi, R^\gamma(X, Y)]\!] - R^\gamma([\rho \circ \xi, X], Y) - R^\gamma(X, [\rho \circ \xi, Y]).$$

Corollary 2.2.6 *The Lie derivative of the curvature may be written as*

$$(\mathcal{L}_\xi R^\gamma)(X, Y) = \mathcal{L}_\xi\gamma \circ [X, Y] - [\![\mathcal{L}_\xi\gamma \circ X, \gamma \circ Y]\!] - [\![\gamma \circ X, \mathcal{L}_\xi\gamma \circ Y]\!].$$

Proof This follows from

$$\begin{aligned} [\![\xi, R^\gamma(X, Y)]\!] &= [\![\xi, \gamma \circ [X, Y]]\!] - [\![\xi, [\![\gamma \circ X, \gamma \circ Y]\!]]\!] \\ R^\gamma([\rho \circ \xi, X], Y) &= \gamma \circ [[\rho \circ \xi, X], Y] - [\![\gamma \circ [\rho \circ \xi, X], \gamma \circ Y]\!] \\ R^\gamma(X, [\rho \circ \xi, Y]) &= \gamma \circ [X, [\rho \circ \xi, Y]] - [\![\gamma \circ X, \gamma \circ [\rho \circ \xi, Y]]\!] \end{aligned}$$

and

$$\begin{aligned} \mathcal{L}_\xi\gamma \circ [X, Y] &= [\![\xi, \gamma \circ [X, Y]]\!] - \gamma \circ [\rho \circ \xi, [X, Y]] \\ [\![\mathcal{L}_\xi\gamma \circ X, \gamma \circ Y]\!] &= [\![[\![\xi, \gamma \circ X]\!], \gamma \circ Y]\!] - [\![\gamma \circ [\rho \circ \xi, X], \gamma \circ Y]\!] \\ [\![\gamma \circ X, \mathcal{L}_\xi\gamma \circ Y]\!] &= [\![\gamma \circ X, [\![\xi, \gamma \circ Y]\!]]\!] - [\![\gamma \circ X, \gamma \circ [\rho \circ \xi, Y]]\!]. \end{aligned}$$ □

All the above applies to infinitesimal connections defined on general transitive Lie algebroids: but if a Lie algebroid is obtained from a locally trivial Lie groupoid with a path connection, we may consider the map γ associated with the path connection as an infinitesimal connection on the Lie algebroid. This correspondence between path connections and infinitesimal connections is a bijection.

Theorem 2.2.7 *If Γ is a path connection on a locally trivial Lie groupoid $\mathcal{G}$ over M then the associated map $\gamma : TM \to A\mathcal{G}$ is an infinitesimal connection on $A\mathcal{G}$. Furthermore, every infinitesimal connection on $A\mathcal{G}$ arises in this way from a unique path connection Γ on $\mathcal{G}$.*

Proof By definition γ is a smooth vector bundle morphism projecting to the identity on M, and from Lemma 2.2.1 it is a section of $\rho = \beta_*$; it is therefore an infinitesimal connection.

Conversely, suppose $\gamma : TM \to A\mathcal{G}$ is an infinitesimal connection. Let $c : [0, 1] \to M$ be a curve; we wish to construct a lifted curve $c^\Gamma : [0, 1] \to \mathcal{G}$ satisfying the conditions for the lift by a path connection and such that $\gamma\big(\dot{c}(0)\big) = \dot{c}^\Gamma(0)$. As $\mathcal{G}$ is locally trivial, for each $t \in [0, 1]$ there is an open neighbourhood U_t of $c(t)$ and a trivialization $\mathrm{T}_t : U_t \times G \times U_t \to \mathcal{G}_{U_t}$ where $G = \mathcal{G}_{c(0)}$ (Lemma 1.1.1). By compactness we may choose finitely many of these neighbourhoods, $U_1, U_2, \ldots, U_m$, covering $c\big([0, 1]\big)$. As in the proof of Lemma 1.1.1 we may assume that each $c\big([0, 1]\big) \cap U_i$ is connected and that the U_i are indexed so that $c(0) \in U_1$ and $c(1) \in U_m$, and so that we may find $c(t_i) \in U_i \cap U_{i+1}$ for $1 \leq i \leq m - 1$ and $0 < t_1 < \cdots < t_{m-1} < 1$; put $t_0 = 0$ and $t_m = 1$.

If c is not smooth on any interval $[t_i, t_{i+1}]$, further subdivide that interval, so that the result is a sequence of subintervals where c is smooth on each subinterval, and where the image of each subinterval lies completely within one of the trivialization neighbourhoods U. We shall demonstrate the existence of a unique lifted curve segment c^Γ satisfying the condition $\gamma\big(\dot{c}(0)\big) = \dot{c}^\Gamma(0)$ and defined on the first of these intervals $[0, a]$ where $a \leq t_1$; this will be the first segment of a lifted curve defined on the whole of $[0, 1]$. Subsequent segments may be defined recursively in a similar way, using the reparametrization property.

Put $U = U_1$ and $\mathrm{T} = \mathrm{T}_1$ so that $c\big([0, a]\big) \subset U$. For each $t \in [0, a]$ we have a tangent vector $\dot{c}(t) \in T_{c(t)}U$, and hence an element $\gamma(\dot{c}(t))$ of the Lie algebroid fibre $A_{c(t)}\mathcal{G}_U \subset T_{1_{c(t)}}\mathcal{G}_U$. Using the trivialization we may write

$$\mathrm{T}_*^{-1}\big(T_{1_{c(t)}}\mathcal{G}_U\big) = T_{c(t)}U \oplus \mathfrak{g}_{\mathrm{R}} \oplus T_{c(t)}U$$

where $\mathfrak{g}_{\mathrm{R}} = T_{1_{\mathcal{G}_{c(0)}}}\mathcal{G}_{c(0)}$ is the opposite Lie algebra of $G = \mathcal{G}_{c(0)}$, and hence we may write, explicitly,

$$\mathrm{T}_*^{-1}\big(A_{c(t)}\mathcal{G}_U\big) = T_{c(t)}U \oplus \mathfrak{g}_{\mathrm{R}} \oplus \{0_{T_{c(t)}U}\}.$$

We may therefore, using $\beta_*\gamma\dot{c}(t) = \rho\gamma\dot{c}(t) = \dot{c}(t)$, put

$$\mathrm{T}_*^{-1}\gamma\dot{c}(t) = \big(\dot{c}(t), v(t), 0_{T_{c(t)}U}\big)$$

where $v : [0, a] \to \mathfrak{g}_{\mathrm{R}}$ is a curve in $\mathfrak{g}_{\mathrm{R}}$.

We now observe that there is a unique curve $g : [0, a] \to G$ with $\dot{g}(t) = r_{g(t)*}v(t)$ and $g(0) = 1_G$, so that v is the right Darboux derivative of g; as the interval $[0, a]$ is one-dimensional, questions of monodromy do not arise and g may be defined on the whole of $[0, a]$. We may therefore define a lifted curve $c^\Gamma : [0, a] \to \mathcal{G}_U$ by

$$c^\Gamma(t) = \mathrm{T}\big(c(t), g(t), c(0)\big).$$

It is clear that $\alpha c^\Gamma(t) = c(0)$ and $\beta c^\Gamma(t) = c(t)$ for all $t \in [0, a]$; we also see that

$$\begin{aligned}
\mathrm{T}_*^{-1}\dot{c}^\Gamma(t) &= \big(\dot{c}(t), \dot{g}(t), 0_{T_{c(0)}U}\big) \qquad (1a)\\
&= \big(\dot{c}(t), r_{g(t)}v(t), 0_{T_{c(0)}U}\big)\\
&= r_{(c(t),g(t),c(0))*}\big(\dot{c}(t), v(t), 0_{T_{c(t)}U}\big)\\
&= r_{\mathrm{T}^{-1}(c^\Gamma(t))*}\mathrm{T}_*^{-1}\gamma\dot{c}(t)\\
&= \mathrm{T}_*^{-1}r_{c^\Gamma(t)*}\gamma\dot{c}(t)
\end{aligned}$$

so that

$$\dot{c}^\Gamma(t) = r_{c^\Gamma(t)*}\gamma\dot{c}(t)$$

for all $t \in [0, a]$, and that in particular $\dot{c}^{\Gamma}(0) = \gamma\dot{c}(0)$. The reparametrization property of c^{Γ} follows from the uniqueness of the solution curve g.

We need to see that this construction does not depend upon the particular choice of local trivializations for $\mathcal{G}$, and it is sufficient to demonstrate this for an alternative trivialization $\tilde{\mathrm{T}} : U \times G \times U \to \mathcal{G}_U$. We may therefore write

$$\tilde{\mathrm{T}}_*^{-1} \gamma\dot{c}(t) = \big(\dot{c}(t), \tilde{v}(t), 0_{T_{c(t)}U}\big)$$

where $\tilde{v} : [0, a] \to \mathfrak{g}_{\mathrm{R}}$ is another curve in $\mathfrak{g}_{\mathrm{R}}$, and therefore obtain another curve $\tilde{g} : [0, a] \to G$ with $\dot{\tilde{g}}(t) = r_{\tilde{g}(t)*}\tilde{v}(t)$ and $\tilde{g}(0) = 1_G$; define $\tilde{c}^{\Gamma} : [0, a] \to \mathcal{G}_U$ by

$$\tilde{c}^{\Gamma}(t) = \tilde{\mathrm{T}}\big(c(t), \tilde{g}(t), c(0)\big),$$

so that as before we will have

$$\dot{\tilde{c}}^{\Gamma}(t) = r_{\tilde{c}^{\Gamma}(t)*}\gamma\dot{c}(t)$$

and therefore

$$r^{-1}_{\tilde{c}^{\Gamma}(t)*}\dot{\tilde{c}}^{\Gamma}(t) = r^{-1}_{c^{\Gamma}(t)*}\dot{c}^{\Gamma}(t). \qquad (2)$$

Now consider the curve $\mathrm{T}^{-1} \circ \tilde{c}^{\Gamma}$ in $\mathrm{T}^{-1}(\mathcal{G}_U)$; this will satisfy

$$\mathrm{T}^{-1}\tilde{c}^{\Gamma}(t) = \big(c(t), \hat{g}(t), c(0)\big)$$

for some curve $\hat{g}$ in G with $\hat{g}(0) = 1_G$, so that

$$\mathrm{T}_*^{-1}\dot{\tilde{c}}^{\Gamma}(t) = \big(\dot{c}(t), \dot{\hat{g}}(t), 0_{T_{c(0)}U}\big). \qquad (1b)$$

It follows from (1a), (1b) and (2) that

$$r^{-1}_{\tilde{c}^{\Gamma}(t)*}\mathrm{T}_*\big(\dot{c}(t), \dot{\hat{g}}(t), 0_{T_{c(0)}U}\big) = r^{-1}_{c^{\Gamma}(t)*}\mathrm{T}_*\big(\dot{c}(t), \dot{g}(t), 0_{T_{c(0)}U}\big)$$

and therefore that

$$r^{-1}_{\hat{g}(t)*}\dot{\hat{g}}(t) = r^{-1}_{g(t)*}\dot{g}(t) = v(t),$$

so that $\hat{g} = g$ by uniqueness; thus $\tilde{c}^{\Gamma} = c^{\Gamma}$.

Using this procedure for each of the successive segments we may, for any curve $c : [0, 1] \to M$, construct a lifted curve $c^{\Gamma} : [0, 1] \to \mathcal{G}$ satisfying the required properties, and in this way we define the path connection Γ whose infinitesimal connection is γ. □

2.3 Connections and Projectability

In Sect. 1.4 we defined full morphisms of Lie groupoids and of Lie algebroids, and indicated that these would be appropriate types of morphism to use when considering whether objects were ‘projectable’. We now use these morphisms to investigate the projectability of path connections on Lie groupoids, and of infinitesimal connections on Lie algebroids.

Let $\mathcal{G}$, $\mathcal{H}$ be Lie groupoids with manifolds M, N of identities, and suppose that $(\mathcal{P}, p)$ is a full morphism from $\mathcal{G}$ to $\mathcal{H}$. We shall say that the path connection Γ on $\mathcal{G}$ is *projectable to* $\mathcal{H}$ if, whenever $c_1, c_2 : [0, 1] \to M$ are curves satisfying $p \circ c_1 = p \circ c_2$, then the lifted curves $c_1^\Gamma, c_2^\Gamma : [0, 1] \to \mathcal{G}$ satisfy $\mathcal{P} \circ c_1^\Gamma = \mathcal{P} \circ c_2^\Gamma$.

Now let $\varkappa : [0, 1] \to N$ be a curve, and suppose initially that there is a neighbourhood $U \subset N$ of $\varkappa(0)$ and a local section $\varsigma : U \to M$ such that $\varkappa(t) \in U$ for all $t \in [0, 1]$. The composite $\varsigma \circ \varkappa$ is then a curve in M satisfying $p \circ (\varsigma \circ \varkappa) = \varkappa$ with a lift $(\varsigma \circ \varkappa)^\Gamma$ in $\mathcal{G}$. Define the lifted curve $\varkappa^{\Gamma^\mathcal{P}} : [0, 1] \to \mathcal{H}$ by $\varkappa^{\Gamma^\mathcal{P}} = \mathcal{P} \circ (\varsigma \circ \varkappa)^\Gamma$; by projectability this does not depend on the choice of local section ς. If there is no such neighbourhood U then split $\varkappa$ into finitely many segments and carry out the procedure described above for each segment individually, combining the results to obtain the lifted curve $\varkappa^{\Gamma^\mathcal{P}}$.

Proposition 2.3.1 *If the path connection Γ on $\mathcal{G}$ is projectable to $\mathcal{H}$ then the lifting operation $\Gamma^\mathcal{P}$ is a well-defined path connection on $\mathcal{H}$, and for any curve $c : [0, 1] \to M$ we have $(p \circ c)^{\Gamma^\mathcal{P}} = \mathcal{P} \circ c^\Gamma$.*

Proof We first describe the lifting procedure in detail for arbitrary curves in N.

Let $\varkappa : [0, 1] \to N$ be a curve. For each $t \in [0, 1]$ let U_t be a neighbourhood of $\varkappa(t)$ which is the domain of a section $\varsigma_t : U_t \to M$; by compactness we may choose finitely many of these neighbourhoods, $U_1, U_2, \ldots, U_m$, covering $\varkappa([0, 1])$, with corresponding local sections ς_i. As in the proof of Lemma 1.1.1 we may assume that each $\varkappa([0, 1]) \cap U_i$ is connected and that the U_i are indexed so that $\varkappa(0) \in U_1$ and $\varkappa(1) \in U_m$, and so that we may find $\varkappa(t_i) \in U_i \cap U_{i+1}$ for $1 \leq i \leq m - 1$ and $0 < t_1 < \cdots < t_{m-1} < 1$; put $t_0 = 0$ and $t_m = 1$. Define curves $\varkappa_i : [0, 1] \to N$, $1 \leq i \leq m$, by

$$\varkappa_i(t) = \varkappa\big(tt_i + (1 - t)t_{i-1}\big),$$

so that $\varkappa_i(t) \in U_i$ for all $t \in [0, 1]$; define the lifted curves $\varkappa_i^{\Gamma^\mathcal{P}}$ by $\varkappa_i^{\Gamma^\mathcal{P}} = \mathcal{P} \circ (\varsigma_i \circ \varkappa_i)^\Gamma$. Finally, noting that

$$\alpha_\mathcal{H} \varkappa_i^{\Gamma^\mathcal{P}}(t) = \varkappa_i(0) = \varkappa(t_{i-1}) = \varkappa_{i-1}(1) = \beta_\mathcal{H} \varkappa_{i-1}^{\Gamma^\mathcal{P}}(1)$$

for $2 \leq i \leq m$ and any $t \in [0, 1]$, define the lifted curve $\varkappa^{\Gamma^\mathcal{P}}$ by

$$\varkappa^{\Gamma^{\mathcal{P}}}(t) = \begin{cases} \varkappa_1^{\Gamma^{\mathcal{P}}}\left(\dfrac{t-t_0}{t_1-t_0}\right) = \varkappa_1^{\Gamma^{\mathcal{P}}}\left(\dfrac{t}{t_1}\right) & t_0 \le t \le t_1 \\ \varkappa_i^{\Gamma^{\mathcal{P}}}\left(\dfrac{t-t_{i-1}}{t_i-t_{i-1}}\right) \cdot \varkappa_{i-1}^{\Gamma^{\mathcal{P}}}(1) & t_{i-1} \le t \le t_i, \quad 2 \le i \le m. \end{cases}$$

We must check that $\varkappa^{\Gamma^{\mathcal{P}}}$ does not depend on the choices made. Suppose that $t_1 < 1$, and that $\hat{U}_1$ is some other neighbourhood of $\varkappa(0)$ with local section $\hat{\varsigma}_1 : \hat{U}_1 : M$ where $\varkappa(t) \in \hat{U}_1$ for $0 \le t \le \hat{t}_1$ where $t_1 < \hat{t}_1$. Let $\varkappa_1, \hat{\varkappa}_1 : [0,1] \to N$ be the corresponding initial curve segments, so that $\varkappa_1(t) = t_1 t$ and $\hat{\varkappa}_1(t) = \hat{t}_1 t$. Then, for $0 \le t \le t_1$, we have

$$\varkappa^{\Gamma^{\mathcal{P}}}(t) = \varkappa_1^{\Gamma^{\mathcal{P}}}\left(\frac{t}{t_1}\right) = \mathcal{P}(\varsigma_1 \circ \varkappa_1)^{\Gamma}\left(\frac{t}{t_1}\right) = \mathcal{P}(\hat{\varsigma}_1 \circ \varkappa_1)^{\Gamma}\left(\frac{t}{t_1}\right)$$

by projectability of Γ. If we now let $\chi : [0,1] \to [0, t_1/\hat{t}_1]$ be given by $\chi(t) = t_1 t/\hat{t}_1$ then χ is a diffeomorphism with the property that $\varkappa_1 = \hat{\varkappa}_1 \circ \chi$, so that

$$\varkappa^{\Gamma^{\mathcal{P}}}(t) = \mathcal{P}(\hat{\varsigma}_1 \circ \varkappa_1)^{\Gamma}\left(\frac{t}{t_1}\right) = \mathcal{P}(\hat{\varsigma}_1 \circ \hat{\varkappa}_1 \circ \chi)^{\Gamma}\left(\frac{t}{t_1}\right) = \mathcal{P}(\hat{\varsigma}_1 \circ \hat{\varkappa}_1)^{\Gamma}\chi\left(\frac{t}{t_1}\right)$$

by the reparametrization property of Γ, using the fact that $(\hat{\varsigma}_1 \circ \hat{\varkappa}_1)^{\Gamma}\chi(0) = (\hat{\varsigma}_1 \circ \hat{\varkappa}_1)^{\Gamma}(0) = 1_{\hat{\varsigma}_1\hat{\varkappa}_1(0)}$. But

$$\mathcal{P}(\hat{\varsigma}_1 \circ \hat{\varkappa}_1)^{\Gamma}\chi\left(\frac{t}{t_1}\right) = \mathcal{P}(\hat{\varsigma}_1 \circ \hat{\varkappa}_1)^{\Gamma}\left(\frac{t}{\hat{t}_1}\right),$$

and this is the first segment of the curve that would be obtained by lifting with the local section $\hat{\varsigma}_1$ rather than the local section ς_1. A similar argument may be applied recursively to show that the whole lifted curve $\varkappa^{\Gamma^{\mathcal{P}}}$ is independent of the particular choice of partition $[0 = t_0, t_1, t_2, \ldots, t_{m-1}, t_m = 1]$ and of the particular choice of local sections ς_i.

We now check that the lifted curve $\varkappa^{\Gamma^{\mathcal{P}}}$ does indeed satisfy the conditions for a path connection.

First, we see that

$$\varkappa^{\Gamma^{\mathcal{P}}}(0) = \varkappa_1^{\Gamma^{\mathcal{P}}}(0) = \mathcal{P}\big((\varsigma_1 \circ \varkappa_1)^{\Gamma}(0)\big) = \mathcal{P}\big(1_{\varsigma_1\varkappa_1(0)}\big) = 1_{p\varsigma_1\varkappa_1(0)} = 1_{\varkappa_1(0)} = 1_{\varkappa(0)}$$

and that

$$\begin{aligned} \alpha_{\mathcal{H}}\varkappa^{\Gamma^{\mathcal{P}}}(t) &= \alpha_{\mathcal{H}}\left(\varkappa_i^{\Gamma^{\mathcal{P}}}\left(\frac{t-t_{i-1}}{t_i-t_{i-1}}\right) \cdot \varkappa_{i-1}^{\Gamma^{\mathcal{P}}}(1) \cdot \varkappa_{i-1}^{\Gamma^{\mathcal{P}}}(1) \cdot \ldots \cdot \varkappa_1^{\Gamma^{\mathcal{P}}}(1)\right) = \alpha_{\mathcal{H}}\varkappa_1^{\Gamma^{\mathcal{P}}}(1) \\ &= \alpha_{\mathcal{H}}\mathcal{P}(\varsigma_1 \circ \varkappa_1)^{\Gamma}(0) = p\alpha_{\mathcal{G}}(\varsigma_1 \circ \varkappa_1)^{\Gamma}(0) = p(\varsigma_1 \circ \varkappa_1)(0) = \varkappa_1(0) = \varkappa(0) \end{aligned}$$

(with the appropriate modification when $0 \le t \le t_1$) whereas

$$\beta_{\mathcal{H}}\varkappa^{\Gamma^{\mathcal{P}}}(t) = \beta_{\mathcal{H}}\left(\varkappa_i^{\Gamma^{\mathcal{P}}}\left(\frac{t-t_{i-1}}{t_i-t_{i-1}}\right)\cdot \varkappa_{i-1}^{\Gamma^{\mathcal{P}}}(1)\cdot \varkappa_{i-1}^{\Gamma^{\mathcal{P}}}(1)\cdot \ldots \cdot \varkappa_1^{\Gamma^{\mathcal{P}}}(1)\right)$$

$$= \beta_{\mathcal{H}}\varkappa_i^{\Gamma^{\mathcal{P}}}\left(\frac{t-t_{i-1}}{t_i-t_{i-1}}\right) = \beta_{\mathcal{H}}\mathcal{P}(\varsigma_i\circ\varkappa_i)^{\Gamma}\left(\frac{t-t_{i-1}}{t_i-t_{i-1}}\right)$$

$$= p\beta_{\mathcal{G}}(\varsigma_i\circ\varkappa_i)^{\Gamma}\left(\frac{t-t_{i-1}}{t_i-t_{i-1}}\right) = p\varsigma_i\varkappa_i\left(\frac{t-t_{i-1}}{t_i-t_{i-1}}\right) = \varkappa_i\left(\frac{t-t_{i-1}}{t_i-t_{i-1}}\right) = \varkappa(t).$$

The reparametrization condition $\varkappa^{\Gamma^{\mathcal{P}}}\circ\chi = r_\varphi\circ(\varkappa\circ\chi)^{\Gamma^{\mathcal{P}}}$, where $\varphi = \varkappa^{\Gamma^{\mathcal{P}}}\chi(0)$, is obtained from the reparametrization condition for Γ using an argument similar to that given when showing that $\Gamma^{\mathcal{P}}$ is well defined.

Now let $c : [0,1] \to M$ be a curve, and suppose that there is a local section $\varsigma : U \to M$ of p such that $pc(t) \in U$ for all $t \in [0,1]$. By definition $(p\circ c)^{\Gamma^{\mathcal{P}}} = \mathcal{P}\circ(\varsigma\circ p\circ c)^{\Gamma}$; but the two curves $\varsigma\circ p\circ c$ and c satisfy the condition $p\circ(\varsigma\circ p\circ c) = p\circ c$, so that $(\varsigma\circ p\circ c)^{\Gamma} = c^{\Gamma}$ by projectability. We therefore see that $(p\circ c)^{\Gamma^{\mathcal{P}}} = \mathcal{P}\circ c^{\Gamma}$. By reparametrization the same result holds if it is necessary to subdivide c into several segments.

Next, suppose that $\varkappa$ is smooth at $t \in [0,1]$, and arrange the subdivision of $[0,1]$ such that $t_{i-1} < t < t_i$ for some $1 \le i \le m$; put $\bar{t} = tt_i + (1-t)t_{i-1}$. Then $\varkappa_i$ is smooth at $\bar{t}$, as is $\varsigma_i\circ\varkappa_i$. So $(\varsigma_i\circ\varkappa_i)^{\Gamma}$ is smooth at $\bar{t}$, and then $\varkappa^{\Gamma^{\mathcal{P}}} = \mathcal{P}\circ(\varsigma_i\circ\varkappa_i)^{\Gamma}$ is also smooth at $\bar{t}$, showing that $\varkappa^{\Gamma^{\mathcal{P}}}$ is smooth at t.

Now suppose that two curves $\varkappa_1, \varkappa_2 : [0,1] \to N$ satisfy $\varkappa_1(0) = \varkappa_2(0)$ and $\dot{\varkappa}_1(0) = \dot{\varkappa}_2(0)$, and that there is a local section $\varsigma : U \to M$ such that $\varkappa_1([0,1]), \varkappa_2([0,1]) \subset U$. Put $c_1 = \varsigma\circ\varkappa_1$ and $c_2 = \varsigma\circ\varkappa_2$, so that $c_1(0) = \varsigma\varkappa_1(0) = \varsigma\varkappa_2(0) = c_2(0)$ and $\dot{c}_1(0) = \varsigma_*\dot{\varkappa}_1(0) = \varsigma_*\dot{\varkappa}_2(0) = \dot{c}_1(0)$; thus

$$\dot{\varkappa}_1^{\Gamma^{\mathcal{P}}}(0) = \mathcal{P}_*\dot{c}_1^{\Gamma}(0) = \mathcal{P}_*\dot{c}_2^{\Gamma}(0) = \dot{\varkappa}_2^{\Gamma^{\mathcal{P}}}(0).$$

Clearly the same result will hold if the domain of ς is simply a neighbourhood of $\varkappa_1(0) = \varkappa_2(0)$.

Now let $\varkappa : [0,1] \to N$ satisfy $\varkappa(0) = y$, $\dot{\varkappa}(0) = w \in T_yN$ and put $\gamma^{\mathcal{P}}(w) = \dot{\varkappa}^{\Gamma^{\mathcal{P}}}(0)$, so that $\gamma^{\mathcal{P}}$ is a well-defined map $T_yN \to T_{1_y}\mathcal{H}$. If $c : [0,1] \to M$ satisfies $p\circ c = \varkappa$ in some neighbourhood of zero then

$$\left.\frac{d(p\circ c)^{\Gamma^{\mathcal{P}}}}{dt}\right|_{t=0} = \left.\frac{d(\mathcal{P}\circ c^{\Gamma})}{dt}\right|_{t=0} = \mathcal{P}_*\left.\frac{dc^{\Gamma}}{dt}\right|_{t=0}$$

so that $\gamma^{\mathcal{P}}\circ p_* = \mathcal{P}_*\circ\gamma$, where $\gamma(v) = \dot{c}^{\Gamma}(0)$, $v = \dot{c}(0)$.

Finally let U be a neighbourhood of y and let $\varsigma : U \to X$ be a local section of p, so that $p_*\circ\varsigma_* = \mathrm{id}_{TU}$. Then

$$\gamma^{\mathcal{P}}|_{TU} = \gamma^{\mathcal{P}}\circ p_*\circ\varsigma_* = \mathcal{P}_*\circ\gamma\circ\varsigma_*$$

so that $\gamma^{\mathcal{P}}$ is a smooth fibre-linear map $TN \to T\mathcal{H}$. □

We now consider the projectability of infinitesimal connections on Lie algebroids. Let $\mathcal{A} \to M$ and $\mathcal{B} \to N$ be Lie algebroids, and let (P, p) be a full Lie algebroid morphism from $\mathcal{A}$ to $\mathcal{B}$. We shall say that the infinitesimal connection γ is *projectable to* $\mathcal{B}$ if, whenever $v_1, v_2 \in TM$ with $p_*(v_1) = p_*(v_2)$, then $P\gamma(v_1) = P\gamma(v_2)$. If γ is projectable, define its projection $\gamma^P : TN \to \mathcal{B}$ by, for $w \in TN$, choosing $v \in TM$ with $p_*(v) = w$ and setting $\gamma^P(w) = P\gamma(v)$.

Proposition 2.3.2 *The map* γ^P *is an infinitesimal connection on* $\mathcal{B}$.

Proof First, take $w_1, w_2 \in T_yN$ for some $y \in N$, and choose $x \in M$ such that $p(x) = y$. As p is a submersion we may find $v_1, v_2 \in T_xM$ such that $p_*(v_1) = w_1$ and $p_*(v_2) = w_2$. Then $\gamma^P(w_1) = P\gamma(v_1)$ and $\gamma^P(w_2) = P\gamma(v_2)$, so that

$$\tau_{\mathcal{B}}\gamma^P(w_1) = \tau_{\mathcal{B}}P\gamma(v_1) = p\tau_{\mathcal{A}}\gamma(v_1) = p(x) = p\tau_{\mathcal{A}}\gamma(v_2) = \tau_{\mathcal{B}}P\gamma(v_2) = \tau_{\mathcal{B}}\gamma^P(w_1),$$

showing that γ^P is well defined. To see that it is fibred over the identity on N, take $w \in T_yN$ and as before choose $x \in M$ such that $p(x) = y$. Taking $v \in T_xM$ such that $p_*(v) = w$ we see that

$$\tau_B\gamma^P(w) = \tau_B P\gamma(v) = p\tau_A\gamma(v) = p(x) = y.$$

To show that γ^P is smooth at $w \in T_yN$, choose a neighbourhood U of y and a local section $\varsigma : U \to M$ of p, so that $\varsigma_* : TU \to TM$ satisfies $p_* \circ \varsigma_* = (\mathrm{id}_U)_* = \mathrm{id}_{TU}$. Then $\gamma^P\big|_{TU} = P \circ \gamma \circ \varsigma_*$, showing that γ^P is smooth.

It also follows from this that, restricted to the fibre T_yN, $\gamma^P\big|_y = P|_{\gamma\varsigma(y)} \circ \gamma|_{\varsigma(y)} \circ \varsigma_{*y}$ is a composition of linear maps and so is linear. Furthermore, the rank of each map in the composition does not depend on the choice of fibre to which it is restricted, so that the composite map γ^P has constant rank on TU. Although the map ς is only a local section of the surjective submersion p, different maps ς_* will all have the same constant rank $\dim M$ (even if M is not connected), and as γ and P have global constant rank it follows that γ^P has constant rank throughout TM and is therefore a vector bundle morphism.

Finally, for any $w \in TN$ take $v \in TM$ with $w = p_*(v)$ so that

$$\rho_B\gamma^P(w) = \rho_B P\gamma(v) = p_*\rho_A\gamma(v) = p_*(v) = w;$$

thus γ^P is a section of the anchor map $\rho_{\mathcal{B}}$, confirming that γ^P is indeed an infinitesimal connection. □

Lemma 2.3.3 *If* γ *is a projectable infinitesimal connection and* X *is a projectable vector field on* M *then the section* $\gamma \circ X$ *of* $\mathcal{A}$ *is projectable; in addition* $(\gamma \circ X)^P = \gamma^P \circ X^p$.

Proof If $x_1, x_2 \in M$ satisfy $p(x_1) = p(x_2)$ then, because $p_*X_{x_1} = p_*X_{x_2}$ by projectability of X, we see that $P\gamma(X_{x_1})) = P\gamma(X_{x_2})$ by projectability of γ, so that the section $\gamma \circ X$ is projectable.

Now let $y \in N$, and let $x \in M$ satisfy $p(x) = y$. By definition the section $(\gamma \circ X)^P$ of $\mathcal{B}$ satisfies

$$(\gamma \circ X)^P(y) = P\big((\gamma \circ X)(x)\big) = P\gamma(X_x);$$

but the section $\gamma^P \circ X^p$ satisfies

$$(\gamma^P \circ X^p)(y) = \gamma^P(X^p_y) = \gamma^P p_*(X_x) = P\gamma(X_x)$$

so $(\gamma \circ X)^P = \gamma^P \circ X^p$. □

Corollary 2.3.4 *The curvature R^{γ^P} of the projected infinitesimal connection γ^P satisfies*

$$R^{\gamma^P}(X^p, Y^p) = \big(R^\gamma(X, Y)\big)^P.$$

Proof A straightforward calculation gives

$$\begin{aligned} R^{\gamma^P}(X^p, Y^p) &= \gamma^P \circ [X^p, Y^p] - [\![\gamma^P \circ X^p, \gamma^P \circ Y^p]\!] \\ &= \gamma^P \circ [X, Y]^p - [\![\gamma^P \circ X^p, \gamma^P \circ Y^p]\!] \\ &= \big(\gamma \circ [X, Y]\big)^P - [\![(\gamma \circ X)^P, (\gamma \circ Y)^P]\!] \\ &= \big(\gamma \circ [X, Y]\big)^P - [\![\gamma \circ X, \gamma \circ Y]\!]^P \\ &= \big(R^\gamma(X, Y)\big)^P. \end{aligned}$$

□

2.4 The Kernel Derivative of an Infinitesimal Connection

In this and the following sections we turn our attention to various notions of covariant derivative associated with an infinitesimal connection on a Lie algebroid.

Let $\gamma : TM \to \mathcal{A}$ be an infinitesimal connection. We define an operator on sections κ of the kernel $\mathcal{K} \to M$, the *kernel derivative*, by $\nabla^\gamma_X \kappa = [\![\gamma \circ X, \kappa]\!]$.

Lemma 2.4.1 *The operator ∇^γ is a covariant derivative.*

Proof The operator is clearly $\mathbb{R}$-linear in both variables, and

$$\nabla^\gamma_{fX}\kappa = [\![\gamma \circ (fX), \kappa]\!] = [\![f(\gamma \circ X), \kappa]\!] = f[\![\gamma \circ X, \kappa]\!] - \big((\rho \circ \kappa)f\big)(\gamma \circ X) = f\nabla^\gamma_X \kappa$$

whereas

$$\nabla^\gamma_X(f\kappa) = [\![\gamma \circ X, f\kappa]\!] = f[\![\gamma \circ X, \kappa]\!] + \big((\rho \circ \gamma \circ X)f\big)\kappa = f\nabla^\gamma_X \kappa + (Xf)\kappa. \qquad \square$$

Note that the formula $[\![\gamma \circ X, \xi]\!]$ does not define a covariant derivative on sections ξ of $\mathcal{A}$ because the condition $\rho \circ \xi = 0$ is needed to ensure that $\nabla^\gamma_{fX} = f\nabla^\gamma_X$.

Lemma 2.4.2 *The kernel derivative* ∇^γ *is a derivation of the Lie algebra bracket* $\{\cdot, \cdot\}$ *on the fibres of* $\mathcal{K}$*:*

$$\nabla^\gamma_X\{\kappa, \lambda\} = \{\nabla^\gamma_X \kappa, \lambda\} + \{\kappa, \nabla^\gamma_X \lambda\}.$$

Proof The Lie algebra bracket is the restriction to $\mathcal{K}$ of the Lie algebroid bracket on sections of $\mathcal{A}$, so that this is just the Jacobi identity again:

$$\begin{aligned}\nabla^\gamma_X\{\kappa, \lambda\} &= [\![\gamma \circ X, \{\kappa, \lambda\}]\!] = [\![\gamma \circ X, [\![\kappa, \lambda]\!]]\!] = [\![[\![\gamma \circ X, \kappa]\!], \lambda]\!] + [\![\kappa, [\![\gamma \circ X, \lambda]\!]]\!] \\ &= [\![\nabla^\gamma_X \kappa, \lambda]\!] + [\![\kappa, \nabla^\gamma_X \lambda]\!] = \{\nabla^\gamma_X \kappa, \lambda\} + \{\kappa, \nabla^\gamma_X \lambda\}. \qquad \square\end{aligned}$$

We may apply the kernel derivative to the curvature of γ.

Lemma 2.4.3 *The curvature* R^γ *of the infinitesimal connection satisfies the identity*

$$\oint \big(\nabla^\gamma_X(R^\gamma(Y, Z)) + R^\gamma(X, [Y, Z])\big) = 0$$

where $\oint$ *indicates the cyclic sum over* X*,* Y *and* Z*.*

Proof We have

$$\begin{aligned}\nabla^\gamma_X(R^\gamma(Y, Z)) &= [\![\gamma \circ X, \gamma \circ [Y, Z]]\!] - [\![\gamma \circ X, [\![\gamma \circ Y, \gamma \circ Z]\!]]\!] \\ &= \gamma \circ [X, [Y, Z]] - R^\gamma(X, [Y, Z]) - [\![\gamma \circ X, [\![\gamma \circ Y, \gamma \circ Z]\!]]\!].\end{aligned}$$

The result now follows by the Jacobi identity for both brackets. $\square$

This single identity generalizes the two Bianchi identities of standard connection theory, and in fact incorporates them both as we shall show later: it is therefore called the *Bianchi identity*.

Lemma 2.4.4 *The curvature* R^∇ *of* ∇^γ *is related to the curvature* R^γ *of* γ *as follows:*

$$R^\nabla(X, Y)\kappa = [\![\kappa, R^\gamma(X, Y)]\!] = \{\kappa, R^\gamma(X, Y)\}.$$

Proof

$$\begin{aligned}
R^{\nabla}(X,Y)\kappa &= \nabla^{\gamma}_{X}\nabla^{\gamma}_{Y}\kappa - \nabla^{\gamma}_{Y}\nabla^{\gamma}_{X}\kappa - \nabla^{\gamma}_{[X,Y]}\kappa \\
&= [\![\gamma\circ X, [\![\gamma\circ Y, \kappa]\!]]\!] - [\![\gamma\circ Y, [\![\gamma\circ X, \kappa]\!]]\!] - [\![\gamma\circ[X,Y], \kappa]\!] \\
&= [\![\kappa, \gamma\circ[X,Y]]\!] - [\![\kappa, [\![\gamma\circ X, \gamma\circ Y]\!]]\!] \\
&= [\![\kappa, R^{\gamma}(X,Y)]\!] = \{\kappa, R^{\gamma}(X,Y)\}. \qquad \square
\end{aligned}$$

We may use the kernel derivative to define a 'differential' on $\mathcal{K}$-valued p-forms on M. Writing $\operatorname{sec}\big(\mathcal{K}\otimes\bigwedge^{p}(T^{*}M)\big)$ for the $C^{\infty}(M)$-module of $\mathcal{K}$-valued p-forms on M, an infinitesimal connection γ defines a map of modules

$$d_{\nabla^{\gamma}} : \operatorname{sec}\big(\mathcal{K}\otimes {\textstyle\bigwedge^{p}}(T^{*}M)\big) \to \operatorname{sec}\big(\mathcal{K}\otimes {\textstyle\bigwedge^{p+1}}(T^{*}M)\big),$$

which we call the exterior kernel differential, as follows:

$$\begin{aligned}
d_{\nabla^{\gamma}}Q(X_1, X_2, \ldots, X_{p+1}) &= \sum_{r=1}^{p+1}(-1)^{r+1}\nabla^{\gamma}_{X_r}(Q(X_1, \ldots \widehat{X_r}\ldots, X_{p+1})) \\
&\quad + \sum_{1\le r,s\le p+1}(-1)^{r+s}Q([X_r, X_s], X_1, \ldots \widehat{X_r}\ldots\widehat{X_s}\ldots, X_{p+1}) \\
&= \sum_{r=1}^{p+1}(-1)^{r+1}[\![\gamma\circ X_r, Q(X_1, \ldots \widehat{X_r}\ldots, X_{p+1})]\!] \\
&\quad + \sum_{1\le r,s\le p+1}(-1)^{r+s}Q([X_r, X_s], X_1, \ldots \widehat{X_r}\ldots\widehat{X_s}\ldots, X_{p+1}).
\end{aligned}$$

The formula has obvious similarities with a well-known one for the exterior derivative of a p-form, and the fact that the exterior kernel differential is a module map can be deduced in the same way as the fact that the exterior derivative has that property can. But $d_{\nabla^{\gamma}}$ is not in general a coboundary operator of course.

Lemma 2.4.5 *The Bianchi identity for* γ*, namely*

$$\oint\big(\nabla^{\gamma}_{X}(R^{\gamma}(Y,Z)) + R^{\gamma}(X,[Y,Z])\big) = 0,$$

may be expressed using the exterior kernel differential as

$$d_{\nabla^{\gamma}}R^{\gamma} = 0. \qquad \square$$

2.5 Covariant Algebroid Derivatives

We now wish to generalize the idea of a covariant derivative, replacing vector fields by sections of a Lie algebroid $\mathcal{A} \to M$. The operators we describe may be defined on any vector bundle over M, and are independent of any infinitesimal connection on $\mathcal{A}$.

As motivation, consider a vector bundle $\pi : E \to M$ with a covariant derivative operator ∇ giving, for every section σ of E and any vector field X on M, a new section $\nabla_X \sigma$ of E. As every Lie algebroid $\mathcal{A} \to M$ supports an anchor map $\rho : \mathcal{A} \to M$, we may also define the derivative of a section of E by a section ξ of $\mathcal{A}$ to be $\nabla_{\rho\circ\xi}\sigma$, its derivative by the vector field $\rho \circ \xi$. This observation suggests that, given a vector bundle and a Lie algebroid over the same base manifold M, we could define a more general type of covariant derivative operator. So a *covariant algebroid derivative on E*, or more specifically an *$\mathcal{A}$-derivative on E*, will be defined to be an $\mathbb{R}$-linear map D from $\mathrm{sec}(\mathcal{A})$ to $\mathbb{R}$-linear operators $\mathrm{sec}(E) \to \mathrm{sec}(E)$, $\xi \mapsto D_\xi$, such that for $f \in C^\infty(M)$ and $\sigma \in \mathrm{sec}(E)$

$$
\begin{aligned}
D_{f\xi}\sigma &= fD_\xi\sigma \\
D_\xi(f\sigma) &= fD_\xi\sigma + (\rho \circ \xi)(f)\sigma.
\end{aligned}
$$

Thus D behaves much like an ordinary covariant derivative operator arising from a linear connection; and indeed a 'TM-derivative' on E is just a covariant derivative in the ordinary sense.

As in the case of an ordinary covariant derivative operator, the first of these conditions ensures that $(D_\xi\sigma)(x)$ depends only on the value of ξ at x, so one may also think of D as defining, for each $\boldsymbol{a} \in \mathcal{A}_x$, an $\mathbb{R}$-linear operator $D_{\boldsymbol{a}}$ from local sections of E defined near x to E_x. From the second condition it follows that if $\rho(\boldsymbol{a}) = 0$ then $D_{\boldsymbol{a}}\sigma$ depends only on the value of σ at x, and $D_{\boldsymbol{a}}$ then defines a linear map of E_x to itself. Likewise, if D and $\hat{D}$ are two $\mathcal{A}$-derivatives on the same vector bundle E then for each $\xi \in \mathrm{sec}(\mathcal{A})$, $\hat{D}_\xi - D_\xi$ is a vector bundle morphism of E.

There is an obvious generalization of the concept of curvature to $\mathcal{A}$-derivatives. For any $\mathcal{A}$-derivative D we set

$$C(\xi, \eta)\sigma = D_\xi(D_\eta\sigma) - D_\eta(D_\xi\sigma) - D_{[\![\xi,\eta]\!]}\sigma.$$

Then $C(\xi, \eta)\sigma \in \mathrm{sec}(E)$; it depends $C^\infty(M)$-linearly on all of its arguments, and it is skew-symmetric in the first two. For each $\xi, \eta \in \mathrm{sec}(\mathcal{A})$, $C(\xi, \eta)$ is a vector bundle morphism of E. The object C so defined is called the *curvature* of D. An $\mathcal{A}$-derivative whose curvature vanishes is said to be *flat*.

For $\mathcal{A}$-derivatives on $\mathcal{A}$ we can also define the *torsion* T of D,

$$T(\xi, \eta) = D_\xi\eta - D_\eta\xi - [\![\xi, \eta]\!];$$

$T(\xi, \eta) \in \sec(\mathcal{A})$, and depends $C^\infty(M)$-linearly and skew-symmetrically on its arguments. Notice that if D is an $\mathcal{A}$-derivative on $\mathcal{A}$ so is D^* defined by

$$D^*_\xi \eta = D_\eta \xi + [\![\xi, \eta]\!];$$

D^* is called the $\mathcal{A}$-derivative on $\mathcal{A}$ *dual* to D. Then

$$T(\xi, \eta) = D_\xi \eta - D^*_\xi \eta.$$

For an $\mathcal{A}$-derivative on $\mathcal{A}$, the operator D extends in the obvious way to tensor-type objects, such as its torsion T, so that

$$D_\xi T(\eta, \zeta) = D_\xi(T(\eta, \zeta)) - T(D_\xi \eta, \zeta) - T(\eta, D_\xi \zeta).$$

Moreover, the curvature and torsion of an $\mathcal{A}$-derivative on $\mathcal{A}$ satisfy *Bianchi identities*, which may be derived in exactly the same way as those for an ordinary covariant derivative on TM: in particular

$$\oint \big(C(\xi, \eta)\zeta + D_\xi T(\eta, \zeta) + T(T(\xi, \eta), \zeta)\big) = 0$$

(cyclic sum)—the first Bianchi identity for D.

2.6 Representations and Semidirect Sums

Let $\mathcal{A} \to M$ be a Lie algebroid, and let D be an $\mathcal{A}$-derivative on $E \to M$. We have a particular interest in $\mathcal{A}$-derivatives with vanishing curvature: flat $\mathcal{A}$-derivatives. A flat $\mathcal{A}$-derivative D satisfies

$$D_{[\![\xi,\eta]\!]} = D_\xi \circ D_\eta - D_\eta \circ D_\xi = [D_\xi, D_\eta]$$

for all $\xi, \eta \in \sec(\mathcal{A})$ and therefore defines a homomorphism from $\sec(\mathcal{A})$ with the Lie algebroid bracket to the space of linear operators on $\sec(E)$ with the commutator bracket; it is therefore called a *representation* of $\mathcal{A}$ on E. A flat $\mathcal{A}$-derivative on $\mathcal{A}$ itself is called a *self-representation* of $\mathcal{A}$.

If $\mathcal{A}$ is transitive and $\mathcal{K}$ is its kernel then for any $\kappa \in \sec(\mathcal{K})$ and $\xi \in \sec(\mathcal{A})$, $[\![\xi, \kappa]\!] \in \sec(\mathcal{K})$; and furthermore if we set $D_\xi \kappa = [\![\xi, \kappa]\!]$ then D is an $\mathcal{A}$-derivative on $\mathcal{K}$, which by the Jacobi identity is flat. The corresponding representation is called the *canonical representation* of $\mathcal{A}$. The kernel $\mathcal{K}$ is a Lie algebra bundle with Lie algebra bracket which is just the restriction of the Lie algebroid bracket. Now for the canonical representation

$$D_\xi [\![\kappa, \lambda]\!] = [\![D_\xi \kappa, \lambda]\!] + [\![\kappa, D_\xi \lambda]\!]$$

for all $\kappa, \lambda \in \sec(\mathcal{K})$, again by the Jacobi identity, which is to say that D is a derivation of the Lie algebra bracket on $\mathcal{K}$. We may regard the canonical representation as a template for the kernel derivatives of infinitesimal connections on $\mathcal{A}$, because for any such connection γ we have

$$\nabla^{\gamma}_{X}\kappa = [\![\gamma \circ X, \kappa]\!] = D_{\gamma \circ X}\kappa.$$

We have already observed that the kernel derivative of an infinitesimal connection does not extend to the whole of $\mathcal{A}$, and for the same reason $\eta \mapsto [\![\xi, \eta]\!]$ does not define an $\mathcal{A}$-derivative on $\mathcal{A}$: there is no canonical self-representation of $\mathcal{A}$. Self-representations of a Lie algebroid which extend the canonical representation, that is, which reduce to it when restricted to acting on $\mathcal{K}$, are therefore of particular interest, as we shall see in Chap. 7.

Now suppose we have a short exact sequence of vector bundles over M,

$$0 \to \mathcal{B} \to \mathcal{V} \xrightarrow{P} \mathcal{A} \to 0,$$

where $\mathcal{A}$ is a Lie algebroid with bracket $[\![\cdot, \cdot]\!]$ and $\mathcal{B}$ is a bundle of Lie algebras with bracket $\{\cdot, \cdot\}$, and suppose also that there is a representation D of $\mathcal{A}$ on $\mathcal{B}$ satisfying the derivation property

$$D_{\xi}\{\kappa, \lambda\} = \{D_{\xi}\kappa, \lambda\} + \{\kappa, D_{\xi}\lambda\}$$

where ξ is a section of $\mathcal{A}$ and κ, λ are sections of $\mathcal{B}$. If it is the case that the short exact sequence of sections

$$0 \to \sec(\mathcal{B}) \to \sec(\mathcal{V}) \to \sec(\mathcal{A}) \to 0$$

splits, so that we can write $\sec(\mathcal{V}) = \sec(\mathcal{A}) \oplus \sec(\mathcal{B})$, then we can define a bracket on sections of $\mathcal{V}$ by the formula

$$[\![(\xi, \kappa), (\eta, \lambda)]\!]^{D} = \Big([\![\xi, \eta]\!], D_{\xi}\lambda - D_{\eta}\kappa + \{\kappa, \lambda\}\Big)$$

Proposition 2.6.1 *The bracket $[\![\cdot, \cdot]\!]^{D}$ is a Lie bracket under which $\mathcal{V}$ becomes a Lie algebroid with anchor map $\rho \circ P$ where ρ is the anchor of $\mathcal{A}$. If $\mathcal{A} = TM$ then $\mathcal{B}$ is the kernel of $\mathcal{V}$.*

Proof The bracket $[\![\cdot, \cdot]\!]^{D}$ is the standard *semidirect sum bracket* on $\sec(\mathcal{V}) = \sec(\mathcal{A}) \oplus \sec(\mathcal{B})$; skew-symmetry and $\mathbb{R}$-bilinearity are obvious, and the Jacobi identity is a straightforward calculation using the facts that D is flat and that D_{ξ} is a derivation. By construction

$$P \circ [\![(\xi, \kappa), (\eta, \lambda)]\!]^{D} = P\Big([\![\xi, \eta]\!], D_{\xi}\lambda - D_{\eta}\kappa + \{\kappa, \lambda\}\Big) = [\![\xi, \eta]\!] = [\![P \circ (\xi, \kappa), P \circ (\eta, \lambda)]\!],$$

so that
$$\rho \circ P \circ [\![(\xi, \kappa), (\eta, \lambda)]\!]^D = [\rho \circ P \circ (\xi, \kappa), \rho \circ P \circ (\eta, \lambda)].$$

Finally if $f \in C^\infty(M)$ then

$$\begin{aligned} [\![(\xi, \kappa), f(\eta, \lambda)]\!]^D &= \Big([\![\xi, f\eta]\!], D_\xi(f\lambda) - D_{f\eta}\kappa + \{\kappa, f\lambda\}\Big) \\ &= \Big(f[\![\xi, \eta]\!] + ((\rho \circ \xi)f)\eta, fD_\xi\lambda + ((\rho \circ \xi)f + f\{\kappa, \lambda\})\lambda - fD_\eta\kappa\Big) \\ &= f\Big([\![\xi, \eta]\!], D_\xi\lambda - D_\eta\kappa + \{\kappa, \lambda\}\Big) + ((\rho \circ \xi)f)(\eta, \lambda) \\ &= f[\![(\xi, \kappa), (\eta, \lambda)]\!]^D + ((\rho \circ P \circ (\xi, \kappa))f)(\eta, \lambda). \end{aligned}$$

It is clear that the restriction of $[\![\cdot, \cdot]\!]^D$ to $\mathcal{B}$ is $\{\cdot, \cdot\}$, so if $\mathcal{A} = TM$ then it is immediate that $\mathcal{B}$ is the kernel of $\mathcal{V}$. □

Corollary 2.6.2 *Suppose that $\mathcal{K} \to M$ is a Lie algebra bundle, with bracket $\{\cdot, \cdot\}$, which as a vector bundle is equipped with a flat TM-connection whose covariant derivative operator ∇ is a derivation of $\{\cdot, \cdot\}$. Then the vector bundle $TM \oplus_M \mathcal{K}$ can be given the structure of a transitive Lie algebroid by defining the bracket of sections by*
$$[\![(X, \kappa), (Y, \lambda)]\!] = \Big([X, Y], \nabla_X\lambda - \nabla_Y\kappa + \{\kappa, \lambda\}\Big)$$
and the anchor by $\rho \circ (X, \kappa) = X$. Furthermore, $X \mapsto (X, 0)$ defines an infinitesimal connection γ, such that $R^\gamma = 0$. □

We have already seen an example of the construction given in the above corollary, namely the Lie algebroid of the trivial Lie groupoid $M \times G \times M$ described in Proposition 1.3.2, where the covariant derivative $\nabla_X\kappa$ of the section κ of the trivial Lie algebra bundle $M \times \mathfrak{g}$ is taken as the Lie derivative of the corresponding $\mathfrak{g}$-valued function $\bar{\kappa}$, namely
$$\nabla_X\kappa = \nabla_X(\mathrm{id}, \bar{\kappa}) = \big(\mathrm{id}, X(\bar{\kappa})\big).$$

An example of the construction in the substantive proposition, where the short exact sequence of vector bundles does not have a canonical splitting but the induced sequence of sections does, comes when we consider the jet bundle of a Lie algebroid $\tau : \mathcal{A} \to M$.

As τ is by definition a vector bundle, its first jet bundle $\tau_1 : J^1\tau \to M$ is also a vector bundle, with the vector space structure on its fibres inherited from the vector space structure on the set of global sections of τ. It is a standard construction that $J^1\tau$ is part of a short exact sequence of vector bundles

$$0 \to \mathcal{A} \otimes T^*M \to J^1\tau \to \mathcal{A} \to 0$$

where the map $\tau_{1,0} : J^1\tau \to \mathcal{A}$ is given by $\tau_{1,0}(j^1_x\xi) = \xi(x)$; the injection $\mathcal{A} \otimes T^*M \to J^1\tau$ arises by observing that the map of sections $\xi \otimes df \mapsto j^1(f\xi) - f(j^1\xi)$, where

$f \in C^\infty(M)$, gives a well-defined pointwise map. A splitting of this sequence amounts to a linear connection on the vector bundle τ, as we shall see in the next section; but the sequence of sections *does* have a canonical splitting $j^1 : \mathrm{sec}(\mathcal{A}) \to \mathrm{sec}(J^1\tau)$, so that any section of $\tau_1 : J^1\tau \to M$ may be written uniquely as a sum $j^1\xi + Q$ where ξ is a section of $\mathcal{A}$ and Q is an $\mathcal{A}$-valued 1-form on M, regarded as a section of $J^1\tau$.

We first define a bracket on the sections of $\mathcal{A} \otimes_M T^*M$, regarded as vector bundle morphisms $TM \to \mathcal{A}$, by

$$\{P, Q\} = Q \circ \rho \circ P - P \circ \rho \circ Q.$$

It is easy to check that this bracket is well-defined pointwise, and defines the structure of a Lie algebra bundle on $\mathcal{A} \otimes T^*M$. If, for instance, P and Q are both infinitesimal connections on $\mathcal{A}$ (that is, $\rho \circ P = \rho \circ Q = \mathrm{id}_M$) then $\{P, Q\}$ is just the difference $Q - P$.

We next define a family of endomorphisms of $\mathrm{sec}(\mathcal{A} \otimes T^*M)$, parametrized by sections of $\mathcal{A}$, by setting

$$D_\xi(\eta \otimes \theta) = [\![\xi, \eta]\!] \otimes \theta + \eta \otimes \mathcal{L}_{\rho\circ\xi}\theta$$

where ξ, η are sections of $\mathcal{A}$ and θ is a 1-form on M, and extending by $\mathbb{R}$-linearity. Each D_ξ is a derivation, and it may easily be shown that D is a representation of $\mathcal{A}$ on $\mathcal{A} \otimes T^*M$, so we conclude that the bracket $[\![\cdot, \cdot]\!]^D$ together with the map $\rho \circ \tau_{1,0} : J^1\tau \to TM$ provide a Lie algebroid structure on $J^1\tau$.

2.7 Linearizing a Connection

We conclude this chapter with some remarks about general connections on vector bundles, with no particular reference to Lie algebroids.

One way of describing a general connection on the vector bundle $\pi : E \to M$ is by specifying a smooth complement H to the vertical sub-bundle $V\pi \subset TE$; this complement may be described equivalently as the image of a horizontal lift operator, whereby a vector field X on M is mapped to a vector field X^h on E. The direct sum decomposition $TE = H \oplus V\pi$ gives rise to a projection operator P_H which, as mentioned in the introduction to Sect. 2.2, may be regarded as a type (1, 1) tensor field on E.

There is no reason why a connection in this general form should be linear (and therefore correspond to a covariant derivative operator). It is, however, always possible to represent such a connection by a map between vector bundles, and hence obtain a derived linear connection on a pullback bundle by differentiation, using jet bundles. If ξ is a section of $\pi : E \to M$ then its jet $j^1_x\xi$ at the point x may be identified with the tangent map $\xi_{*x} : T_xM \to T_{\xi(x)}E$; using this identification we may construct a canonical tensor field h along the map $\pi_{1,0} : J^1\pi \to E$ by setting $\mathrm{h}_{j^1_x\xi}(v) = (\xi_x \circ \pi)_* v$ for any $v \in T_{\xi(x)}E$. A connection may then be described by a section $\sigma : E \to J^1\pi$, so that $\mathrm{h} \circ \sigma$ is a genuine tensor field on E (rather than along $\pi_{1,0}$) giving rise to the decomposition $TE = H \oplus V\pi$.

We are now able to say that the connection is a linear connection if the map σ is linear: that is, if it provides a splitting of the short exact sequence

$$0 \to E \otimes T^*M \to J^1\pi \to E \to 0,$$

as we mentioned earlier. Recalling that the corresponding exact sequence of spaces of sections

$$0 \to \sec(E \otimes T^*M) \to \sec(J^1\pi) \to \sec(E) \to 0$$

has the canonical splitting $\xi \mapsto j^1\xi$, we may define a covariant differential

$$\nabla : \sec(E) \to \sec(E \otimes T^*M), \qquad \nabla\xi = j^1\xi - \sigma \circ \xi$$

and hence, for any vector field X on M, a covariant derivative ∇_X acting on sections of $E \to M$.

If the connection σ is not linear, we proceed in the following way.

For any $x \in M$ let σ_x denote the restriction of σ to the vector space E_x, so that $\sigma_x : E_x \to J^1_x\pi$. The derivative σ'_x is a map $E_x \times E_x \to J^1_x\pi$, linear in the second variable, so we may construct the fibre derivative

$$\mathcal{F}\sigma : \pi^*E \to J^1\pi, \qquad \mathcal{F}\sigma(y, z) = \sigma'_{\pi(y)}(y, z)$$

where $\pi^*(\pi) : \pi^*(E) = E \times_M E \to M$ is the pullback bundle. There is also a canonical inclusion $J^1\pi \subset J^1\big(\pi^*(\pi)\big)$ as a vector sub-bundle, given by $j^1_x\xi \mapsto j^1_x(\xi, 0_\pi)$ where $0_\pi : M \to E$ is the zero section, and in this way $\mathcal{F}\sigma$ becomes a linear connection on the pullback bundle $\pi^*(\pi)$.

This construction of a derived linear connection will be used in Chap. 12, and it will be convenient to describe it using coordinates. If (x^i, y^α) are fibred coordinates on E then the horizontal lift $X \mapsto X^\mathrm{h}$ is given by

$$\frac{\partial}{\partial x^i} \mapsto \frac{\partial}{\partial x^i} - \Gamma^\alpha_i(x, y)\frac{\partial}{\partial y^\alpha}$$

(the minus sign is conventional) and the projection operator P_H is given by

$$dx^i \otimes \left(\frac{\partial}{\partial x^i} - \Gamma^\alpha_i(x, y)\frac{\partial}{\partial y^\alpha}\right).$$

For the jet bundle description we let y^α_i be the jet coordinates, so that

$$\mathrm{h} = dx^i \otimes \left(\frac{\partial}{\partial x^i} + y_i^\alpha \frac{\partial}{\partial y^\alpha} \right);$$

then $\Gamma_i^\alpha = -\sigma_i^\alpha = -y_i^\alpha \circ \sigma$. If in addition $(x^i; y^\alpha, z^\alpha)$ are fibred coordinates on $\pi^*(E)$ then

$$y_i^\alpha \circ \mathcal{F}\sigma = -\frac{\partial \Gamma_i^\alpha}{\partial y^j} z^j, \qquad z_i^\alpha \circ \mathcal{F}\sigma = 0.$$

Chapter 3
Groupoids of Fibre Morphisms

Our main concern in this work will not be with general Lie groupoids and Lie algebroids, but with those arising when we are given a fibre bundle and consider diffeomorphisms from one fibre to another that respect the group action in a suitable sense. Whereas the maps from a single fibre *to itself* satisfying such a condition will form a group, this will obviously not be the case when we consider maps with different domains and codomains: we will obtain, instead, a groupoid. This is exactly analogous to the relationship between the fundamental group of a topological space with a given basepoint, and the fundamental groupoid of the space.

We shall see that the groupoid obtained from such maps—fibre morphisms, we shall call them—is indeed a Lie groupoid, and so gives rise to an accompanying Lie algebroid. We shall then see that we can represent sections of this Lie algebroid by projectable vector fields on the total space of the original fibre bundle, and elements of the Lie algebroid by vector fields 'along a fibre' of the bundle.

We shall also compare this with a more traditional approach. Every fibre bundle is an associated bundle of some principal G-bundle using an action of the group G on the standard fibre of the fibre bundle. In this way the Lie groupoid of fibre morphisms (and its corresponding Lie algebroid) may be identified with the gauge groupoid of the principal bundle (and its corresponding Atiyah algebroid).

3.1 Lie Groupoids and Fibre Bundles

Let $\pi : E \to M$ be a fibre bundle with connected base manifold M, structure group G and standard fibre F; we shall sometimes write $\pi^G : E \to M$ to emphasize that this is a G-bundle. We denote a fibre of this bundle by $E_x = \pi^{-1}(x)$. For our applications in later chapters the action of G on F will be effective, and often it will also be transitive, so that F may be regarded as a homogeneous space of G; the results in the present section do not, though, depend on these restrictions.

Let $\varphi : E_{\tilde{x}} \to E_x$ be a diffeomorphism and let $U_{\tilde{\lambda}}$, U_λ be neighbourhoods of $\tilde{x}$, x respectively where $\mathrm{T}_{\tilde{\lambda}} : U_{\tilde{\lambda}} \times F \to \pi^{-1}(U_{\tilde{\lambda}})$ and $\mathrm{T}_\lambda : U_\lambda \times F \to \pi^{-1}(U_\lambda)$

M. Crampin and D. Saunders, *Cartan Geometries and their Symmetries*,
Atlantis Studies in Variational Geometry 4, DOI 10.2991/978-94-6239-192-5_3

are local bundle trivializations. We shall say that φ *respects the action of* G if the representation of φ in these two trivializations, the map

$$\mathrm{T}_\lambda^{-1}\Big|_{E_x} \circ \varphi \circ \mathrm{T}_{\tilde{\lambda}}\Big|_{\{\tilde{x}\}\times F} : \{\tilde{x}\} \times F \to \{x\} \times F,$$

corresponds to the left action l_g of some element $g \in G$ on F. Choosing two different trivializations $\mathrm{T}_{\hat{\tilde{\lambda}}}$, $\mathrm{T}_{\hat{\lambda}}$ will give the left action of some other element $\hat{g} \in G$ as the transition functions of the bundle take their values in G.

Theorem 3.1.1 *Let $\mathcal{G}$ be the collection of maps $\varphi : E_{\tilde{x}} \to E_x$ that are diffeomorphisms respecting the action of G. Then $\mathcal{G}$ is a locally trivial Lie groupoid.*

Proof Let $\mathcal{G}_{x,\tilde{x}} \subset \mathcal{G}$ denote the subset of maps from $E_{\tilde{x}}$ to E_x. If $\varphi \in \mathcal{G}_{x,\tilde{x}}$ put $\alpha(\varphi) = \tilde{x}$ and $\beta(\varphi) = x$; this defines source and target maps $\alpha, \beta : \mathcal{G} \to M$. Let $1_x : E_x \to E_x$ denote the identity diffeomorphism; then $1 : x \mapsto 1_x$ maps M to $\mathcal{G}$. Finally, let the partial multiplication be composition of maps, and the inverse be the usual inverse of maps. It is immediate that, with these definitions, $\mathcal{G}$ is a groupoid.

We now give $\mathcal{G}$ a smooth structure. Let $\{U_\mu, \mathrm{T}_\mu\}$ be a family of local trivializations of the bundle $\pi : E \to M$ where the sets U_μ are coordinate patches covering M and where, writing $\mathrm{T}_{\tilde{\lambda},\tilde{x}} : F \to E_{\tilde{x}}$ and $\mathrm{T}_{\lambda,x} : F \to E_x$ for the maps given by $\mathrm{T}_{\tilde{\lambda},\tilde{x}}(q) = \mathrm{T}_{\tilde{\lambda}}(\tilde{x}, q), \mathrm{T}_{\lambda,x}(q) = \mathrm{T}_\lambda(x, q)$ where $x \in U_\lambda$ and $\tilde{x} \in U_{\tilde{\lambda}}$, each composition $\mathrm{T}_{\lambda,x}^{-1} \circ \mathrm{T}_{\tilde{\lambda},\tilde{x}} : F \to F$ may be represented by the left action of an element of G on F. Let

$$\mathcal{G}_{\lambda,\tilde{\lambda}} = \bigcup_{x\in U_\lambda,\, \tilde{x}\in U_{\tilde{\lambda}}} \mathcal{G}_{x,\tilde{x}}$$

and define

$$\psi_{\lambda,\tilde{\lambda}} : \mathcal{G}_{\lambda,\tilde{\lambda}} \to U_\lambda \times G \times U_{\tilde{\lambda}},$$

as follows. If $\varphi \in \mathcal{G}_{\lambda,\tilde{\lambda}}$ then $\varphi \in \mathcal{G}_{x,\tilde{x}}$ for some $\tilde{x} \in U_{\tilde{\lambda}}$ and $x \in U_\lambda$. Thus $\left(\mathrm{T}_{\lambda,x}\right)^{-1} \circ \varphi \circ \mathrm{T}_{\tilde{\lambda},\tilde{x}}$ is a map $l_g : F \to F$ corresponding to the left action of a unique element $g \in G$. We may therefore put

$$\psi_{\lambda,\tilde{\lambda}}(\varphi) = (x, g, \tilde{x}),$$

and it is straightforward to see that this defines a bijection.

To show that $\mathcal{G}$ has a smooth structure, we need to show that if $\psi_{\lambda_2,\tilde{\lambda}_2} \circ \left(\psi_{\lambda_1,\tilde{\lambda}_1}\right)^{-1}$ has a nonempty domain then it is smooth. The nonempty domain condition is that there is some point common to the image of $\psi_{\lambda_1,\tilde{\lambda}_1}$ and the domain of $\psi_{\lambda_2,\tilde{\lambda}_2}$, so that

$$\left(\psi_{\lambda_1,\tilde{\lambda}_1}\right)^{-1}(U_{\lambda_1} \times G \times U_{\tilde{\lambda}_1}) \cap \mathcal{G}_{\lambda_2,\tilde{\lambda}_2} \neq \varnothing,$$

in other words that

$$\mathcal{G}_{\lambda_1,\tilde{\lambda}_1} \cap \mathcal{G}_{\lambda_2,\tilde{\lambda}_2} \neq \varnothing.$$

The condition is therefore that

$$U_{\bar{\lambda}_1} \cap U_{\bar{\lambda}_2} \neq \varnothing, \qquad U_{\lambda_1} \cap U_{\lambda_2} \neq \varnothing.$$

If this condition holds then the domain of $\psi_{\lambda_2,\bar{\lambda}_2} \circ \left(\psi_{\lambda_1,\bar{\lambda}_1}\right)^{-1}$ is

$$(U_{\lambda_1} \cap U_{\lambda_2}) \times G \times (U_{\bar{\lambda}_1} \cap U_{\bar{\lambda}_2}),$$

and if $(x, g, \tilde{x})$ is in the domain then $\left(\psi_{\lambda_1,\bar{\lambda}_1}\right)^{-1}(x, g, \tilde{x})$ is the diffeomorphism

$$\mathrm{T}_{\lambda_1,x} \circ l_g \circ \left(\mathrm{T}_{\bar{\lambda}_1,\tilde{x}}\right)^{-1} : E_{\tilde{x}} \to E_x .$$

Thus

$$\begin{aligned}\psi_{\lambda_2,\bar{\lambda}_2} \circ \left(\psi_{\lambda_1,\bar{\lambda}_1}\right)^{-1}(x, g, \tilde{x}) &= \psi_{\lambda_2,\bar{\lambda}_2}\left(\mathrm{T}_{\lambda_1,x} \circ l_g \circ \left(\mathrm{T}_{\bar{\lambda}_1,\tilde{x}}\right)^{-1}\right)\\ &= \left(x, \left(\left(\mathrm{T}_{\lambda_2,x}\right)^{-1} \circ \mathrm{T}_{\lambda_1,x}\right) \circ l_g \circ \left(\left(\mathrm{T}_{\bar{\lambda}_1,\tilde{x}}\right)^{-1} \circ \mathrm{T}_{\bar{\lambda}_2,\tilde{x}}\right), \tilde{x}\right).\end{aligned}$$

Note that $\left(\mathrm{T}_{\bar{\lambda}_1,\tilde{x}}\right)^{-1} \circ \mathrm{T}_{\bar{\lambda}_2,\tilde{x}} = l_{g_{\bar{\lambda}}}$, and also $\left(\mathrm{T}_{\lambda_2,x}\right)^{-1} \circ \mathrm{T}_{\lambda_1,x} = l_{g_\lambda}$, where $g_{\bar{\lambda}}, g_\lambda \in G$, so the central term above is the left action of an element of G, as required. Moreover, $\tilde{x} \mapsto \left(\mathrm{T}_{\bar{\lambda}_1,\tilde{x}}\right)^{-1} \circ \mathrm{T}_{\bar{\lambda}_2,\tilde{x}}$ is the transition function $U_{\bar{\lambda}_1} \cap U_{\bar{\lambda}_2} \to G$ for the bundle structure on $\pi : E \to M$, and likewise $x \mapsto \left(\mathrm{T}_{\lambda_1,x}\right)^{-1} \circ \mathrm{T}_{\lambda_2,x}$ is the transition function $U_{\lambda_1} \cap U_{\lambda_2} \to G$. So $\psi_{\lambda_2,\bar{\lambda}_2} \circ \left(\psi_{\lambda_1,\bar{\lambda}_1}\right)^{-1}$ is constructed from smooth maps by smooth operations, and therefore is smooth.

Now $U_{\bar{\lambda}}, U_\lambda \subset M$ are coordinate patches, by assumption. Taking a coordinate patch $V \in G$ we obtain a coordinate patch $\left(\psi_{\lambda,\bar{\lambda}}\right)^{-1}(U_\lambda \times V \times U_{\bar{\lambda}})$ on $\mathcal{G}$. The transition functions for such patches are smooth as a consequence of the immediately preceding result about $\psi_{\lambda_2,\bar{\lambda}_2} \circ \left(\psi_{\lambda_1,\bar{\lambda}_1}\right)^{-1}$. Thus $\mathcal{G}$ is a differentiable manifold of dimension $2n + d$, $n = \dim M$, $d = \dim G$.

It is straightforward to show that the structure maps of the groupoid are smooth and that the source and target projections are surjective submersions, so we conclude that $\mathcal{G}$ is a Lie groupoid of dimension $2n + d$.

It is evident that the vertex groups $\mathcal{G}_x$ are all isomorphic to G, and that each map $\psi_{\lambda,\bar{\lambda}} : \mathcal{G}_{\lambda,\bar{\lambda}} \to U_\lambda \times G \times U_{\bar{\lambda}}$ is a basepoint-preserving isomorphism, so that $\mathcal{G}$ is a locally trivial Lie groupoid by Lemma 1.1.1, using the connectedness of M. □

We shall call $\mathcal{G}$ the *Lie groupoid of fibre morphisms* of $\pi^G : E \to M$.

We see from this argument that we may take $(x^i, z^A, \tilde{x}^i)$ as local coordinates in a neighbourhood of $\varphi \in \mathcal{G}$ where $\tilde{x}^i$ are coordinates in M of the source $\alpha(\varphi)$, x^i are coordinates of the target $\beta(\varphi)$, and z^A are coordinates on the group G of the bundle $\pi^G : E \to M$.

Corollary 3.1.2 *The Lie algebroid of a Lie groupoid of fibre morphisms is transitive.*

Proof From Proposition 1.3.1, as the Lie groupoid is locally trivial. □

An important special case of this construction arises when the fibre bundle $\pi^G : E \to M$ is trivial, so that $E = M \times F$ and π is projection on the first factor. Let $\mathrm{T} : M \times F \to E$ be the canonical trivialization (so that T is just the identity map); then if $\varphi \in \mathcal{G}$ and $q \in F$ we have[1]

$$\mathrm{T}^{-1}\varphi\mathrm{T}\big(\alpha(\varphi), q\big) = \big(\beta(\varphi), gq\big)$$

for a unique group element $g \in G$. The map

$$\mathcal{G} \to M \times G \times M, \qquad \varphi \mapsto \big(\beta(\varphi), g, \alpha(\varphi)\big)$$

is a bijection with inverse $(x, g, \tilde{x}) \mapsto \varphi$ where $\mathrm{T}^{-1}\varphi\mathrm{T}(\tilde{x}, q) = (x, gq)$, and the argument in the above proof shows that it is a diffeomorphism. In this case, therefore, $\mathcal{G}$ may be identified with a trivial groupoid.

We now return to the general case. Suppose that $o \in F$ is a distinguished element of the standard fibre, and let G_o be the isotropy group of o, so that $G_o \subset G$ is a closed Lie subgroup.

Proposition 3.1.3 *Let $\mathcal{G}_o$ be the collection of diffeomorphisms $\varphi : E_u \to E_x$ respecting the action of $G_o \subset G$. Then $\mathcal{G}_o$ is a wide closed Lie subgroupoid of $\mathcal{G}$.*

Proof It is clear that $\mathcal{G}_o$ satisfies the algebraic conditions to be a wide subgroupoid of $\mathcal{G}$; we must show that $\mathcal{G}_o \subset \mathcal{G}$ is a closed submanifold, that the restricted source and target maps are surjective submersions, and that the subset $\mathcal{G}_o^{(2)} \subset \mathcal{G}_o \times \mathcal{G}_o$ of admissible pairs is a submanifold.

Let $\{U_\mu, \mathrm{T}_\mu\}$ be a family of local trivializations of the bundle $\pi : E \to M$ and put $\psi_{\lambda,\tilde{\lambda}} : \mathcal{G}_{\lambda,\tilde{\lambda}} \to U_\lambda \times G \times U_{\tilde{\lambda}}$, as in the proof of Theorem 3.1.1, so that the family $\{\mathcal{G}_{\lambda,\tilde{\lambda}}, \psi_{\lambda,\tilde{\lambda}}\}$ is an atlas for $\mathcal{G}$. By definition

$$\mathcal{G}_o \cap \mathcal{G}_{\lambda,\tilde{\lambda}} = \psi_{\lambda,\tilde{\lambda}}^{-1}\big(U_\lambda \times G_o \times U_{\tilde{\lambda}}\big)$$

and $G_o \subset G$ is closed, so that $\mathcal{G}_o \subset \mathcal{G}$ is a closed submanifold. The remaining conditions hold because, from Theorem 3.1.1, $\mathcal{G}_o$ is a Lie groupoid in its own right under the restricted maps. □

If there is a reduction of the G-bundle $\pi^G : E \to M$ to a G_o-bundle $\pi^{G_o} : E \to M$ then there is a global section ς of π, defined in any local G_o-trivialization $\mathrm{T} : U \times F \to \pi^{-1}(U)$ by $\varsigma(x) = \mathrm{T}(x, o)$; the definition is consistent on overlaps because $go = o$ for any $g \in G_o$.

[1] Where there will be no confusion we write gq instead of $l_g q$ for the left action of $g \in G$ on $q \in F$.

3.2 Lie Algebroids and Fibre Bundles

We continue with $\mathcal{G}$ as the Lie groupoid of fibre morphisms of some G-bundle $\pi^G : E \to M$, and we assume that G acts effectively on the standard fibre F of π. We now show that we may regard elements of the transitive Lie algebroid $A\mathcal{G}$ as vector fields along fibres of π, and that consequently we may regard sections of $A\mathcal{G} \to M$ as vector fields on E. (There is also a finite version of this infinitesimal correspondence which will be given in Theorem 5.1.4.)

To see how this relationship arises, take $x \in M$ and let $i_x : E_x \to E$ be the inclusion map of the fibre E_x in the total space of the bundle. The infinite-dimensional vector space $\mathfrak{X}(i_x)$ consists of vector fields along the map i_x, in other words sections of the pullback bundle $i_x^*(TE) \to E_x$, and we shall let $\overline{\mathfrak{X}}(i_x)$ denote the subspace of projectable vector fields along i_x: that is of those $\mathcal{Y} \in \mathfrak{X}(i_x)$ satisfying $\pi_*(\mathcal{Y}_{y_1}) = \pi_*(\mathcal{Y}_{y_2}) \in T_xM$ for all $y_1, y_2 \in E_x$. In local coordinates x^i on M (around x) and y^α on E_x we would write a vector field $\mathcal{Y} \in \overline{\mathfrak{X}}(i_x)$ as

$$\mathcal{Y} = \lambda^i \frac{\partial}{\partial x^i} + \mathcal{Y}^\alpha \frac{\partial}{\partial y^\alpha}$$

where $\lambda^i \in \mathbb{R}$ and $\mathcal{Y}^\alpha \in C^\infty(E_x)$.

Now let $\overline{\mathfrak{X}}(E)$ denote the vector space of projectable vector fields on E, so that if $Y \in \overline{\mathfrak{X}}(E)$ then $Y|_{E_x} \in \overline{\mathfrak{X}}(i_x)$ for each $x \in M$. If ψ is the flow of Y, defined on an open subset of $E \times \mathbb{R}$, and if $y_1, y_2 \in E$ with $\pi(y_1) = \pi(y_2)$ then $\pi\psi(y_1, t) = \pi\psi(y_2, t)$ for any t where both sides are defined. We shall derive the flow of such a vector field on E from the flow of a right-invariant vector field on $\mathcal{G}$, and it will be clear from our construction that any vector field on E obtained in this way must have the property that, for each $x \in M$, there must be some value $\varepsilon_x > 0$ such that $\psi(y, t)$ is defined for all $(y, t) \in E_x \times (-\varepsilon_x, \varepsilon_x)$. We shall say that a vector field with this property has *vertically uniform flow boxes*.

Let $\overline{\mathfrak{X}}^G(E) \subset \overline{\mathfrak{X}}(E)$ be the subset of projectable vector fields having vertically uniform flow boxes, whose flows, when restricted to complete fibres in this way, consist of fibre morphisms of the G-bundle $\pi^G : E \to M$. Next, let $\overline{\mathfrak{X}}^G(i_x) \subset \overline{\mathfrak{X}}(i_x)$ denote the subset containing those $\mathcal{Y} \in \overline{\mathfrak{X}}(i_x)$ such that $\mathcal{Y} = Y|_{E_x}$ for some $Y \in \overline{\mathfrak{X}}^G(E)$. Finally, let $\mathcal{A}^G = \bigcup_{x \in M} \overline{\mathfrak{X}}^G(i_x)$.

Theorem 3.2.1 *Each $\overline{\mathfrak{X}}^G(i_x)$ is a vector space satisfying $\overline{\mathfrak{X}}^G(i_x) \cong A_x\mathcal{G}$, and $\mathcal{A}^G \to M$ is a Lie algebroid isomorphic to $A\mathcal{G} \to M$.*

Proof Take $\boldsymbol{a} \in A_x\mathcal{G}$, and let ξ be a section of $A\mathcal{G} \to M$ satisfying $\xi(x) = \boldsymbol{a}$. Let ξ^{R} be the corresponding right-invariant vector field on $\mathcal{G}$, and let Ψ_t be the flow of ξ^{R} in a neighbourhood V of 1_x, so that $t \mapsto \Psi_t(1_x)$ is a curve in $\mathcal{G}$ and $d\Psi_t(1_x)/dt|_{t=0} = \boldsymbol{a}$. By right invariance

$$\alpha\Psi_t(1_x) = \alpha\Psi_0(1_x) = x,$$

so that $\Psi_t(1_x) : E_x \to E_{\beta\Psi_t(1_x)}$, and of course $\Psi_0(1_x) = 1_x : E_x \to E_x$.

Now let U be a neighbourhood of x such that $1_U \subset V$, and let $\varepsilon > 0$ be such that $\Psi_t(1_x)$ is defined for $t \in (-\varepsilon, \varepsilon)$. For each such t let the map $\psi_t : E_U \to E$ be given by

$$\psi_t(y) = \Psi_t(1_{\pi(y)})(y). \tag{1}$$

We show first that ψ_t satisfies the pseudogroup property. We have

$$\psi_{s+t}(y) = \Psi_{s+t}(1_{\pi(y)})(y) = \Psi_s\Psi_t(1_{\pi(y)})(y);$$

but $\Psi_s\Psi_t(1_{\pi(y)}) = \Psi_s(1_{\beta\Psi_t(1_{\pi(y)})}) \cdot \big(\Psi_t(1_{\pi(y)})$ by right invariance, so that

$$\begin{aligned}
\Psi_s\Psi_t(1_{\pi(y)})(y) &= \Psi_s(1_{\beta\Psi_t(1_{\pi(y)})})\big(\Psi_t(1_{\pi(y)})(y)\big) \\
&= \Psi_s(1_{\pi(\Psi_t(1_{\pi(y)})(y))})\big(\Psi_t(1_{\pi(y)})(y)\big) \\
&= \psi_s\big(\Psi_t(1_{\pi(y)})(y)\big) \\
&= \psi_s\psi_t(y)
\end{aligned}$$

as required. As ψ_0 is the identity on E_U, we see that ψ_t is the flow of a vector field Y on E_U. In addition,

$$\pi\psi_t(y) = \pi\big(\Psi_t(1_{\pi(y)})(y)\big) = \beta\Psi_t(1_{\pi(y)})$$

is independent of the choice of y in any given fibre E_x, so that Y is projectable. By construction the flow of Y comprises fibre morphisms. Putting $\Lambda_x(\boldsymbol{a}) = Y|_{E_x}$, we see that the correspondence $\boldsymbol{a} \mapsto \Lambda_x(\boldsymbol{a})$ is well defined as it depends only on the derivative of Ψ_t at $t = 0$, and hence is independent of the choice of section ξ. By construction $\Lambda_x(\boldsymbol{a}) \in \overline{\mathfrak{X}}^G(i_x)$; let $\Lambda : A\mathcal{G} \to \mathcal{A}^G$ be defined by $\Lambda|_{A_x\mathcal{G}} = \Lambda_x$.

Now suppose instead we are given $\mathcal{Y} \in \overline{\mathfrak{X}}^G(i_x)$; we may, by definition, choose a projectable vector field $Y \in \overline{\mathfrak{X}}^G(E)$ with $\mathcal{Y} = Y|_{E_x}$. Let ψ be the flow of Y and ϕ be the flow of the projection of Y to M, so that

$$\pi\psi(y, t) = \phi(\pi(y), t) = \phi(x, t)$$

for every $y \in E_x$ and every $t \in (-\varepsilon_x, \varepsilon_x)$, where $\varepsilon_x > 0$ exists because Y has vertically uniform flow boxes; thus $\psi(y, t) \in E_{\phi(x,t)}$. Furthermore, for each such t the map $y \mapsto \psi(y, t)$ is a fibre morphism $\varphi_t : E_x \to E_{\phi(x,t)}$, so that we may define a curve $c : (-\varepsilon_x, \varepsilon_x) \to \mathcal{G}$ by $c(t) = \varphi_t$. Put $\boldsymbol{a} = \dot{c}(0) \in T_{1_x}\mathcal{G}$; then $\alpha c(t) = x$ for all $t \in (-\varepsilon_x, \varepsilon_x)$, so that $\alpha_*(\boldsymbol{a}) = 0$, and therefore that $\boldsymbol{a} \in A_x\mathcal{G}$. A different choice of Y will result in a different flow ψ, but for any $y \in E_x$ the derivative $\partial\psi/\partial t|_{(y,0)}$ will be unchanged. As $c(t) = \big(y \mapsto \psi(y, t)\big)$, it follows that $\boldsymbol{a} = \dot{c}(0)$ will be unchanged. It is clear that this method of obtaining $\boldsymbol{a}$ from $\mathcal{Y}$ is just the inverse of the map $\Lambda_x : \boldsymbol{a} \mapsto \mathcal{Y}$, so that Λ_x is a bijection $A_x\mathcal{G} \leftrightarrow \overline{\mathfrak{X}}^G(i_x) \subset \overline{\mathfrak{X}}(i_x)$.

In order to consider linearity, we use the pullback bundle $\alpha^* E \to \mathcal{G}$ and the evaluation map $\mu : \alpha^* E \to E$, $\mu(\varphi, y) = \varphi(y)$ with $\alpha(\varphi) = \pi(y) = x$. We use this to rewrite the definition (1) of ψ_t as

$$\psi_t(y) = \mu\big(\Psi_t(1_x), y\big)$$

so that, for each $y \in E_x$, we obtain a linear map $\mu_{*y} : T_{(1_x,y)}(\alpha^* E) \to T_y E$. As

$$A_x\mathcal{G} \cong A_x\mathcal{G} \oplus \{0_y\} \subset T_{(1_x,y)}(\alpha^* E) \subset T_{1_x}\mathcal{G} \oplus T_y E,$$

we see that the restriction of μ_{*y} to $A_x\mathcal{G}$ is just the linear map $\boldsymbol{a} \mapsto \Lambda_x(\boldsymbol{a})|_y$. It follows that $\Lambda_x : A_x\mathcal{G} \to \overline{\mathfrak{X}}(i_x)$ itself is linear, so that the image space $\Lambda_x(A_x\mathcal{G}) = \overline{\mathfrak{X}}^G(i_x)$ is a vector subspace of $\overline{\mathfrak{X}}(i_x)$, and that $\Lambda_x : A_x\mathcal{G} \to \overline{\mathfrak{X}}^G(i_x)$ is a linear isomorphism. It is now evident that $\mathcal{A}^G \to M$ is a vector bundle, and that $\Lambda : A\mathcal{G} \to \mathcal{A}^G$ is a vector bundle isomorphism projecting to the identity on M; we also see that the anchor maps of the two bundles are the same.

Finally we must show that Λ is an isomorphism of Lie algebroids, in other words that $\Lambda \circ [\![\xi, \eta]\!] = [\Lambda \circ \xi, \Lambda \circ \eta]$ for sections $\xi, \eta : M \to A\mathcal{G}$, and to do this we again use the flow relation (1). If Ψ_t is the flow of ξ^{R} and Φ_t is the flow of η^{R} then

$$\Phi_{-\sqrt{t}} \circ \Psi_{-\sqrt{t}} \circ \Phi_{\sqrt{t}} \circ \Psi_{\sqrt{t}}$$

is the flow of $[\xi^{\mathsf{R}}, \eta^{\mathsf{R}}] = [\![\xi, \eta]\!]^{\mathsf{R}}$. On the other hand, if ψ_t is the flow of $\Lambda \circ \xi$ and ϕ_t is the flow of $\Lambda \circ \eta$ then

$$\phi_{-\sqrt{t}} \circ \psi_{-\sqrt{t}} \circ \phi_{\sqrt{t}} \circ \psi_{\sqrt{t}}$$

is the flow of $[\Lambda \circ \xi, \Lambda \circ \eta]$. We demonstrate the relation between these two formulæ using the right invariance of the vector fields ξ^{R} and η^{R}, so that for any $\varphi \in \mathcal{G}$ and any valid parameter t we have

$$\Psi_t(\varphi) = \Psi_t(1_{\beta(\varphi)}) \cdot \varphi, \qquad \Phi_t(\varphi) = \Phi_t(1_{\beta(\varphi)}) \cdot \varphi.$$

Thus

$$\begin{aligned}
\phi_{\sqrt{t}}\big(\psi_{\sqrt{t}}(y)\big) &= \phi_{\sqrt{t}}\big(\Psi_{\sqrt{t}}(1_{\pi(y)})(y)\big) \\
&= \Phi_{\sqrt{t}}(1_{\pi(\Psi_{\sqrt{t}}(1_{\pi(y)})(y))})\big(\Psi_{\sqrt{t}}(1_{\pi(y)})(y)\big) \\
&= \Phi_{\sqrt{t}}(1_{\beta(\Psi_{\sqrt{t}}(1_{\pi(y)}))})\big(\Psi_{\sqrt{t}}(1_{\pi(y)})(y)\big) \\
&= \Phi_{\sqrt{t}}\big(\Psi_{\sqrt{t}}(1_{\pi(y)})(y)\big),
\end{aligned}$$

and applying this procedure twice more we see that

$$(\phi_{-\sqrt{t}} \circ \psi_{-\sqrt{t}} \circ \phi_{\sqrt{t}} \circ \psi_{\sqrt{t}})(y) = (\Phi_{-\sqrt{t}} \circ \Psi_{-\sqrt{t}} \circ \Phi_{\sqrt{t}} \circ \Psi_{\sqrt{t}})(1_{\pi(y)})(y)$$

from which the result follows. □

We should make a remark about exactly what is happening here. Under ordinary circumstances one regards a projectable vector field Y on E as a section of the tangent bundle $TE \to E$, which may be evaluated at a point $y \in E$ to give a tangent vector $Y_y \in T_yE$. When regarding $Y = \Lambda \circ \xi$ as a section of the Lie algebroid $\mathcal{A}^G$, on the other hand, we 'evaluate' it at a point $x \in M$ to give $Y|_{E_x}$, a vector field along the inclusion map $i_x : E_x \to E$.

It is of interest to consider the expression of Λ in coordinates. Let $(x^i, z^A, \tilde{x}^i)$ be local coordinates in a neighbourhood of $1_x \in \mathcal{G}$; as each element $\boldsymbol{a} \in A\mathcal{G}$ may be regarded as an element of the kernel of α_*, we may write $\boldsymbol{a}$ as

$$\boldsymbol{a} = \boldsymbol{a}^i \left.\frac{\partial}{\partial x^i}\right|_{1_x} + \boldsymbol{a}^A \left.\frac{\partial}{\partial z^A}\right|_{1_x}.$$

If $\boldsymbol{a} = \xi(x)$ then in terms of the flow Ψ_t of ξ^{R} we have

$$\boldsymbol{a}^i = \left.\frac{d\Psi_t^i(1_x)}{dt}\right|_{t=0}, \qquad \boldsymbol{a}^A = \left.\frac{d\Psi_t^A(1_x)}{dt}\right|_{t=0}$$

where $\Psi_t^i = x^i \circ \Psi_t$ and $\Psi_t^A = z^A \circ \Psi_t$ (note that $\tilde{x}^i \circ \Psi_t$ is constant by right invariance). We may also write the projectable vector field Y on E using general fibred local coordinates (x^i, y^α) as

$$Y = X^i \frac{\partial}{\partial x^i} + Y^\alpha \frac{\partial}{\partial y^\alpha}$$

where

$$X^i(\pi(y)) = \left.\frac{d\psi_t^i(y)}{dt}\right|_{t=0}, \qquad Y^\alpha(y) = \left.\frac{d\psi_t^\alpha(y)}{dt}\right|_{t=0}$$

with

$$\psi_t^i = x^i \circ \psi_t, \qquad \psi_t^\alpha = y^\alpha \circ \psi_t.$$

Thus, using the evaluation map $\mu : \alpha^* E \to E$, we have

$$X^i(x) = \left.\frac{\partial \mu^i}{\partial x^j}\right|_{(1_x, y)} \left.\frac{d\Psi_t^j(1_x)}{dt}\right|_{t=0} = \boldsymbol{a}^i$$

because $\partial \mu^i / \partial x^j = \delta^i_j$, and

$$
\begin{aligned}
Y^\alpha(y) &= \left.\frac{\partial\mu^\alpha}{\partial x^j}\right|_{(1_x, y)} \left.\frac{d\Psi_t^j(1_x)}{dt}\right|_{t=0} + \left.\frac{\partial\mu^\alpha}{\partial z^A}\right|_{(1_x, y)} \left.\frac{d\Psi_t^A(1_x)}{dt}\right|_{t=0} \\
&= \boldsymbol{a}^j \left.\frac{\partial\mu^\alpha}{\partial x^j}\right|_{(1_x, y)} + \boldsymbol{a}^A \left.\frac{\partial\mu^\alpha}{\partial z^A}\right|_{(1_x, y)}
\end{aligned}
$$

where there is no term in $\partial\mu^\alpha/\partial u^j$ because $u^i \circ \Psi_s$ is constant.

Now we have used general local fibred coordinates (x^i, y^α) on E, and in those coordinates the above expression for $Y^\alpha(y)$ will depend on both $\boldsymbol{a}^i$ and $\boldsymbol{a}^A$. We could, however, use a bundle trivialization $\mathrm{T} : U \times F \to \pi^{-1}(U)$ where $\pi(y) \in U$ and specify that (x^i, y^α) should be local coordinates on $U \times F$ around $\mathrm{T}^{-1}(y)$; with such a choice of coordinates we would have $\partial\mu^\alpha/\partial x^j = 0$, so that $Y^\alpha(y)$ would depend only on $\boldsymbol{a}^A$. Indeed, each such local trivialization allows us to identify a fibre of the Lie algebroid with the direct sum of the corresponding tangent space to M and the opposite Lie algebra $\mathfrak{g}_\mathsf{R}$ of G.

Proposition 3.2.2 *If $\mathrm{T} : U \times F \to \pi^{-1}(U)$ is a local trivialization then $x \in U$, $v \in T_x M$ and $Z \in \mathfrak{X}(F)$ determine the vector field $\mathrm{T}_{x\dagger}(v, Z)$ along the inclusion map $i_x : E_x \to E$ by, for $q \in F$,*

$$
\mathrm{T}_{x\dagger}(v, Z)\big|_{\mathrm{T}(x,q)} = \mathrm{T}_*(v, Z_q) \in T_{\mathrm{T}(x,q)} E
$$

and in this way they determine an element of $\mathcal{A}_x^G$.

In particular, taking an element[2] $a \in \mathfrak{g}_\mathsf{R}$ and using its fundamental vector field $a^\sharp \in \mathfrak{X}(F)$ gives an isomorphism $\mathrm{T}_{x\sharp} : T_x M \oplus \mathfrak{g}_\mathsf{R} \to \mathcal{A}_x^G$ satisfying

$$
\mathrm{T}_{x\sharp}(v, a)\big|_{\mathrm{T}(x,q)} = \mathrm{T}_{x\dagger}(v, a^\sharp)\big|_{\mathrm{T}(x,q)} = \mathrm{T}_*(v, a_q^\sharp).
$$

Furthermore, the restriction of $\mathrm{T}_{x\sharp}$ to $\{0_x\} \oplus \mathfrak{g}_\mathsf{R}$ defines a map $\mathrm{T}_{x\sharp} : \mathfrak{g}_\mathsf{R} \to \mathcal{K}_x^G$ which is an isomorphism of Lie algebras.

Proof For each $q \in F$ we may identify $T_{(x,q)}(M \times F)$ with $T_x M \oplus T_q F$, so that we may regard $\mathrm{T}_* : T_{(x,q)}(M \times F) \to T_{\mathrm{T}(x,q)} E$ as a map $T_x M \oplus T_q F \to T_{\mathrm{T}(x,q)} E$; with this understanding $\mathrm{T}_*(v, Z_q)$ is a tangent vector to E at $\mathrm{T}(x, q)$. Furthermore,

$$
\pi_* \mathrm{T}_*(v, Z_q) = \left(\pi \circ \mathrm{T}\right)_*(v, Z_q) = \mathrm{pr}_{1*}(v, Z_q) = v \in T_x M
$$

where $\mathrm{pr}_1 : M \times F \to M$, so that $\mathrm{T}_{x\dagger}(v, Z) : \mathrm{T}(x, q) \to \mathrm{T}_{(x,q)*}(v, Z_q)$ is indeed an element of $\mathcal{A}_x^G$.

The correspondence $T_x M \oplus \mathfrak{g}_\mathsf{R} \to \mathcal{A}_x^G$ given by $(v, a) \mapsto \mathrm{T}_{x\sharp}(v, a) = \mathrm{T}_{x\dagger}(v, a^\sharp)$ is evidently a linear map. If $\mathrm{T}_{x\dagger}(v, a^\sharp) = 0$ then $\mathrm{T}_*(v, a_q^\sharp) = 0$ for all $q \in F$, and as T_* is a fibrewise isomorphism it follows that $v = 0$ and that $a^\sharp = 0$; as

[2] Recall that we write elements of a Lie algebra as $a, b, \ldots$ whereas elements of a Lie algebroid are written as $\boldsymbol{a}, \boldsymbol{b}, \ldots$.

the action of G is effective we may conclude that $a = 0$. Thus the correspondence $\mathrm{T}_{x\sharp} : T_xM \oplus \mathfrak{g}_{\mathsf{R}} \to \mathcal{A}^G_x$ is injective; it is surjective because both spaces have the same dimension $\dim M + \dim G$.

For any $v \in T_xU$ and $w \in T_qF$, $\mathrm{T}_*(v, w) \in \mathcal{K}^G_x$ exactly when $v = 0$, so that $\mathrm{T}_{x\sharp}(\mathfrak{g}_{\mathsf{R}}) = \mathcal{K}^G_x$. If $a, b \in \mathfrak{g}_{\mathsf{R}}$ then the bracket formula

$$\mathrm{T}_{x\sharp}\big(\{a, b\}_{\mathsf{R}}\big) = \mathrm{T}_{x\dagger}\big(0, [a^\sharp, b^\sharp]\big) = [\![\mathrm{T}_{x\dagger}(0, a^\sharp), \mathrm{T}_{x\dagger}(0, b^\sharp)]\!] = [\![\mathrm{T}_{x\sharp}(a), \mathrm{T}_{x\sharp}(b)]\!]$$

follows from Proposition 1.3.2 and Corollary 1.3.3. □

In the remainder of this volume we shall often identify the Lie algebroids $A\mathcal{G}$ and $\mathcal{A}^G$, and use symbols such as ξ to represent projectable vector fields on E, writing $[\xi, \eta]$ instead of $[\![\xi, \eta]\!]$. When using this convention there is a potential ambiguity concerning the argument of ξ which we resolve as follows: if the argument is an element $y \in E$ then we use a subscript, so that $\xi_y \in T_yE$ is a tangent vector to the manifold E, whereas if the argument is $x \in M$ then we use parentheses, so that $\xi(x) \in \overline{\mathfrak{X}}(i_x)$ is a vector field along the fibre E_x.

Now let ξ be a section of $\mathcal{A}^G \to M$ (a projectable vector field on E) and set $X = \rho \circ \xi$. We consider the representation of ξ with respect to the local trivialisation $\mathrm{T} : U \times F \to \pi^{-1}(U)$. We have

$$\xi(x) = \mathrm{T}_{x\sharp}(X_x, \bar{\xi}(x))$$

where $\bar{\xi}$ is a $\mathfrak{g}_{\mathsf{R}}$-valued function on U. We may write

$$\xi|_U = \mathrm{T}_\sharp \circ (X, \bar{\xi})$$

where the map $\mathrm{T}_\sharp : TU \oplus \mathfrak{g}_{\mathsf{R}} \to \mathcal{A}^G|_U$ is given by $\mathrm{T}_\sharp(x, v, a) = \mathrm{T}_{x\sharp}(v, a)$; it is an isomorphism of vector bundles over the identity. We compute the Lie algebroid bracket on $\mathcal{A}^G$ in terms of the trivialization. We have, over U,

$$\begin{aligned} [\![\xi, \eta]\!] = [\xi, \eta] &= \big[\mathrm{T}_\sharp \circ (X, \bar{\xi}), \mathrm{T}_\sharp \circ (Y, \bar{\eta})\big] \\ &= \big[\mathrm{T}_*(X, \bar{\xi}^\sharp), \mathrm{T}_*(Y, \bar{\eta}^\sharp)\big] = \mathrm{T}_*\big[(X, \bar{\xi}^\sharp), (Y, \bar{\eta}^\sharp)\big], \end{aligned}$$

where $(X, \bar{\xi}^\sharp)$ is the vector field on $U \times F$ given on the fibre $\{x\} \times F$ by $(X_x, \bar{\xi}(x)^\sharp)$. So $[\![\xi, \eta]\!]$ corresponds, via the trivialisation, to $[(X, \bar{\xi}^\sharp), (Y, \bar{\eta}^\sharp)]$, the bracket of vector fields on $U \times F$. Now

$$\big[(X, \bar{\xi}^\sharp), (Y, \bar{\eta}^\sharp)\big] = \big([X, Y], (X(\bar{\eta}) - Y(\bar{\xi}) + \{\bar{\xi}, \bar{\eta}\}_{\mathsf{R}})^\sharp\big)$$

where $\{\cdot, \cdot\}_{\mathsf{R}}$ is the Lie algebra bracket on $\mathfrak{g}_{\mathsf{R}}$, and $\{\bar{\xi}, \bar{\eta}\}_{\mathsf{R}}$ is the $\mathfrak{g}_{\mathsf{R}}$-valued function given by $\{\bar{\xi}, \bar{\eta}\}_{\mathsf{R}}(x) = \{\bar{\xi}(x), \bar{\eta}(x)\}_{\mathsf{R}}$. To confirm that this is the case one can proceed as follows. Let $\{e_\alpha\}$ be a fixed basis of $\mathfrak{g}_{\mathsf{R}}$. Then

$$[(X, 0), (0, e^\sharp_\alpha)] = (0, 0), \quad [(0, e^\sharp_\alpha), (0, e^\sharp_\beta)] = (0, \{e_\alpha, e_\beta\}^\sharp),$$

and of course $e^\sharp_\alpha(f) = 0$ for any $f \in C^\infty(U)$. Now $\bar{\xi} = \xi^\alpha e_\alpha$ with $\xi^\alpha \in C^\infty(U)$. Thus

$$\begin{aligned}\big[(X, \bar{\xi}^\sharp), (Y, \bar{\eta}^\sharp)\big] &= \big[(X, \xi^\alpha e^\sharp_\alpha), (Y, \eta^\beta e^\sharp_\beta)\big] \\ &= \big([X, Y], X(\eta^\alpha)e^\sharp_\alpha - Y(\xi^\alpha)e^\sharp_\alpha + \xi^\beta\eta^\gamma\{e_\beta, e_\gamma\}^\sharp_{\mathsf{R}}\big) \\ &= \big([X, Y], (X(\bar{\eta}) - Y(\bar{\xi}) + \{\bar{\xi}, \bar{\eta}\}_{\mathsf{R}})^\sharp\big).\end{aligned}$$

Now $TM \oplus \mathfrak{g}_{\mathsf{R}}$ is a Lie algebroid, with bracket

$$[\![(X, \bar{\xi}), (Y, \bar{\eta})]\!] = \big([X, Y], X(\bar{\eta}) - Y(\bar{\xi}) + \{\bar{\xi}, \bar{\eta}\}_{\mathsf{R}}\big),$$

where for simplicity we have ignored the distinction between a section of $M \times \mathfrak{g}_{\mathsf{R}} \to M$ and a $\mathfrak{g}_{\mathsf{R}}$-valued function on M. It is in fact the Lie algebroid of the trivial Lie groupoid $M \times G \times M$, where G is a Lie group whose opposite Lie algebra is $\mathfrak{g}_{\mathsf{R}}$, as we showed in Proposition 1.3.2. We have proved the following result.

Proposition 3.2.3 *The map $\mathrm{T}_\sharp : TU \oplus \mathfrak{g}_{\mathsf{R}} \to \mathcal{A}^G|_U$ is an isomorphism of Lie algebroids.*

3.3 Projectability of Bundle Structures

When considering morphisms between Lie groupoids or Lie algebroids we have emphisized the use of 'full morphisms' so that features of the domain of the morphism, such as certain bisections and path connections in the case of Lie groupoids, or certain sections and infinitesimal connections in the case of Lie algebroids, may be regarded as projectable to the codomain. We now review these constructions in the specific context of Lie groupoids of fibre morphisms and their Lie algebroids. We shall need a specification of those morphisms between fibre bundles which will be both full and also compatible with the two group actions, and so we start by considering a manifold F with a left action $\mu : G \times F \to F$ of a Lie group G; we write $gq = \mu_q(g) = \mu(g, q)$.

Let $\varpi : F \to K$ be a surjective submersion. We shall say that ϖ and μ are *compatible* if the action μ preserves the fibres of ϖ: that is, if $q_1, q_2 \in F$ and $\varpi(q_1) = \varpi(q_2)$ then, for each $g \in G$, we have $\varpi(gq_1) = \varpi(gq_2)$.

Put

$$S_k = \{g \in G : g(F_k) \subset F_k\}, \qquad F_k = \varpi^{-1}(k), \qquad k \in K,$$

and put $S = \bigcap_{k\in K} S_k$.

Proposition 3.3.1 *If ϖ and μ are compatible then the set S is a closed normal subgroup of G, the quotient G/S is a Lie group with projection $\theta : G \to G/S$, and the projected action μ^ϖ of G/S on K given by*

$$\mu^\varpi(gS, k) = \mu^\varpi_{gS}(k) = \varpi(gq), \qquad \varpi(q) = k$$

is a well-defined left action which is effective, whether or not the original action χ was effective.

Proof It is evident that each S_k is a subgroup of G, so that S is a subgroup of G.
If $h \in G$ then

$$\begin{aligned} hS_kh^{-1} &= \{hgh^{-1}, g \in G : g(F_k) \subset F_k\} \\ &= \{g \in G : h^{-1}gh(F_k) \subset F_k\} \\ &= \{g \in G : gh(F_k) \subset h(F_k)\} \\ &= \{g \in G : g(F_{\bar{k}}) \subset F_{\bar{k}}\} \\ &= S_{\bar{k}} \end{aligned}$$

where $\bar{k} = \varpi(hq)$ for any $q \in F_k$. But every element of K is of the form $\varpi(hq)$ for some $q \in F$, so that if $g \in S$ then $hgh^{-1} \in S_{\bar{k}}$ for every $\bar{k} \in K$, showing that $hSh^{-1} \subset S$.

If $g \in S_k$ then $g \in \mu_q^{-1}(F_k)$ for every $q \in F_k$, and conversely, so that $S_k = \bigcap_{q \in F_k} \mu_q^{-1}(F_k)$; but each F_k is closed in F and each μ_q is continuous, so that S_k is closed in G and hence S is closed in G.

Thus S is a closed normal subgroup of G, so that G/S is a Lie group.

Next, if $q_1, q_2 \in F_k$ then $\varpi(gq_1) = \varpi(gq_2)$, and if $g_1, g_2 \in G$ with $g_1S = g_2S$ then

$$\varpi(g_2q) = \varpi\left(g_1g_1^{-1}g_2q\right) = \varpi(g_1q)$$

because $g_1^{-1}g_2 \in S$, so that the map $\mu^\varpi : G/S \times K \to K$ is well defined.

Now take $gS \in G/S$ and $k \in K$. As ϖ is a surjective submersion there is a neighbourhood $U \subset K$ of k and a local section $\varsigma : U \to F$; there is also [44, Theorem 3.58] a neighbourhood $V \subset G/S$ of gS and a local section $s : V \to G$. Thus

$$\mu^\varpi|_{V \times U} = \varpi \circ \mu \circ (s, \varsigma),$$

showing that μ^ϖ is a smooth map. In addition, if $g, h \in G$ then

$$\mu^\varpi_{(gS)(hS)}(k) = \mu^\varpi_{(gh)S}(k) = \varpi(ghq) = \mu^\varpi_{gS}(hq) = \mu^\varpi_{gS}\mu^\varpi_{hS}(k),$$

and also $\mu^\varpi_S(k) = \varpi(1_Gq) = \varpi(q) = k$, so that μ^ϖ is a left action of G/S on K.

Finally, suppose that the coset $gS \in G/S$ satisfies $\mu^\varpi_{gS}(k) = k$ for every $k \in K$. Then for every $q \in F$ we have

$$\mu(gq) = \mu^\varpi_{gS}\varpi(q) = \varpi(q)$$

so that $gq \in F_{\varpi(q)}$, showing that $g \in S$; thus the action μ^ϖ is effective. □

Now let $\pi_E^G : E \to M$ be a fibre bundle with structure group G and standard fibre F, and let $\pi_L^H : L \to N$ be another fibre bundle with structure group H and standard fibre K. Let (P, p) be a fibred map, so that $\mathrm{P} : E \to L$, $p : M \to N$ and $p \circ \pi_E = \pi_L \circ \mathrm{P}$, and let $\varpi : F \to K$ be compatible with the action of G on F; we shall say that (P, p) is a *compatible fibre bundle map* if

- (P, p) is a full map of fibred manifolds;
- there is an isomorphism of Lie groups $G/S \leftrightarrow H$ such that, under this isomorphism, the action of H on K is the same as the projection μ^{ϖ} of the action μ of G on F; and
- there is a covering of E by *projectable local trivializations*: that is, each $x_0 \in M$ has a neighbourhood U and a local trivialization $\mathrm{T} : U \times F \to E_U$ such that
  - if $x_1, x_2 \in U$ and $q \in F$ then $\mathrm{PT}(x_1, q) = \mathrm{PT}(x_2, q)$ whenever $p(x_1) = p(x_2)$;
  - if $x \in U$ and $q_1, q_2 \in F$ then $\mathrm{PT}(x, q_1) = \mathrm{PT}(x, q_2)$ if, and only if, $\varpi(q_1) = \varpi(q_2)$; and
  - the map $\mathrm{T}^{\mathrm{P}} : p(U) \times K \to L_{p(U)}$ given by $\mathrm{T}^{\mathrm{P}}\big(p(x), \varpi(q)\big) = \mathrm{PT}(x, q)$, well defined by the above, is a local trivialization of π_L^H.

We now show that any compatible bundle map gives rise to a full morphism between the corresponding Lie groupoids of fibre morphisms.

Theorem 3.3.2 *Let (P, p) be a compatible fibre bundle map from $E \to M$ to $L \to N$, and let $\mathcal{G}$ and $\mathcal{H}$ be the corresponding Lie groupoids of fibre morphisms. Define a map $\mathcal{P} : \mathcal{G} \to \mathcal{H}$ by setting*

$$\mathcal{P}(\varphi) : L_{p\alpha_{\mathcal{G}}(\varphi)} \to L_{p\beta_{\mathcal{G}}(\varphi)}, \qquad \mathcal{P}(\varphi)(z) = \mathrm{P}\varphi(y)$$

for $z \in L_{p\alpha_{\mathcal{G}}(\varphi)}$, where $y \in E_{\alpha_{\mathcal{G}}(\varphi)}$ is any element satisfying $\mathrm{P}(y) = z$; then $(\mathcal{P}, p)$ is a full morphism of Lie groupoids.

Proof We show first that $\mathcal{P}$ is well defined. Put $\tilde{x} = \alpha_{\mathcal{G}}(\varphi)$ and $x = \beta_{\mathcal{G}}(\varphi)$, and let $(U_{\tilde{\lambda}}, \mathrm{T}_{\tilde{\lambda}})$ and $(U_\lambda, \mathrm{T}_\lambda)$ be projectable local trivializations of π_E^G around $\tilde{x}$ and x. Let $g \in G$ be such that, for any $(\tilde{x}, q) \in \{\tilde{x}\} \times F$, $\varphi\mathrm{T}_{\tilde{\lambda}}(\tilde{x}, q) = \mathrm{T}_\lambda\big(x, gq\big)$.

If $y_1, y_2 \in E_{\tilde{x}}$, put $q_1 = \mathrm{pr}_2 \mathrm{T}_{\tilde{\lambda}}^{-1}(y_1)$ and $q_2 = \mathrm{pr}_2 \mathrm{T}_{\tilde{\lambda}}^{-1}(y_2)$, so that if $\mathrm{P}(y_1) = \mathrm{P}(y_2)$ then $\mathrm{PT}_{\tilde{\lambda}}(\tilde{x}, q_1) = \mathrm{PT}_{\tilde{\lambda}}(\tilde{x}, q_2)$. Thus $\varpi(q_1) = \varpi(q_2)$ as (P, p) is full, giving $\varpi(gq_1) = \varpi(gq_2) \in K$ by compatibility of ϖ, and then $\mathrm{PT}_\lambda\big(x, gq_1\big) = \mathrm{PT}_\lambda\big(x, gq_2\big)$ as (P, p) is full again, finally giving $\mathrm{P}\varphi\mathrm{T}_{\tilde{\lambda}}(\tilde{x}, q_1) = \mathrm{P}\varphi\mathrm{T}_{\tilde{\lambda}}(\tilde{x}, q_2)$ so that $\mathrm{P}\varphi(y_1) = \mathrm{P}\varphi(y_2)$ as required.

We must next show that the map $\mathcal{P}(\varphi) : L_{p(\tilde{x})} \to L_{p(x)}$ is a fibre morphism of π_L^H. Let $y = \mathrm{T}_{\tilde{\lambda}}(\tilde{x}, q)$, so that $\varphi(y) = \mathrm{T}_\lambda\big(x, l_g(q)\big)$ and therefore that

$$\mathrm{P}(y) = \mathrm{PT}_{\tilde{\lambda}}(\tilde{x}, q) = \mathrm{T}_{\tilde{\lambda}}^{\mathrm{P}}\big(p(\tilde{x}), \varpi(q)\big) \tag{1}$$
$$\mathcal{P}(\varphi)\big(\mathrm{P}(y)\big) = \mathrm{P}\varphi(y) = \mathrm{PT}_\lambda\big(x, gq\big) = \mathrm{T}_\lambda^{\mathrm{P}}\big(p(x), \varpi(gq)\big) = \mathrm{T}_\lambda^{\mathrm{P}}\big(p(x), \mu_{gS}^{\varpi}\varpi(q)\big),$$

showing that $\mathcal{P}(\varphi)$ corresponds to the projected action of $gS \in G/S \cong H$ and is therefore an element of $\mathcal{H}$.

We now show that $\mathcal{P}$ is a smooth map $\mathcal{G} \to \mathcal{H}$ in a neighbourhood of φ. Let $\psi_{\lambda,\bar{\lambda}} : \mathcal{G}_{\lambda,\bar{\lambda}} \to U_\lambda \times G \times U_{\bar{\lambda}}$ be the map constructed in the proof of Theorem 3.1.1; then $\psi_{\lambda,\bar{\lambda}}$ is a diffeomorphism as it was used to define the smooth structure on $\mathcal{G}$. Similarly the map $\psi^{\mathrm{P}}_{\lambda,\bar{\lambda}} : \mathcal{H}_{\lambda,\bar{\lambda}} \to p(U_\lambda) \times H \times p(U_{\bar{\lambda}})$ is a diffeomorphism. But now the restriction of $\mathcal{P}$ to $\mathcal{G}_{\lambda,\bar{\lambda}}$ is given by

$$\mathcal{P}|_{\mathcal{G}_{\lambda,\bar{\lambda}}} = \left(\psi^{\mathrm{P}}_{\lambda,\bar{\lambda}}\right)^{-1} \circ (p, \theta, p) \circ \psi_{\lambda,\bar{\lambda}}$$

and as p and $\theta : G \to G/S \cong H$ are both smooth it follows that $\mathcal{P}|_{\mathcal{G}_{\lambda,\bar{\lambda}}}$ is smooth; thus $\mathcal{P}$ is smooth.

By construction $\alpha_{\mathcal{H}}\mathcal{P}(\varphi) = p\alpha_{\mathcal{G}}(\varphi)$ and $\beta_{\mathcal{H}}\mathcal{P}(\varphi) = p\beta_{\mathcal{G}}(\varphi)$ for any $\varphi \in \mathcal{G}$, so that $(\mathcal{P}, p)$ is a Lie groupoid morphism.

Finally we must show that, for each $\tilde{x} \in M$, the restriction of $\mathcal{P}$ to a map $\alpha_{\mathcal{G}}^{-1}(\tilde{x}) \to \alpha_{\mathcal{H}}^{-1}\big(p(\tilde{x})\big)$ is a surjective submersion. The fact that it is a submersion is a local property and follows from

$$\mathcal{P}|_{\mathcal{G}_{\lambda,\tilde{x}}} = \left(\psi^{\mathrm{P}}_{\lambda,\tilde{x}}\right)^{-1} \circ (p, \theta) \circ \psi_{\lambda,\tilde{x}}$$

where $\psi_{\lambda,\tilde{x}} : \mathcal{G}_{\lambda,\tilde{x}} \to U_\lambda \times G \times \{\tilde{x}\}$ and $\psi^{\mathrm{P}}_{\lambda,\tilde{x}} : \mathcal{H}_{\lambda,\tilde{x}} \to p(U_\lambda) \times \theta(G) \times \{p(\tilde{x})\}$ because both p and θ are submersions (the latter following from the existence of local sections as mentioned in the proof of Proposition 3.3.1).

To show that the restriction of $\mathcal{P}$ is surjective, suppose $\tilde{\varphi} \in \alpha_{\mathcal{H}}^{-1}\big(p(\tilde{x})\big)$, and choose any $x \in M$ such that $p(x) = \beta_{\mathcal{H}}(\tilde{\varphi})$. There is then an element $gS \in G/S \cong H$ such that, for any $z \in L_{p(\tilde{x})}$, if $k \in K$ satisfies $z = \mathrm{T}^{\mathrm{P}}_{\bar{\lambda}}(p(\tilde{x}), k)$ then $\tilde{\varphi}(z) \in L_{p(x)}$ satisfies $\tilde{\varphi}(z) = \mathrm{T}^{\mathrm{P}}_{\lambda}\big(p(x), \mu^{\varpi}_{gS}(k)\big)$. Choose a fixed representative g of the coset gS.

Now define $\varphi \in \alpha_{\mathcal{G}}^{-1}(u)$ by setting $\varphi(y) = \mathrm{T}_\lambda\big(x, gq\big)$ for any $y \in \alpha_{\mathcal{G}}^{-1}(\tilde{x})$, where $q \in F$ satisfies $y = \mathrm{T}_{\bar{\lambda}}(\tilde{x}, q)$. By construction $\beta_{\mathcal{G}}(\varphi) = x$. If $z = \mathrm{P}(y)$ then we have

$$z = \mathrm{P}(y) = \mathrm{P}\mathrm{T}_{\bar{\lambda}}(\tilde{x}, q) = \mathrm{T}^{\mathrm{P}}_{\bar{\lambda}}\big(p(\tilde{x}), \varpi(q)\big)$$

so that $\tilde{\varphi}(z) = \mathrm{T}^{\mathrm{P}}_{\lambda}\big(p(x), \mu^{\varpi}_{gS}\varpi(q)\big)$. But then, using (1),

$$\mathcal{P}(\varphi)(z) = \mathcal{P}(\varphi)\big(\mathrm{P}(y)\big) = \mathrm{T}^{\mathrm{P}}_{\lambda}\big(p(x), \chi^{\varpi}_{gS}\varpi(q)\big) = \tilde{\varphi}(z),$$

so that $\mathcal{P}(\varphi) = \tilde{\varphi}$. □

We sometimes consider a fibre bundle $\pi^G : E \to N$ with standard fibre F where there is, in addition, some other fibred manifold $p : M \to N$. It is then natural to study the pullback bundle $p^*\pi : p^*E \to M$, and we may check that this, too, is a G-bundle with the same standard fibre. So let $x \in M$ and let $U \subset N$ be a neighbourhood of $p(x)$ such that $\mathrm{T} : U \times F \to \pi^{-1}(U)$ is a local trivialization; define the map T^p by

$$\mathrm{T}^p : p^{-1}(U) \times F \to (p^*\pi)^{-1}p^{-1}(U), \qquad \mathrm{T}^p(\tilde{x}, z) = \big(\tilde{x}, \mathrm{T}(p(\tilde{x}), z)\big).$$

The collection of such maps T^p forms a family of local trivializations for $p^*\pi$ whose transition functions take their values in G.

Proposition 3.3.3 *If $\mathcal{G}$ is the Lie groupoid of fibre morphisms of $\pi : E \to N$ then the Lie groupoid $\mathcal{G}^p$ of fibre morphisms of $p^*\pi : p^*E \to M$ may be identified with the pullback Lie groupoid $p^{**}\mathcal{G}$, and the induced map $\mathcal{P} : \mathcal{G}^p \to \mathcal{G}$ may be identified with the map $p^{**}\mathcal{G} \to \mathcal{G}$ given by $(x, \varphi, \tilde{x}) \mapsto \varphi$. The map $\mathrm{P} : p^*E \to E$ given by $\mathrm{P}(x, y) = y$ defines a compatible map (P, p) of fibre bundles.*

Proof Take $\varphi^p \in \mathcal{G}^p$, and suppose $\varphi^p : (p^*E)_{\tilde{x}} \to (p^*E)_x$. As

$$(p^*E)_{\tilde{x}} = \{(\tilde{x}, z) : z \in E_{p(\tilde{x})}\}, \qquad (p^*E)_x = \{(x, y) : y \in E_{p(x)}\}$$

we see that φ^p determines a unique map $\varphi : E_{p(\tilde{x})} \to E_{p(x)}$ which is evidently a fibre morphism of π and hence an element of $\mathcal{G}$. As $\alpha(\varphi) = p(\tilde{x})$ and $\beta(\varphi) = p(x)$ we may identify φ^p with $(x, \varphi, \tilde{x}) \in p^{**}\mathcal{G}$, and it is clear that this correspondence is a bijection preserving the groupoid operations. The identification of the two maps $\mathcal{G}^p \to \mathcal{G}$ and $p^{**}\mathcal{G} \to \mathcal{G}$ is then immediate.

The map $\mathrm{P} : p^*E \to E$ defines a fibred map (P, p) because $(p^*\pi)(x, y) = x$ so that

$$(p \circ p^*\pi)(x, y) = p(x) = \pi(y) = (\pi \circ \mathrm{P})(x, y).$$

Also, the restriction of P to a map of fibres $(p^*E)_x \to E_{p(x)}$ is given by $(x, y) \mapsto y$ and is a diffeomorphism, so that P is a fibrewise surjective submersion, and therefore (P, p) is full. We also see that the fibre projection ϖ is the identity map on F and so is compatible with the action.

We now see that the local trivializations T^p are projectable. First, if $x_1, x_2 \in p^{-1}(U)$ and $q \in F$, and if $p(x_1) = p(x_2)$, then

$$\begin{aligned}\mathrm{P}\mathrm{T}^p(x_1, q) &= \mathrm{P}\big(x_1, \mathrm{T}(p(x_1), q)\big) = \mathrm{T}(p(x_1), q) = \mathrm{T}(p(x_2), q) = \mathrm{P}\big(x_2, \mathrm{T}(p(x_2), q)\big) \\ &= \mathrm{P}\mathrm{T}^p(x_2, q).\end{aligned}$$

Next, if $x \in p^{-1}(U)$ and $q_1, q_2 \in F$ then

$$\begin{aligned}\mathrm{P}\mathrm{T}^p(x, q_1) &= \mathrm{P}\big(x, \mathrm{T}(p(x), q_1)\big) = \mathrm{T}(p(x), q_1), \\ \mathrm{P}\mathrm{T}^p(x, q_2) &= \mathrm{P}\big(x, \mathrm{T}(p(x), q_2)\big) = \mathrm{T}(p(x), q_2)\end{aligned}$$

and $\mathrm{T}(p(x), q_1) = \mathrm{T}(p(x), q_2)$ if, and only if, $q_1 = q_2$ if, and only if, $\varpi(q_1) = \varpi(q_2)$.

Finally, defining $\mathrm{T}^{p\mathrm{P}} : U \times F \to \pi^{-1}(U)$ by $\mathrm{T}^{p\mathrm{P}}\big(p(x), \varpi(q)\big) = \mathrm{P}\mathrm{T}^p(x, q)$, we see that

$$\mathrm{T}^{p\mathrm{P}}\big(p(x), q\big) = \mathrm{P}\mathrm{T}^p(x, q) = \mathrm{P}\big(x, \mathrm{T}(p(x), q)\big) = \mathrm{T}(p(x), q),$$

so we see that (P, p) is indeed a compatible map of fibre bundles. □

We now return to the general case of two fibre bundles $\pi_E^G : E \to M$ and $\pi_L^H : L \to N$, and a compatible fibre bundle map (P, p) between them. The induced full morphism $\mathcal{P} : \mathcal{G} \to \mathcal{H}$ of fibre morphism groupoids gives rise to a full morphism of Lie algebroids $P : A\mathcal{G} \to A\mathcal{H}$; we shall be interested in the appearance of this latter morphism when we consider the corresponding Lie algebroids $\mathcal{A}^G$, $\mathcal{A}^H$ of vector fields along the bundle fibres.

A general full morphism $\mathcal{A}^G \to \mathcal{A}^H$ will of course be defined as a map of vector fields along fibres, and need not be given in terms of a pointwise map of the underlying vector bundles $TE \to TL$. Indeed, if (P, p) is an arbitrary fibred map from $E \to M$ to $L \to N$ then there is no reason why an element $\boldsymbol{a} \in \mathcal{A}^G$ should be projectable under P_* to give an element of $\mathcal{A}^H$: the key condition is that (P, p) should be a compatible fibre bundle map. To see the reason for this, we look at the infinitesimal action of a Lie algebra on a manifold.

Lemma 3.3.4 *Let F be a manifold, let $\mu : G \times F \to F$ be an effective left action of a Lie group G, and let $\varpi : F \to K$ be a surjective submersion compatible with μ. If $\mathfrak{g}_{\mathrm{R}}$ denotes the opposite Lie algebra of G then, for any $a \in \mathfrak{g}_{\mathrm{R}}$, the fundamental vector field $a^\sharp$ on F is projectable to K; its projection is the fundamental vector field $\big(\theta_*(a)\big)^\sharp$ of the action μ^{P} of G/S on K, where $\theta : G \to G/S$ is the canonical projection.*

Proof Let c be a curve in G with $\dot{c}(0) = a$. For any $q \in F$ we have

$$a^\sharp|_q = \frac{d}{dt}\mu\big(c(t), q\big)\Big|_{t=0}$$

so that, using Proposition 3.3.1,

$$\varpi_*\big(a_q^\sharp\big) = \frac{d}{dt}\varpi\mu\big(c(t), q\big)\Big|_{t=0} = \frac{d}{dt}\mu^{\varpi}\big(\theta c(t), \varpi(q)\big)\Big|_{t=0} = \big(\theta_*(a)\big)^\sharp_{\varpi(q)}.$$ □

Theorem 3.3.5 *Let $\pi_E^G : E \to M$ and $\pi_L^H : L \to N$ be fibre bundles, and let* (P, p) *be a compatible fibre bundle map between them. Let $\mathcal{G}$ and $\mathcal{H}$ be the corresponding fibre morphism groupoids, with induced map $\mathcal{P} : \mathcal{G} \to \mathcal{H}$, and let $P : A\mathcal{G} \to A\mathcal{H}$ be the resulting map of Lie algebroids.*

If we identify $A\mathcal{G} \cong \mathcal{A}^G$ and $A\mathcal{H} \cong \mathcal{A}^H$ as Lie algebroids of vector fields along fibres then P may be identified with the map $\mathrm{P}_* : \mathcal{A}^G \to \mathcal{A}^H$, and a projectable (under P) section of $A\mathcal{G}$ corresponds to a doubly projectable (under both π_{E*}^G and P_*) vector field on E.

Proof Let $\Lambda_G : A\mathcal{G} \to \mathcal{A}^G$ and $\Lambda_H : A\mathcal{H} \to \mathcal{A}^H$ denote the Lie algebroid isomorphisms. If $\boldsymbol{a} \in A_x\mathcal{G}$ then, as in the proof of Theorem 3.2.1, there is a local flow Ψ_t on $\mathcal{G}$ such that $\boldsymbol{a} = d\Psi_t(1_x)/dt|_{t=0}$, and so

$$P(\boldsymbol{a}) = \mathcal{P}_*(\boldsymbol{a}) = \mathcal{P}_*\frac{d\Psi_t(1_x)}{dt}\Big|_{t=0} = \frac{d}{dt}\mathcal{P}\Psi_t(1_x)\Big|_{t=0}.$$

We also know from the proof of that theorem that, for each $y \in E_x$,

$$\Lambda_G(\boldsymbol{a})\big|_y = d\Psi_t(1_{\pi(y)})(y)/dt\big|_{t=0},$$

so that

$$\begin{aligned}\mathrm{P}_*\big(\Lambda_G(\boldsymbol{a})|_y\big) &= \mathrm{P}_*\frac{d\Psi_t(1_{\pi(y)})(y)}{dt}\bigg|_{t=0} \\ &= \frac{d}{dt}\mathrm{P}\big(\Psi_t(1_{\pi(y)})(y)\big)\Big|_{t=0} = \frac{d}{dt}\big(\mathcal{P}\Psi_t(1_{\pi(y)})\big)\big(\mathrm{P}(y)\big)\Big|_{t=0} = \Lambda_H P(\boldsymbol{a})|_{\mathrm{P}(y)}\,.\end{aligned}$$

The fact that a section of $A\mathcal{G}$ projectable to $A\mathcal{H}$ corresponds to a vector field on E projectable to L is a consequence of Lemma 3.3.4 using local trivializations. □

3.4 Connections on Fibre Morphism Groupoids and Algebroids

We continue with a fibre bundle $\pi^G : E \to M$ with standard fibre F, but now suppose that we have a path connection Γ on the Lie groupoid $\mathcal{G}$ of fibre morphisms of π^G. We also consider the corresponding infinitesimal connection as a map $\gamma : TM \to \mathcal{A}^G \cong A\mathcal{G}$, so that a section of TM (a vector field X on M) will map to its *horizontal lift*, a section of $\mathcal{A}^G$ (a projectable vector field $X^{\mathsf{h}} = \gamma \circ X$ on E).

We now wish to see how a curve in M might give rise to a lifted curve of a new kind, namely a 'vertical' curve in the total space E of the bundle π, taking its values in a single fibre of the bundle. In order to construct such a curve, we assume that π has a global section $\varsigma : M \to E$; we think of E as attached to M along the image of ς, and call ς the *attachment section.*

The path connection associates with each curve $c : [0, 1] \to M$ a curve c^Γ in $\mathcal{G}$, so that for each $t \in [0, 1]$, $c^\Gamma(t)$ is a fibre morphism $E_{c(0)} \to E_{c(t)}$. We now use the attachment section to construct the *development* $c_\Gamma : [0, 1] \to E_{c(0)}$: this is the curve in the fibre over $c(0)$ given by

$$c_\Gamma(t) = \big(c^\Gamma(t)\big)^{-1}\varsigma c(t).$$

On the right-hand side, $\varsigma c(t) \in E_{c(t)}$ and $c^\Gamma(t) : E_{c(0)} \to E_{c(t)}$, so $c_\Gamma(t) \in E_{c(0)}$ as claimed. Note that $c_\Gamma(0) = \varsigma c(0)$, so the development starts on the image of M under the attachment section, at the point $c(0)$ if one were to identify the image with M itself.

Lemma 3.4.1 *If $c : [0, 1] \to M$ is a curve and $T_{c(0)}M \to T_{\varsigma c(0)}E$, $v \mapsto v^{\mathsf{h}}$ is the horizontal lift determined by the path connection Γ then*

$$\dot{c}_\Gamma(0) = \varsigma_*\dot{c}(0) - \dot{c}(0)^{\mathsf{h}} = \mathrm{v}(\varsigma_*\dot{c}(0))$$

where $\mathrm{v} : T_y E \to V_y E$ *is the vertical projection determined by the horizontal lift.*

Proof For the development we have $c^\Gamma(t)(c_\Gamma(t)) = \varsigma(c(t))$; moreover $c^\Gamma(0) = \mathrm{id}_{E_x}$ and $c_\Gamma(0) = \varsigma(x)$. We therefore have

$$\begin{aligned} \varsigma_* \dot{c}(0) &= \frac{d}{dt}\big(c^\Gamma(t)(c_\Gamma(t))\big)_{t=0} \\ &= \frac{d}{dt}\big(c^\Gamma(t)(\varsigma(x))\big)_{t=0} + \dot{c}_\Gamma(0) \\ &= \frac{d}{dt}\big(c^{\mathsf{h}}_{\varsigma(x)}(t)\big)_{t=0} + \dot{c}_\Gamma(0) \\ &= \dot{c}(0)^{\mathsf{h}} + \dot{c}_\Gamma(0), \end{aligned}$$

so that

$$\dot{c}_\Gamma(0) = \varsigma_* \dot{c}(0) - \dot{c}(0)^{\mathsf{h}} = \mathrm{v}(\varsigma_* \dot{c}(0)). \qquad \square$$

We shall now work towards the establishment of a technical result, Lemma 3.4.2 below, which we shall need in order to compute the development of a curve in particular cases. Recall first that, when discussing path connections, we defined lifted curves c^Γ where c was defined on $\mathbb{R}$ (or on a subinterval of $\mathbb{R}$) rather than on $[0, 1]$; we then showed in Proposition 2.1.8 that if χ is a reparametrization preserving 0 then $(c \circ \chi)^\Gamma = c^\Gamma \circ \chi$. Thus, for the development,

$$(c \circ \chi)_\Gamma(t) = \big((c \circ \chi)^\Gamma(t)\big)^{-1} \varsigma c \chi(t) = \big(c^\Gamma \chi(t)\big)^{-1} \varsigma c \chi(t) = c_\Gamma \chi(t),$$

so that developments also respect reparametrization.

Now let $c : [0, 1] \to M$ be, in particular, an injective curve. We may, by the remarks above, regard c as taking its values in $U \subset M$, where $\mathrm{T} : U \times F \to \pi^{-1}(U)$ is some given local trivialization of $\pi^G : E \to M$. We may use this trivialization to identify $c^* E$, the pull-back of E to $[0, 1]$ defined by c, with $[0, 1] \times F$. We may then regard $\gamma(\dot{c})$ as a vector field C on $[0, 1] \times F$; and since $\gamma(\dot{c})$ projects onto $\partial/\partial t$ on $[0, 1]$, we may write

$$C(t, q) = \frac{\partial}{\partial t} - \bar{C}(t, q)$$

where for each t, $\bar{C}(t, q) \in T_q F$.

The flow of C maps fibres to fibres, and consists of fibre morphisms; c^Γ is the flow of C acting on $E_{c(0)} \cong \{0\} \times F$. Let $s \mapsto \phi_s$ be the flow of C. It takes the form

$$\phi_s(t, q) = (t + s, \Phi(s, t)q), \quad q \in F,$$

where for each s, t, $\Phi(s, t) \in G$, with

$$\left.\frac{\partial \Phi}{\partial s}\right|_{(0,t)} \in T_{\Phi(0,t)} G \cong \mathfrak{g}, \qquad \left(\left.\frac{\partial \Phi}{\partial s}\right|_{(0,t)}\right)^\sharp_q = -\bar{C}(t, q),$$

where $a^\sharp$ is the fundamental vector field on F corresponding to $a \in \mathfrak{g}$ under the action of G. We may define a curve in $\mathfrak{g}$, denoted by $t \mapsto C_\mathfrak{g}(t)$, by $\bar{C}(t, q) = C_\mathfrak{g}^\sharp|_q$; then

$$\left.\frac{\partial \Phi}{\partial s}\right|_{(0,t)} = -C_\mathfrak{g}(t).$$

The attachment section ς corresponds to some $o \in F$. The development of c into $E_{c(0)} \sim \{0\} \times F$ is

$$c_\Gamma(t) = \left(c^\Gamma(t)\right)^{-1} \varsigma c(t) = \phi_{-t}(t, o) = (0, \Phi(-t, t)o),$$

so that $c_\Gamma(t) = (0, \bar{c}_\Gamma(t))$ where $\bar{c}_\Gamma$ is the curve in F given by

$$\bar{c}_\Gamma(t) = \Psi(t)o$$

where $\Psi(t) = \Phi(-t, t)$; note that Ψ is a curve in G, with $\Psi(0) = 1_G$. We denote by l the left action of G on itself: then

$$t \mapsto l_{\Psi(t)^{-1}*} \left.\frac{d\Psi}{dt}\right|_t$$

is a curve in $\mathfrak{g}$, as is $t \mapsto C_\mathfrak{g}(t)$.

Lemma 3.4.2

$$l_{\Psi(t)^{-1}*} \left.\frac{d\Psi}{dt}\right|_t = C_\mathfrak{g}(t);$$

that is to say, the curve $C_\mathfrak{g}$ in $\mathfrak{g}$ is the left Darboux derivative of the curve Ψ in G.

Proof As $s \mapsto \phi_s$ has the one-parameter group property, it follows that

$$\Phi(s_1 + s_2, t) = \Phi(s_1, t + s_2) \cdot \Phi(s_2, t).$$

Differentiating this equation with respect to s_2 and evaluating the result at $s_1 = -t$, $s_2 = 0$ gives

$$\left.\frac{\partial \Phi}{\partial s}\right|_{(-t,t)} = \left.\frac{\partial \Phi}{\partial t}\right|_{(-t,t)} + l_{\Phi(-t,t)*} \left.\frac{\partial \Phi}{\partial s}\right|_{0,t} \in T_{\Phi(-t,t)}G,$$

and then using

$$\left.\frac{d\Psi}{dt}\right|_t = -\left.\frac{\partial\Phi}{\partial s}\right|_{(-t,t)} + \left.\frac{\partial\Phi}{\partial t}\right|_{(-t,t)}$$

we obtain the required result. □

We continue the study of developments, and investigate the way in which they may be used to define geodesics, in Chap. 11.

3.5 The Comparison with Principal Bundles

To conclude this chapter we look briefly at the way in which the basic structures of our approach relate to those of the principal bundle theory.[3]

As before we take a general fibre bundle $\pi^G : E \to M$ with standard fibre F and a left action $y \mapsto gy$ of G on F. Let $\Pi^G : P \to M$ be a principal bundle for which π^G is an associated bundle, so that $E = (P \times F)/G$ where the right action of G on $P \times F$ is given by $(p, y) \mapsto (pg, g^{-1}y)$. Writing the equivalence class of (p, y) under this right action as $[(p, y)]$, we see that each $p \in P$ determines a map $\hat{p} : F \to E_{\Pi(p)}$ by $\hat{p}(y) = [(p, y)]$, and that

$$\widehat{pg}(y) = [(pg, y)] = [(p, gy)] = \hat{p}(gy).$$

Now recall that we have defined the groupoid $\mathcal{G}$ of fibre morphisms to be the collection of maps $\varphi : E_{\tilde{x}} \to E_x$ that are diffeomorphisms respecting the action of G. We see that such a diffeomorphism φ is one satisfying the condition that, for $p, q \in P$ with $\Pi(p) = \tilde{x}$ and $\Pi(q) = x$,

$$\hat{q}^{-1} \circ \varphi \circ \hat{p} \in G$$

so that, when the fibres $E_{\Pi(p)}$, $E_{\Pi(q)}$ are identified with F using $\hat{p}$ and $\hat{q}$, the map becomes an element of G. If this condition holds for some p, q then it holds for all p, q (in the same fibres), because $\widehat{pg} = \hat{p} \circ l_g$. In fact we can always choose p, q such that $\hat{q}^{-1} \circ \varphi \circ \hat{p} = 1_G$, so that we may assume without loss of generality that $\varphi = \hat{q} \circ \hat{p}^{-1}$.

Let $\mathcal{P}$ be the gauge groupoid of the principal bundle $P \to M$, so that elements of $\mathcal{P}$ are equivalence classes $[(p, q)]$, where $(p_1, q_1) \sim (p_2, q_2)$ if there is some $g \in G$ with $(p_1 g, q_1 g) = (p_2, q_2)$. As $\widehat{pg} \circ \widehat{qg}^{-1} = \hat{q} \circ \hat{p}^{-1}$ for any $g \in G$ we see that the map

$$\mathcal{P} \to \mathcal{G}, \qquad [(p, q)] \mapsto \hat{q} \circ \hat{p}^{-1}$$

[3] In earlier sections we have used the symbols $P, \mathrm{P}, \mathcal{P}, p$ to denote projection maps of various kinds; in the present section we instead use these symbols for objects associated with principal bundles. We also use q to denote a second element of P rather than of the standard fibre F, so in this section we shall take y to be an element of F rather than of E.

is well defined, and indeed it is an isomorphism of Lie groupoids. We may therefore identify the fibre morphism groupoid $\mathcal{G}$ as the gauge groupoid $\mathcal{P}$ of any principal bundle for which the fibre bundle is an associated bundle.

Now take a curve $t \mapsto p(t)$ in P with $p(0) = p$, $\Pi(p) = x$, and define the corresponding curve

$$c : t \mapsto \widehat{p(t)} \circ \hat{p}^{-1}$$

in $\mathcal{G}$, so that $\alpha(c(t)) = x$ and therefore that the tangent vector $\dot{c}(0)$ is an element of $A_x\mathcal{G}$. This procedure clearly depends only on the initial tangent vector to $t \mapsto p(t)$, so for any $u \in T_pP$ we may let $\tilde{p}(u)$ be the element of $A_x\mathcal{G}$ obtained by carrying out this construction on any curve through p which has u as its initial tangent vector. The map $\tilde{p} : T_pP \to A_x\mathcal{G}$, $x = \Pi(p)$, is then an isomorphism of vector spaces, and it is easy to see that, for $g \in G$,

$$\widetilde{pg} \circ r_{g*} = \tilde{p},$$

so that the construction is invariant under the right action of G. It follows from this that we may identify $A\mathcal{G}$ with TP/G, the Atiyah algebroid of P.

If, in particular, $u \in T_pP$ is vertical then $\tilde{p}(u) \in K_x\mathcal{G}$. In such a case u takes the form $a^\sharp_p$, the fundamental vector at p corresponding to some $a \in \mathfrak{g}$, so that p determines a vector space isomorphism $\mathfrak{g} \to K_x\mathcal{G}$ given by $a \mapsto \tilde{p}(a^\sharp_p)$. But

$$r_{g*}a^\sharp_p = (\mathrm{Ad}_{g^{-1}}(a))^\sharp_{pg},$$

so that

$$\widetilde{pg}(a^\sharp_{pg}) = \tilde{p}\big((\mathrm{Ad}_g(a))^\sharp_p\big).$$

Chapter 4
Four Case Studies

In this chapter we see how the approach of considering the Lie groupoid $\mathcal{G}$ of fibre morphisms may be applied to four standard cases, where the morphisms are linear maps, affine maps, projective maps and Euclidean maps; the fibres of the bundle $\pi^G : E \to M$ will, of course, need to carry an appropriate structure so that each class of map can be defined. If the example supports a canonical global section then the subgroup G_o of the bundle group G may be interpreted as the isotropy group of an 'origin' in the standard fibre F: we shall see the significance of this when we investigate Cartan geometries in later chapters. In order to carry out this task for projective maps we shall need to define a bundle over M whose fibres are projective spaces of dimension n, and in Sect. 4.3 we describe this construction in two different ways.

4.1 Vector Bundles and Linear Maps

We first consider linear maps between the fibres of a vector bundle $\pi : E \to M$ of rank k, which we regard as a fibre bundle with group $\mathsf{GL}(k)$, the group of nonsingular matrices (A^α_β), $\alpha = 1, 2, \ldots, k$. We note that in this case there is indeed a canonical global section, namely the zero section of the vector bundle, but that the isotropy group is trivial.

Proposition 4.1.1 *Let $\pi : E \to M$ be a vector bundle, and let $\mathcal{G}$ be the set of linear maps $\varphi : E_{\tilde{x}} \to E_x$ for all $x, \tilde{x} \in M$. Then $\mathcal{G}$ is a locally trivial Lie groupoid. We shall call such maps φ* linear morphisms *of π (or of E).*

Proof Immediate from Theorem 3.1.1. □

An element of the Lie algebroid $A\mathcal{G}$ may therefore be written in the coordinates $(x^i, z^\alpha_\beta, \tilde{x}^i)$ as

M. Crampin and D. Saunders, *Cartan Geometries and their Symmetries*,
Atlantis Studies in Variational Geometry 4, DOI 10.2991/978-94-6239-192-5_4

$$a = a^i \left.\frac{\partial}{\partial x^i}\right|_{1_x} + a^\alpha_\beta \left.\frac{\partial}{\partial z^\alpha_\beta}\right|_{1_x} \in A_x\mathcal{G} = T_{1_x}\mathcal{G}\,.$$

We now recall the result of Theorem 3.2.1, that a section of the Lie algebroid $A\mathcal{G}$ may be composed with the map Λ to give a projectable vector field on E, so that elements of the image $\mathcal{A}^G = \Lambda(A\mathcal{G})$ are vector fields along the fibres of $E \to M$. We indicated earlier that we use symbols such as $\xi, \eta, \ldots$ to represent sections of $A\mathcal{G}$, but we shall now use the same symbols for the corresponding projectable vector fields on E. We shall therefore say that a projectable vector field ξ with a vertically uniform flow box, where the restriction $\psi_{t;x} = \psi_t|_{E_x}$ of its flow is a linear map between fibres of E, is a *fibre-linear projectable vector field*.

Proposition 4.1.2 *If ξ is a projectable vector field on E then ξ corresponds to a section of $A\mathcal{G}$ if, and only if, its expression in coordinates (x^i, y^α) on E is*

$$\xi = X^i \frac{\partial}{\partial x^i} + Y^\alpha_\beta y^\beta \frac{\partial}{\partial y^\alpha}$$

where the functions X^i and Y^α_β are projectable to M.

Proof Let $\pi^{-1}(U) \subset E$ be the domain of the coordinates (x^i, y^α). Let ψ_t be the flow of ξ, and put $\phi^i_t = x^i \circ \psi_t$ and $\psi^\alpha_t = y^\alpha \circ \psi_t$.

Suppose first that ξ corresponds to a section of $A\mathcal{G}$. In these coordinates we may write ξ as

$$\xi = X^i \frac{\partial}{\partial x^i} + Y^\alpha \frac{\partial}{\partial y^\alpha}$$

where the functions X^i are projectable to U, and where

$$X^i = \left.\frac{d\phi^i_t}{dt}\right|_{t=0}, \qquad Y^\alpha = \left.\frac{d\psi^\alpha_t}{dt}\right|_{t=0}.$$

Fix $x \in U$. For each $y \in E_x$ there is a nonempty interval $(-\varepsilon_y, \varepsilon_y)$ such that

$$y^\alpha\big(\psi_{t;x}(y)\big) = c^\alpha_\beta(t) y^\beta(y)$$

for $t \in (-\varepsilon_y, \varepsilon_y)$ and for some functions $c^\alpha_\beta : (-\varepsilon_y, \varepsilon_y) \to \mathbb{R}$. Thus, putting $a^\alpha_\beta = dc^\alpha_\beta/dt|_{t=0}$, we see that

$$\xi(x) = \xi|_{E_x} = X^i(x) \left.\frac{\partial}{\partial x^i}\right|_{E_x} + a^\alpha_\beta y^\beta \left.\frac{\partial}{\partial y^\alpha}\right|_{E_x}$$

so that $Y^\alpha(y) = a^\alpha_\beta y^\beta(y)$, showing that ξ takes the required coordinate form.

Suppose conversely that ξ takes the appropriate form in coordinates on E_U, so that $Y^\alpha = Y^\alpha_\beta y^\beta$ and therefore that

$$\frac{d\psi_t^\alpha}{dt} = Y^\alpha_\beta \psi_t^\beta . \tag{1}$$

Let $(-\varepsilon_x, \varepsilon_x)$ be a nonempty interval such that $0_x \in E_x$ is in the domain of ψ_t for each $t \in (-\varepsilon_x, \varepsilon_x)$; then the entire fibre E_x is in the domain of ψ_t for each $t \in (-\varepsilon_x, \varepsilon_x)$ because the equations (1) are linear in the fibre coordinates y^α. For the same reason, the restriction $\psi_{t;x}$ is a linear map $E_x \to E_{\phi_t(x)}$. Thus ξ corresponds to a section of $A\mathcal{G}$. □

A path connection on the Lie groupoid of linear morphisms of the vector bundle $\pi : E \to M$ will be called a *linear connection*, and the same terminology will be used for the infinitesimal connection on the Lie algebroid. We may regard the latter as a map $\gamma : E \to \mathcal{A}^G$, and then for any vector field X on M the composition $X^{\mathsf{h}} = \gamma \circ X$ is a projectable vector field on E; the map $X \to X^{\mathsf{h}}$ is then a 'horizontal lift' operation which defines a connection (in the classical sense) on E. Each vector field X^{h} is a fibre-linear projectable vector field as described above, given in terms of a local basis by

$$\frac{\partial}{\partial x^i} \mapsto \left(\frac{\partial}{\partial x^i}\right)^{\mathsf{h}} = \frac{\partial}{\partial x^i} - \Gamma^\alpha_{i\beta} y^\beta \frac{\partial}{\partial y^\alpha}$$

where $\Gamma^\alpha_{i\beta}$ are functions defined locally on M. We shall see in the next section that linear connections may also arise in the context of affine bundles, and we shall also see the reason for the choice of a minus sign in the coordinate formula above.

4.2 Affine Bundles and Affine Maps

For any positive integer k let $\mathsf{A}(k)$ be the subgroup of matrices $(A^a_b) \in \mathsf{GL}(k+1)$, $a, b = 0, 1, 2, \ldots, k$, where $A^0_0 = 1$ and $A^0_\alpha = 0$ for $\alpha = 1, 2, \ldots, k$. As model affine space of dimension k we take the number space $\mathbb{R}^k$ and the action of the group $\mathsf{A}(k)$ on it by $(x^\alpha) \mapsto (A^\alpha_0 + A^\alpha_\beta x^\beta)$. By an *affine bundle* of rank k we mean a fibre bundle with the affine space $\mathbb{R}^k$ as standard fibre, and with group $\mathsf{A}(k)$. We now consider affine maps between the fibres of such an affine bundle $\pi : E \to M$ of rank k.

Proposition 4.2.1 *Let $\pi : E \to M$ be an affine bundle, and let $\mathcal{G}$ be the set of affine maps $\varphi : E_{\tilde{x}} \to E_x$ for all $x, \tilde{x} \in M$. Then $\mathcal{G}$ is a locally trivial Lie groupoid. We shall call such maps φ* affine morphisms *of π (or of E).*

Proof Immediate from Theorem 3.1.1. □

An element of this Lie algebroid $A\mathcal{G}$ may therefore be written in the coordinates $(x^i, (z^\alpha_0, z^\alpha_\beta), \tilde{x}^i)$ as

$$a = a^i \left.\frac{\partial}{\partial x^i}\right|_{1_x} + a_0^\alpha \left.\frac{\partial}{\partial z_0^\alpha}\right|_{1_x} + a_\beta^\alpha \left.\frac{\partial}{\partial z_\beta^\alpha}\right|_{1_x} \in A_x\mathcal{G} = T_{1_x}\mathcal{G}\,.$$

Once again we regard sections of $A\mathcal{G}$ as projectable vector fields on E, and we say that a projectable vector field ξ with a vertically uniform flow box, where the restriction $\psi_{t;x} = \psi_t|_{E_x}$ of its flow is an affine map between fibres of E, is a *fibre-affine projectable vector field.*

Proposition 4.2.2 *If ξ is a projectable vector field on E then ξ corresponds to a section of $A\mathcal{G}$ if, and only if, its expression in coordinates (x^i, y^α) on E is*

$$\xi = X^i \frac{\partial}{\partial x^i} + (Y_0^\alpha + Y_\beta^\alpha y^\beta)\frac{\partial}{\partial y^\alpha}$$

where the functions X^i, Y_0^α and Y_β^α are projectable to M.

Proof A straightforward modification of the proof of 4.1.2. □

If an affine bundle admits a global section, which we can identify pointwise with the origin o of $\mathbb{R}^k$, then its group reduces to $G_o = \mathsf{GL}(k)$, and the bundle with the reduced group is a vector bundle. Conversely, any vector bundle can be considered as an affine bundle by as it were forgetting about the zero section, or in other words enlarging its group to $\mathsf{A}(k)$ where k is its rank. In such a case we may take the y^α in Proposition 4.2.2 to be linear (rather than affine) fibre coordinates. With respect to such coordinates it makes sense to require Y_0^α to vanish: the vector fields ξ with $Y_0^\alpha = 0$ are then fibre-linear projectable vector fields on E, and are those with flow ψ_t such that each $\psi_{t;x}$ is in fact a linear map.

We shall be most interested in the case of the tangent bundle $\tau_M : TM \to M$ considered as an affine bundle. We now specialize to this case, and consider path connections on $\mathcal{G}$ and their associated infinitesimal connections $\gamma : TM \to A\mathcal{G}$. We note that the tangent bundle has an attachment section, namely its zero section; thus, given any path connection Γ on $\mathcal{G}$, we may define the development of the curve $c : I \to M$ to a curve in the tangent space $c_\Gamma : I \to T_{c(0)}M$.

We shall give specific names to certain types of connection: our classification is chosen to match that of Kobayashi and Nomizu [29] in the standard approach. Any such connection, not necessarily having any further properties, will be called a *generalized affine connection* on M. If Γ satisfies the property that, for each curve $c : I \to M$ and each $t \in I$ the map $c(t) : T_{c(0)}M \to T_{c(t)}M$ is a linear map, rather than merely being affine, then we say that Γ is a *linear connection* on M, and it is indeed a linear connection on the vector bundle $TM \to M$ in the sense of the previous section. If, instead, Γ satisfies the property that the tangent vector $\dot{c}_\Gamma(0) \in T_0T_{c(0)}M$ to the developed curve at $t = 0$ is always the vertical lift of the tangent vector $\dot{c}(0) \in T_{c(0)}M$ to the original curve at $t = 0$, then we say that Γ is an *affine connection* on M. We note that, for any generalized affine connection Γ, there is an associated linear connection $\bar{\Gamma}$ given by

$$c^{\bar{\Gamma}}(t)(v) = c^{\Gamma}(t)(v) - c^{\Gamma}(t)(0) \qquad (v \in T_{c(0)}M).$$

Now suppose that Γ is a generalized affine connection on M, and regard the corresponding infinitesimal connection γ as a map $TM \to \mathcal{A}^G$. For any vector field X on M, the composition $X^{\mathrm{h}} = \gamma \circ X$ is a projectable vector field on TM; the map $X \to X^{\mathrm{h}}$ is then a horizontal lift operation which defines a connection (in the classical sense) on TM. Each vector field X^{h} is a fibre-affine projectable vector field as described above, given in adapted coordinates (x^i, y^i) on TM by

$$\left(\frac{\partial}{\partial x^i}\right)^{\mathrm{h}} = \frac{\partial}{\partial x^i} - (A_i^k + \Gamma_{ij}^k y^j)\frac{\partial}{\partial y^k}$$

where A_i^k, Γ_{ij}^k are functions defined locally on M, and where we have chosen a minus sign for reasons which will become apparent.

Lemma 4.2.3 *If the generalized affine connection Γ is, in particular, an affine connection then $A_i^k = \delta_i^k$; if instead it is a linear connection then $A_i^k = 0$.*

Proof Let $c : I \to M$ be a curve with coordinate representation c^i; if $\boldsymbol{a}^i = dc^i/dt|_{t=0}$ then the tangent vector $\dot{c}(0) \in T_{c(0)}M$ is given by

$$\dot{c}(0) = \boldsymbol{a}^i \left.\frac{\partial}{\partial x^i}\right|_{c(0)} .$$

Now suppose that, for suitably small t, the affine map $c^{\Gamma}(t) : T_{c(0)}M \to T_{c(t)}M$ is given by

$$c^{\Gamma}(t)\left(v^i \left.\frac{\partial}{\partial x^i}\right|_{c(0)}\right) = \left(c_0^i(t) + c_j^i(t)v^j\right) \left.\frac{\partial}{\partial x^i}\right|_{c(t)}$$

so that the developed curve c_{Γ} is given by

$$c_{\Gamma}^i(t) = -b_j^i(t)c_0^j(t) \left.\frac{\partial}{\partial x^i}\right|_{c(0)}$$

where $b_j^i(t)$ is the matrix inverse of $c_j^i(t)$. Thus if $\boldsymbol{a}_0^i = dc_0^i/dt|_{t=0}$ and $\boldsymbol{a}_j^i = dc_j^i/dt|_{t=0}$ the tangent vector $\dot{c}_{\Gamma}(0) \in T_0T_{c(0)}M$ is given by

$$\dot{c}_{\Gamma}^i(0) = -\left.\frac{d}{dt}b_j^i(t)c_0^j(t)\right|_{t=0} \left.\frac{\partial}{\partial y^i}\right|_{0_{c(0)}} = -\boldsymbol{a}_0^i \left.\frac{\partial}{\partial y^i}\right|_{0_{c(0)}}$$

using $c_0^i(0) = 0$ and $c_j^i(0) = \delta_j^i$. The condition that Γ be an affine connection (rather than a generalized affine connection) is that $\dot{c}_{\Gamma}(0)$ must be the vertical lift of $\dot{c}(0)$, so that we must have $\boldsymbol{a}_0^i = -\boldsymbol{a}^i$. The condition that Γ be a linear connection is that the map $c^{\Gamma}(t)$ is linear for each t, so that $c_0^i(t) = 0$ and therefore that $\boldsymbol{a}_0^i = 0$.

We now use the definition of γ to see that $\gamma(\dot{c}(0)) = \dot{c}^\Gamma(0)$, and as γ is a section of the anchor map ρ we have, in terms of the groupoid coordinates $\left(x^i, (z_0^i, z_j^i), \tilde{x}^i\right)$,

$$\gamma\left(\boldsymbol{a}^i \left.\frac{\partial}{\partial x^i}\right|_{c(0)}\right) = \boldsymbol{a}^i \left.\frac{\partial}{\partial x^i}\right|_{1_{c(0)}} + \boldsymbol{a}_0^i \left.\frac{\partial}{\partial z_0^i}\right|_{1_{c(0)}} + \boldsymbol{a}_j^i \left.\frac{\partial}{\partial z_j^i}\right|_{1_{c(0)}} \in A_x\mathcal{G} = T_{1_x}\mathcal{G}\,.$$

Writing this as an element of $\mathcal{A}_x^G$ we have, for an affine connection,

$$\begin{aligned}\left(\boldsymbol{a}^i \left.\frac{\partial}{\partial x^i}\right|_{c(0)}\right)^{\mathsf{h}} &= \boldsymbol{a}^i \left.\frac{\partial}{\partial x^i}\right|_{T_{c(0)}M} + (\boldsymbol{a}_0^i + \boldsymbol{a}_j^i y^j) \left.\frac{\partial}{\partial y^i}\right|_{T_{c(0)}M} \\ &= \boldsymbol{a}^i \left.\frac{\partial}{\partial x^i}\right|_{T_{c(0)}M} + (-\boldsymbol{a}^i + \boldsymbol{a}_j^i y^j) \left.\frac{\partial}{\partial y^i}\right|_{T_{c(0)}M}\,;\end{aligned}$$

as this holds for every curve c it follows that $A_j^i = \delta_j^i$. In the same way we see that for a linear connection we have $A_j^i = 0$. □

Corollary 4.2.4 *The map $\Gamma \mapsto \bar{\Gamma}$, giving the linear connection associated to an affine connection (rather than a generalized affine connection) is a bijection. If γ and $\bar{\gamma}$ are the corresponding infinitesimal connections, and if X is a vector field on M and*

$$X^{\mathsf{h}} = \gamma \circ X\,, \qquad X^{\bar{\mathsf{h}}} = \bar{\gamma} \circ X,$$

then

$$X^{\bar{\mathsf{h}}} = X^{\mathsf{h}} + X^{\mathsf{v}}$$

where X^{v}, a vector field on TM, is the vertical lift of X.

Proof For suitably small t we have, using the definition $c^{\bar{\Gamma}}(t)(v) = c^\Gamma(t)(v) - c^\Gamma(t)(0)$,

$$\begin{aligned}c^\Gamma(t)\left(v^i \left.\frac{\partial}{\partial x^i}\right|_{c(0)}\right) &= \left(c_0^i(t) + c_j^i(t)v^j\right) \left.\frac{\partial}{\partial x^i}\right|_{c(t)}\,, \\ c^{\bar{\Gamma}}(t)\left(v^i \left.\frac{\partial}{\partial x^i}\right|_{c(0)}\right) &= c_j^i(t)v^j \left.\frac{\partial}{\partial x^i}\right|_{c(t)}\end{aligned}$$

so that, using the fact that Γ is an affine connection,

$$\begin{aligned}X^{\mathsf{h}} &= X^i\left(\frac{\partial}{\partial x^i} - (\delta_i^k + \Gamma_{ij}^k y^j)\frac{\partial}{\partial y^k}\right), \\ X^{\bar{\mathsf{h}}} &= X^i\left(\frac{\partial}{\partial x^i} - \Gamma_{ij}^k y^j \frac{\partial}{\partial y^k}\right).\end{aligned}$$

It follows that

$$X^{\mathsf{h}} - X^{\bar{\mathsf{h}}} = X^i \frac{\partial}{\partial y^i} = X^{\mathsf{v}} .$$

Thus the map from infinitesimal affine connections to the associated infinitesimal linear connections is a bijection, and so the same must hold for the corresponding path connections. □

As we might expect, the kernel derivative of an affine connection is related to the (ordinary) covariant derivative of its associated linear connection. We may obtain an expression for this relationship in coordinates.

Lemma 4.2.5 *If*

$$\kappa = (Y^i + Y^i_j y^j) \frac{\partial}{\partial y^i}$$

is a section of the kernel of $\mathcal{A}^G$ then

$$\nabla^{\gamma}_{\partial/\partial x^i} \kappa = \left(Y^j_{|i} - Y^j_i + Y^j_{k|i} y^k\right) \frac{\partial}{\partial y^j}$$

where the rule denotes covariant differentiation with respect to the linear connection.

Proof A straightforward calculation gives

$$\begin{aligned}
\nabla^{\gamma}_{\partial/\partial x^i} \kappa &= [(\partial/\partial x^i)^{\mathsf{h}}, \kappa] \\
&= \left[\frac{\partial}{\partial x^i} - (\delta^k_i + \Gamma^k_{il} y^l) \frac{\partial}{\partial y^k}, (Y^j + Y^j_m y^m) \frac{\partial}{\partial y^j} \right] \\
&= \left(\frac{\partial Y^j}{\partial x^i} - Y^j_i + \Gamma^j_{ik} Y^k + \left(\frac{\partial Y^j_k}{\partial x^i} - \Gamma^l_{ik} Y^j_l + \Gamma^j_{il} Y^l_k \right) y^k \right) \frac{\partial}{\partial y^j} \\
&= \left(Y^j_{|i} - Y^j_i + Y^j_{k|i} y^k\right) \frac{\partial}{\partial y^j} .
\end{aligned}$$

□

We may now say that an affine connection, or its associated linear connection, is *symmetric* (or *torsion-free*) if, for any two vector fields X, Y on M, we have

$$[X^{\bar{\mathsf{h}}}, Y^{\bar{\mathsf{h}}}] = [X^{\mathsf{h}}, Y^{\mathsf{h}}] + [X, Y]^{\mathsf{v}} .$$

Lemma 4.2.6 *An affine connection is symmetric if, and only if,* $\Gamma^k_{ij} = \Gamma^k_{ji}$.

Proof Put $H_i = (\partial/\partial x^i)^{\mathsf{h}}$ and $\bar{H}_i = (\partial/\partial x^i)^{\bar{\mathsf{h}}}$, so that

$$H_i = \frac{\partial}{\partial x^i} - (\delta^k_i + \Gamma^k_{ij} y^j) \frac{\partial}{\partial y^k} , \qquad \bar{H}_i = \frac{\partial}{\partial x^i} - \Gamma^k_{ij} y^j \frac{\partial}{\partial y^k} = H_i + \frac{\partial}{\partial y^i} .$$

Then

$$[\bar{H}_i, \bar{H}_j] = \left[H_i + \frac{\partial}{\partial y^i}, H_j + \frac{\partial}{\partial y^j} \right] = [H_i, H_j] + (\Gamma^k_{ij} - \Gamma^k_{ji}) \frac{\partial}{\partial y^k}$$

and, taking $X = X^i \partial/\partial x^i$ and $Y = Y^i \partial/\partial x^i$, we have

$$[X^{\bar{\mathsf{h}}}, Y^{\bar{\mathsf{h}}}] = [X^i \bar{H}_i, Y^j \bar{H}_j] = X^i Y^j [\bar{H}_i, \bar{H}_j] + \left(X^i \frac{\partial Y^j}{\partial x^i} - Y^i \frac{\partial X^j}{\partial x^i} \right) \bar{H}_j$$

whereas

$$[X^{\mathsf{h}}, Y^{\mathsf{h}}] = [X^i H_i, Y^j H_j] = X^i Y^j [H_i, H_j] + \left(X^i \frac{\partial Y^j}{\partial x^i} - Y^i \frac{\partial X^j}{\partial x^i} \right) H_j.$$

Thus

$$\begin{aligned} [X^{\bar{\mathsf{h}}}, Y^{\bar{\mathsf{h}}}] - [X^{\mathsf{h}}, Y^{\mathsf{h}}] &= X^i Y^j ([\bar{H}_i, \bar{H}_j] - [H_i, H_j]) + \left(X^i \frac{\partial Y^j}{\partial x^i} - Y^i \frac{\partial X^j}{\partial x^i} \right) (\bar{H}_j - H_j) \\ &= X^i Y^j (\Gamma^k_{ij} - \Gamma^k_{ji}) \frac{\partial}{\partial y^k} + \left(X^i \frac{\partial Y^j}{\partial x^i} - Y^i \frac{\partial X^j}{\partial x^i} \right) \frac{\partial}{\partial y^j} \\ &= X^i Y^j (\Gamma^k_{ij} - \Gamma^k_{ji}) \frac{\partial}{\partial y^k} + [X, Y]^{\mathsf{v}}. \end{aligned}$$

□

An affine connection is therefore symmetric as defined above if its associated linear connection is symmetric in the classical sense: that is, if its torsion tensor vanishes. Indeed the torsion tensor is closely related to the curvatures of the two connections.

Lemma 4.2.7 *For any vector fields X, Y on M the curvatures R^γ, $R^{\bar{\gamma}}$ satisfy*

$$R^\gamma(X, Y) = R^{\bar{\gamma}}(X, Y) + \big(T(X, Y)\big)^{\mathsf{v}}$$

where T denotes the torsion tensor of $\bar{\gamma}$, so that $R^\gamma = R^{\bar{\gamma}}$ if, and only if, the connections are symmetric.

Proof By definition

$$R^\gamma(X, Y) = [X, Y]^{\mathsf{h}} - [X^{\mathsf{h}}, Y^{\mathsf{h}}]$$

so that, putting $R^\gamma_{ij} = R^\gamma(\partial/\partial x^i, \partial/\partial x^j)$ and $R^{\bar{\gamma}}_{ij} = R^{\bar{\gamma}}(\partial/\partial x^i, \partial/\partial x^j)$,

$$\begin{aligned} R^\gamma_{ij} &= -[H_i, H_j] = -\left[\bar{H}_i - \frac{\partial}{\partial y^i}, \bar{H}_j - \frac{\partial}{\partial y^j} \right] \\ &= -[\bar{H}_i, \bar{H}_j] + (\Gamma^k_{ij} - \Gamma^k_{ji}) \frac{\partial}{\partial y^k} = R^{\bar{\gamma}}_{ij} + \left(T^k_{ij} \frac{\partial}{\partial x^k} \right)^{\mathsf{v}}; \end{aligned}$$

the result follows as all the maps are tensorial. □

Corollary 4.2.8 *If γ is a symmetric affine connection then*

$$R^{\gamma}_{ij} = R^{k}_{lij} y^{l} \frac{\partial}{\partial y^{k}}$$

where R^{k}_{lij} are components of the curvature tensor of the associated linear connection.

Proof This follows from

$$\begin{aligned}
R^{\gamma}_{ij} = R^{\bar{\gamma}}_{ij} &= -[\bar{H}_i, \bar{H}_j] \\
&= -\left[\frac{\partial}{\partial x^i} - \Gamma^{k}_{ih} y^{h} \frac{\partial}{\partial y^{k}}, \frac{\partial}{\partial x^{j}} - \Gamma^{\,l}_{jm} y^{m} \frac{\partial}{\partial y^{l}}\right] \\
&= y^{m} \frac{\partial \Gamma^{\,l}_{jm}}{\partial x^{i}} \frac{\partial}{\partial y^{l}} - y^{h} \frac{\partial \Gamma^{k}_{ih}}{\partial x^{j}} \frac{\partial}{\partial y^{k}} - y^{h} \Gamma^{k}_{ih} \Gamma^{\,l}_{jk} \frac{\partial}{\partial y^{l}} + y^{m} \Gamma^{\,l}_{jm} \Gamma^{k}_{il} \frac{\partial}{\partial y^{k}} \\
&= R^{k}_{lij} y^{l} \frac{\partial}{\partial y^{k}} \,.
\end{aligned}$$

□

We shall continue with the study of affine connections on TM in Sect. 5.5.

4.3 Interlude: Densities and Projective Structures

We now consider fibre bundles $\pi^{G} : E \to M$ where the typical fibre F is real projective space P^{n} of the same dimension as the base manifold M, and where the group G of the bundle is the projective group $\mathsf{PGL}(n+1)$, the quotient of $\mathsf{GL}(n+1)$ by nonzero multiples of the identity.

When n is even, $\mathsf{PGL}(n+1) \cong \mathsf{PSL}(n+1) \cong \mathsf{SL}(n+1)$. When n is odd, though, $\mathsf{PGL}(n+1)$ is not connected; in this case $\mathsf{PGL}(n+1)$ may be identified with a group whose elements are equivalence classes containing pairs of matrices $\pm g$ where $\det g = \pm 1$, and which has an identity component $\mathsf{PSL}(n+1)$ containing those equivalence classes where $\det g = 1$. Although elements of $\mathsf{PGL}(n+1)$ are often represented by matrices g with $|\det g| = 1$, it should be remembered that for n odd such a matrix is determined only up to sign. For any value of n, even or odd, the Lie algebra of $\mathsf{PGL}(n+1)$ may be identified with $\mathfrak{sl}(n+1)$, as indeed may the Lie algebra of $\mathsf{PSL}(n+1)$.

We shall consider a canonical choice of such a fibre bundle π. This will not be the simplest choice: we could, for example, consider $\mathrm{P}(TM \times \mathbb{R}) \to M$, but we shall see that this alternative choice has a little too much structure for our immediate need. We shall describe two approaches to the construction of this new bundle π, one being close in spirit to the classical approach taken by T.Y. Thomas [42] and the other following the lines of 'tractor calculus' described in, for example, [2]. In this section we shall describe both constructions in some detail, and in a form which can be adapted for use in other contexts; we shall consider the Lie groupoid of projective transformations, its Lie algebroid, and the corresponding connections, in the following section.

The first approach to constructing the bundle π starts with the manifold of *unoriented volume elements* $\mathcal{V}M$. This is obtained by taking the nonzero volume elements $\mathsf{v} \in \bigwedge^n T^*M$; each element of $\mathcal{V}M$ will be an equivalence class $z = [\mathsf{v}] = \{\pm\mathsf{v}\}$. Clearly $\mathcal{V}M$ is a manifold, and it is a bundle over M with projection $\nu : \mathcal{V}M \to M$ defined by $\nu(z) = x$ whenever $\mathsf{v}, -\mathsf{v} \in \bigwedge^n T^*_x M$. Indeed it is a principal $\mathbb{R}_+$ bundle with (right) action $\mu^1(z, s) = [s\mathsf{v}]$, and as such is may be regarded as a reduction of the frame bundle on M.

We can, however, use a different right action $\mu^p(z, s) = [s^{1/p}\,\mathsf{v}]$ for any $p \neq 0$, and write $\mu^p_s : \mathcal{V}M \to \mathcal{V}M$ for the map defined by $\mu^p_s(z) = \mu^p(z, s)$. For projective geometry we shall choose $p = 1/(n+1)$; other values of p will be appropriate in other contexts (for example, in conformal geometry).

We also consider the tangent bundle $\tau_{\mathcal{V}M} : T\mathcal{V}M \to \mathcal{V}M$. Let $\mu^p_{s*} : T\mathcal{V}M \to T\mathcal{V}M$ be the derivative of the action μ^p_s on the fibres of $\nu : \mathcal{V}M \to M$, and let $\delta_t : T\mathcal{V}M \to T\mathcal{V}M$ denote the action of scaling in the fibres of $T\mathcal{V}M$ by $t \in \mathbb{R}_+$. We shall write u^0, u^i for fibre coordinates on $T\mathcal{V}M$, reserving y^i for fibre coordinates on the projective bundle we shall define shortly.

To see the expression of these two actions in coordinates, let x^i be coordinates on M. The fibre coordinate on $\nu : \mathcal{V}M \to M$ corresponding to the action μ^1 is $|v|$, where the linear coordinate v satisfies

$$z = \left[v(\mathsf{v})dx^1 \wedge \ldots dx^n|_x\right]$$

for any nonzero $\mathsf{v} \in \bigwedge^n T^*M$; with the action μ^p we take a new fibre coordinate $x^0 = |v|^p$ instead. We may check that

$$\mu^p(z, s) = \left[s^{1/p}v(\mathsf{v})dx^1 \wedge \ldots dx^n|_x\right] = \left[v(s^{1/p}\mathsf{v})dx^1 \wedge \ldots dx^n|_x\right]$$

so that $x^0(\mu^p(z, s)) = sx^0(z)$: we may write this as $\mu^p_s(x^0, x^i) = (sx^0, x^i)$. Thus, on $T\mathcal{V}M$, we have

$$\mu^p_{s*}(x^0, x^i, u^0, u^i) = (sx^0, x^i, su^0, u^i)$$

and also

$$\delta_t(x^0, x^i, u^0, u^i) = (x^0, x^i, tu^0, tu^i)$$

where δ_t is multiplication by t in the fibres of $T\mathcal{V}M \to \mathcal{V}M$. It is evident that $\mu^p_{s*} \circ \delta_t = \delta_t \circ \mu^p_{s*}$ for all $(s, t) \in \mathbb{R}^2_+$, and that

$$(\mu^p_{s*} \circ \delta_{1/s}) \circ (\mu^p_{t*} \circ \delta_{1/t}) = \mu_{st*} \circ \delta_{1/st},$$

so that the map $\lambda^p_s : T\mathcal{V}M \times \mathbb{R}_+ \to T\mathcal{V}M$ given by $\lambda^p_s = \mu^p_{s*} \circ \delta_{1/s}$ defines an action of $\mathbb{R}_+$ on $T\mathcal{V}M$ given in coordinates by

$$\lambda^p_s(x^0, x^i, u^0, u^i) = (sx^0, x^i, u^0, s^{-1}u^i)\,.$$

If we write Υ^p for the infinitesimal generator of the action μ^p_s on $\mathcal{V}M$, and $(\Upsilon^p)^{\mathrm{c}}$, $\tilde{\Delta}$ and $\tilde{\Upsilon}^p$ for the infinitesimal generators of the actions μ^p_{s*}, δ_t and λ^p_s on $T\mathcal{V}M$ (so that $(\Upsilon^p)^{\mathrm{c}}$ is the complete lift of Υ^p), then in coordinates

$$\Upsilon^p = x^0 \frac{\partial}{\partial x^0}$$

on $\mathcal{V}M$, and

$$(\Upsilon^p)^{\mathrm{c}} = x^0 \frac{\partial}{\partial x^0} + u^0 \frac{\partial}{\partial u^0}\,, \qquad \tilde{\Delta} = u^0 \frac{\partial}{\partial u^0} + u^i \frac{\partial}{\partial u^i}$$

and

$$\tilde{\Upsilon}^p = (\Upsilon^p)^{\mathrm{c}} - \tilde{\Delta} = x^0 \frac{\partial}{\partial x^0} - u^i \frac{\partial}{\partial u^i}$$

on $T\mathcal{V}M$.

We now consider the orbit spaces of the actions μ^p_{s*} and λ^p_s, which we will denote by $\mathcal{W}M$ and $\widetilde{\mathcal{W}}^pM$. Both these spaces are vector bundles over M; as the orbits of μ^p_{s*} do not depend upon the choice of p we do not use a superscript in the notation $\mathcal{W}M$.

Although these two bundles are related, they behave in slightly different ways. As $\nu_* \circ \mu^p_{s*} = (\nu \circ \mu^p_s)_* = \nu_*$ we see that there is a well-defined projection $\rho : \mathcal{W}M \to TM$ satisfying $\rho \circ \chi = \nu_*$ where $\chi : T\mathcal{V}M \to \mathcal{W}M$ is the quotient map, so that $\mathcal{W}M$ is an anchored vector bundle. Furthermore, the bracket of vector fields on $\mathcal{V}M$ induces a bracket of sections under which $\tau : \mathcal{W}M \to M$ becomes a Lie algebroid: it is in fact the Atiyah algebroid of the principal bundle $\nu : \mathcal{V}M \to M$. By contrast, it is not the case that $\nu_* \circ \lambda^p_s = \nu_*$, so that $\widetilde{\mathcal{W}}^pM$ does not have a similar structure. Nevertheless, the projectivized bundles $\mathrm{P}\mathcal{W}M$ and $\mathrm{P}\widetilde{\mathcal{W}}^pM$ are equal, because both actions μ^p_{s*} and λ^p_s are projectable to $\mathrm{P}T\mathcal{V}M$ and their projections are equal.

Starting with coordinates (x^0, x^i, u^0, u^i) on $T\mathcal{V}M$, the induced coordinates on $\mathcal{W}M$ will be written as (x^i, w^0, w^i) where $w^0 = u^0/x^0$ and $w^i = u^i$, and those on $\widetilde{\mathcal{W}}^pM$ as $(x^i, \tilde{w}^0, \tilde{w}^i)$ where $\tilde{w}^0 = u^0$ and $\tilde{w}^i = x^0 u^i$. We therefore see that homogeneous fibre coordinates on $\mathrm{P}\mathcal{W}M$ are either $w^i/w^0 = x^0u^i/u^0$, or else w^i/w^j and $w/u^j = u^0/x^0u^j$, whereas homogeneous fibre coordinates on $\mathrm{P}\widetilde{\mathcal{W}}^pM$ are either $\tilde{w}^i/\tilde{w}^0 = x^0u^i/u^0$, or else $\tilde{w}^i/\tilde{w}^j = u^i/u^j$ and $\tilde{w}^0/\tilde{w}^j = u^0/x^0u^j$, so that the two families of coordinates on the projective bundles are the same.

We should remark that the coordinate w^0, which may be used as a fibre coordinate on the line bundle $\ker\rho \to M$, *does* depend on the choice of p, even though the manifold $\mathcal{W}M$ does not.

We also observe that the vector field Υ^p is projectable to $\mathcal{W}M$, and defines a global section $e^p : M \to \mathcal{W}M$ (depending on p, as the notation indicates) with coordinates $w^0 = 1$, $w^i = 0$ and spanning $\ker\rho$; if we write $\ker\rho$ as $\langle e^p \rangle$ then we obtain a global section $x \mapsto [e^p(x)]$ of $\mathrm{P}\mathcal{W}M$ which now does not depend on p. On the other hand, Υ^p does not project to a section of $\widetilde{\mathcal{W}}^p M \to M$; its image is instead a line bundle with coordinates $\tilde{w}^i = 0$ whose projection to $\mathrm{P}\widetilde{\mathcal{W}}^p M = \mathrm{P}\mathcal{W}M$ defines the same global section as $x \mapsto [e^p(x)]$.

We shall let π be the $\mathsf{PGL}(n+1)$-bundle $\mathrm{P}\widetilde{\mathcal{W}}^p M = \mathrm{P}\mathcal{W}M \to M$.

We now consider the second approach, using weighted vector bundles. Let $\nu^p : \mathcal{D}^p M \to M$ be the vector bundle associated with the frame bundle on M by the action of $\mathsf{GL}(n)$ on $\mathbb{R}$ given by $(A, t) \mapsto |\det A|^{-p}\, t$. We shall call $\mathcal{D}^p M$ the *bundle of scalar densities of weight* p. An element $\varrho \in \mathcal{D}^p_x M$ is an equivalence class $\varrho = [(\boldsymbol{v}, A)]$ where $\boldsymbol{v} = (v_i)$, $v_i \in T_x M$, is a frame on M and $A \in \mathsf{GL}(n)$; the equivalence relation is given by

$$(\boldsymbol{v}A, t) \sim (\boldsymbol{v}, |\det A|^{-p}\, t)\,.$$

Our use of the symbol ν^p indicates a relationship between weighted vector bundles and the principal $\mathbb{R}_+$-bundle $\nu : \mathcal{V}M \to M$ we have defined above, and indeed we may equally consider $\mathcal{D}^p M$ to be the vector bundle associated with the principal bundle $\mathcal{V}M \to M$ by the action of $\mathbb{R}_+$ on $\mathbb{R}$ given by $(s, t) \mapsto s^p t$: with this second interpretation an element $\varrho \in \mathcal{D}^p_x M$ may be regarded as an equivalence class $\varrho = \big[(z, t)\big]$ where the equivalence relation on $\mathcal{V}M \times \mathbb{R}$ is given by $(\mu^p(z, s), t) \sim (z, s^p t)$. The difference in sign between $|\det A|^{-p}\, t$ and $s^p t$ arises because elements of $\mathcal{V}M$ are equivalence classes of cotangent n-vectors rather than tangent n-vectors. The map $\iota^p : \mathcal{V}M \to \mathcal{D}^p M$ given by $\iota^p(z) = \big[(z, 1)\big]$ satisfies the condition that $\iota^p(\mu^p(z, s)) = s\iota^p(z)$, so that we may regard $\mathcal{V}M$, with scalar multiplication given by μ^p, as the 'positive half' of $\mathcal{D}^p M$, and regard $\mathcal{D}^p M$ as an oriented vector bundle.

Yet another interpretation would be to regard $\varrho \in \mathcal{D}^p_x M$ as a map $\varrho : \mathcal{V}_x M \to \mathbb{R}$ satisfying the homogeneity condition $\varrho(\mu^{-p}(z, s)) = s\varrho(z)$. It is evident from this third point of view that we may regard $\mathcal{D}^{-p} M$ as the dual bundle $\mathcal{D}^{p*} M$.

The bundle $\mathcal{D}^p M \to M$ may be used to form weighted versions of other vector bundles over M by taking tensor products.

Lemma 4.3.1 *The bundle $\widetilde{\mathcal{W}}^p M \to M$ is canonically isomorphic to the weighted bundle $\mathcal{W}M \otimes \mathcal{D}^p M \to M$.*

Proof Take $v \in T_z \mathcal{V}M$ where $\nu(z) = x$, so that the orbit of v under the action λ^p is

$$[v]_\lambda = \{\lambda^p_s(v) : s \in \mathbb{R}_+\} \in \widetilde{\mathcal{W}}^p_x M\,.$$

Define a map $\mathcal{D}^{-p}_x M \to \mathcal{W}_x M$ by

$$\varrho \mapsto [\varrho(z)v]_{\mu *} = \{\mu^p_{s*}(\varrho(z)v) : s \in \mathbb{R}_+\} = \{\varrho(z)\mu^p_{s*}(v) : s \in \mathbb{R}_+\}\,.$$

If we choose a different representative $\hat{v} \in [v]_\lambda$ then, for some $s \in \mathbb{R}_+$,

$$\hat{v} = \lambda^p_s(v) = s^{-1}\mu^p_{s*}(v) \in T_{\hat{z}}\mathcal{V}M\,, \qquad \hat{z} = \mu^p_s(z)\,.$$

We then see that

$$\varrho(\hat{z})\hat{v} = \varrho(\mu^p_s(z))s^{-1}\mu^p_{s*}(v) = \varrho(z)\mu^p_{s*}(v)$$

so that $[\varrho(\hat{z})\hat{v}]_{\mu*} = [\varrho(z)v]_{\mu*}$, showing that the map depends on $[v]_\lambda$ rather than any representative vector v. The map is clearly linear, and hence an element of $\mathcal{W}_x M \otimes \mathcal{D}^p_x M$; the correspondence $\widetilde{\mathcal{W}}^p_x M \to \mathcal{W}_x M \otimes \mathcal{D}^p_x M$ is easily seen to be a linear isomorphism. □

This relationship between $\widetilde{\mathcal{W}}^p M$ and $\mathcal{W}M$ may also be seen in coordinates. The fibre coordinates on $\mathcal{W}M$ are (w^0, w^i) where $w^0 = u^0/x^0$ and $w^i = u^i$; multiplying these by the weight coordinate x^0 gives $(x^0 w, x^0 u^i) = (\tilde{w}^0, \tilde{w}^i)$, the fibre coordinates on $\widetilde{\mathcal{W}}^p M$.

We now consider the first jet bundle $J^1\nu^p \to M$. This too is a vector bundle, and forms part of a short exact sequence of vector bundles over M given by

$$0 \to T^*M \otimes \mathcal{D}^p M \to J^1\nu^p \to \mathcal{D}^p M \to 0$$

where $T^*M \otimes \mathcal{D}^p M$ is the bundle of weighted cotangent vectors. The induced coordinates on the jet manifold $J^1\nu^p$ are (x^i, x^0, x_i) where

$$x_i(j^1_x\sigma) = \left.\frac{\partial \sigma^0}{\partial x^i}\right|_x$$

where $\sigma^0 = x^0 \circ \sigma$ for any section (or local section) σ of $\nu^p : \iota^p(\mathcal{V}M) \to M$.

Lemma 4.3.2 *There is a canonical identification of $J^1\nu^p \to M$ with the vector bundle dual to $\widetilde{\mathcal{W}}^{-p} M \to M$.*

Proof We first construct a canonical nondegenerate fibre-bilinear map $\mathcal{W}M \times_M J^1\nu^p \to \mathcal{D}^p M$.

Take an element $[v] \in \mathcal{W}_x M$, and a representative tangent vector $v \in T_z \mathcal{V}M$ for some $z \in \mathcal{V}_x M$. Take an element $j^1_x\sigma \in J^1_x\nu^p$, and a representative section $\sigma : M \to \mathcal{D}^p M$.

Let c be a curve in $\mathcal{V}M$ such that $c(0) = z$ and $\dot{c}(0) = v$. Regarding $\mathcal{V}M$ as the positive half of $\mathcal{D}^p M$, we may regard both c and $\sigma \circ \nu \circ c$ as curves in $\mathcal{D}^p M$ such that, for t in the domain of c, both $c(t)$ and $\sigma\nu c(t)$ lie in the same fibre over M. As $c(t)$ cannot be the zero element of the fibre $\mathcal{D}^p_{\nu c(t)} M$, we may write $\sigma\nu c(t) = f(t)c(t)$ for some real-valued function f defined on the domain of c.

Define the element of $\mathcal{D}^p_x M$ corresponding to the pair $\big([v], j^1_x\sigma\big)$ to be $f'(0)c(0)$. Choosing a different section σ to represent $j^1_x\sigma$, and choosing a different curve c to represent the tangent vector v, will not affect the derivative $f'(0)$. We must, however, confirm that choosing a different representative tangent vector $\hat{v}$ instead of v will not affect the result. By construction we must have $\hat{v} = \mu_{s*}(v)$ for some $s > 0$, so that $\hat{v}$ will be represented by a new curve $\hat{c}$ given by $\hat{c}(t) = sc(t)$. As $\sigma\nu\hat{c}(t) = \sigma\nu c(t)$, we must have $\hat{f}(t)\hat{c}(t) = f(t)c(t)$, so that $\hat{f}(t) = s^{-1}f(t)$; we therefore see that $\hat{f}'(0)\hat{c}(0) = f'(0)c(0)$ as required, so that the pairing is well defined.

Let the curve c in $\mathcal{V}M$ have coordinates (c^0, c^i), so that $\dot{c}^i(0) = u^i(v) = u^i([v])$ and $\dot{c}^0(0) = u^0(v) = w([v])x^0(v)$. Let the section σ have coordinate $\sigma^0 = x^0 \circ \sigma$. Then

$$f(t) = \frac{\sigma^0 \nu c(t)}{c^0(t)}, \qquad f'(0) = \frac{1}{\bigl(c^0(0)\bigr)^2}\left(c^0(0)\left.\frac{\partial \sigma^0}{\partial x^i}\right|_x \dot{c}^i(0) - \sigma^0(x)\dot{c}^0(0)\right)$$

so that, simplifying,

$$f'(0) = \frac{1}{c^0(0)}\Bigl(x_i(j^1_x\sigma)w^i([v]) - x^0(j^1_x\sigma)w^0([v])\Bigr);$$

we therefore see that the pairing is given in coordinates by[1] $\bigl((w^0, w^i), (x^0, x_i)\bigr) \mapsto x_i w^i - x^0 w^0$, showing that it is bilinear and nondegenerate as required.

We may therefore regard $J^1\nu^p$ as a bundle of fibre-linear maps $\mathcal{W}M \to \mathcal{D}^p M$, and hence as a bundle of fibre-linear maps

$$\widetilde{\mathcal{W}}^{-p}M \cong \mathcal{W}M \otimes_M \mathcal{D}^{-p}M \quad \to \quad \mathcal{D}^p M \otimes_M \mathcal{D}^{-p}M \cong \mathbb{R}. \qquad \square$$

To see the effect of coordinate changes on these various manifolds, let $\hat{x}^j$ be some other coordinates on M. Then

$$z = \bigl[\hat{v}(\mathrm{v})d\hat{x}^1 \wedge \ldots d\hat{x}^n|_x\bigr] = \bigl[\hat{v}(\mathrm{v})J\, dx^1 \wedge \ldots dx^n|_x\bigr]$$

where $J = \det J^i_j$ is the Jacobian determinant[2] of the coordinate transformation, $J^i_j = \partial\hat{x}^i/\partial x^j$, so that $\hat{v} = J^{-1}v$ and therefore that $\hat{x}^0 = |J|^{-p}x^0$; as the choice of coordinate x^0 implies a particular choice of p, it is no surprise that the transformation rule for x^0 depends upon p. The transformation rules for the additional coordinates on $T\mathcal{V}M$ are then

$$\hat{u}^i = J^i_j u^j, \qquad \hat{u}^0 = |J|^{-p}u^0 + u^i\frac{\partial|J|^{-p}}{\partial x^i}x^0 = |J|^{-p}\bigl(u^0 - p\,u^i\psi_i x^0\bigr)$$

where $\psi_i = \partial \log|J|/\partial x^i$. The transformation rules for the coordinates on $\mathcal{W}M$ are

$$\hat{w}^i = J^i_j w^j, \qquad \hat{w}^0 = \hat{u}^0/\hat{x}^0 = w^0 - p\,w^i\psi_i.$$

[1] The product $x^0 w^0$ in this formula may seem unusual as both indices are superscripts. In fact for full consistency the coordinate x^0 on $\mathcal{V}M$ should have its index as a subscript, as it is effectively the coordinate of a weighted cotangent n-vector. It is, however, more convenient to write $(x^a) = (x^0, x^i)$ as in many cases we can deal with all $n + 1$ indices together.

[2] There should be no confusion between the use of J^1 here to denote a jet manifold, and J with any other indices (or without any decoration at all) to denote something related to the Jacobian.

We also see that

$$\hat{x}_i(j^1_x\sigma) = \left.\frac{\partial\hat{\sigma}^0}{\partial\hat{x}^i}\right|_x = \bar{J}^j_i(x)\left(|J|^{-p}\frac{\partial\sigma^0}{\partial x^j} + \sigma^0\frac{\partial(|J|^{-p})}{\partial x^j}\right)(x)$$

so that $\hat{x}_i = |J|^{-p}\bar{J}^j_i\big(x_j - p\,x^0\psi_j\big)$. Similarly, the transformation rules for the fibre coordinates on $\widetilde{\mathcal{W}}^p M$ are

$$\hat{\tilde{w}}^i = \hat{x}^0\hat{u}^i = |J|^{-p}J^i_j x^0 u^j = |J|^{-p}J^i_j\tilde{w}^j,$$
$$\hat{\tilde{w}}^0 = \hat{u}^0 = |J|^{-p}\big(u^0 - p\,u^i\psi_i x^0\big) = |J|^{-p}\big(\tilde{w}^0 - p\,\tilde{w}^i\psi_i\big).$$

The pairing between $\mathcal{W}M$ and $J^1\nu^p$ is, as we have seen, given in coordinates by $x_i w^i - x^0 w^0$, so that

$$\begin{aligned}\hat{x}_i\hat{w}^i - \hat{x}^0\hat{w}^0 &= |J|^{-p}\bar{J}^j_i\big(x_j - p\,x^0\psi_j\big)J^i_k w^k - |J|^{-p}x^0\big(w^0 - p\,w^i\psi_i\big)\\ &= |J|^{-p}(x_j w^j - x^0 w^0)\,,\end{aligned}$$

confirming that the result is an element of $\mathcal{D}^p M$.

4.4 Projective Bundles and Projective Maps

We now take $p = 1/(n+1)$. (The reason for choosing this particular value of p will not, though, become apparent until Chap. 9.) We shall call $\pi : \mathrm{P}\mathcal{W}M \to M$ the *Cartan projective bundle*, and we note that it has an attachment section $x \mapsto [e(x)]$, where for simplicity we write $e = e^p = e^{1/(n+1)}$. We consider projective maps between the fibres of this bundle.

Proposition 4.4.1 *Let $\mathcal{G}$ be the set of projective maps $\varphi : \mathrm{P}\mathcal{W}_{\tilde{x}}M \to \mathrm{P}\mathcal{W}_x M$ for all $x, \tilde{x} \in M$. Then $\mathcal{G}$ is a locally trivial Lie groupoid.*

Proof Immediate from Theorem 3.1.1. □

Let $\varphi : \mathrm{P}\mathcal{W}_{\tilde{x}}M \to \mathrm{P}\mathcal{W}_x M$ be such a map, and suppose that both x and $\tilde{x}$ are in the domain of a single chart on M with coordinates (x^i). The linear fibre coordinates (w^0, w^i) on $\mathcal{W}M$ may be used to define local projective fibre coordinates on $\mathrm{P}\mathcal{W}M$, by taking $y^i_0 = w^i/w^0$ where $w^0 \neq 0$, and by taking

$$y^i_j = \begin{cases} w^i/w^j & (i \neq j)\\ w^0/w^j & (i = j)\end{cases}$$

where $w^j \neq 0$. Let (y^i) denote any of these sets of fibre coordinates; then

$$y^i \circ \varphi = \frac{\varphi^i_j y^j + \varphi^i_0}{\varphi^0_k y^k + \varphi^0_0}$$

for some matrix

$$\begin{pmatrix} \varphi^0_0 & \varphi^0_j \\ \varphi^i_0 & \varphi^i_j \end{pmatrix}$$

determined up to an overall scalar factor.

Proposition 4.4.2 *If ξ is a projectable vector field on* $\mathcal{PWM}$ *then ξ corresponds to a section of $A\mathcal{G}$ if, and only if, its expression in coordinates (x^i, y^i) on* $\mathcal{PWM}$ *is*

$$\xi = X^i \frac{\partial}{\partial x^i} + \left(Y^i_0 + (Y^i_j - \delta^i_j Y^0_0)y^j - (Y^0_j y^j)y^i\right)\frac{\partial}{\partial y^i}$$

where the functions X^i, Y^i_0, Y^i_j, Y^0_0 and Y^0_j are projectable to M.

Proof Let ψ_t be the flow of ξ. As in the affine case (Proposition 4.2.2) we may write

$$\xi = X^i \frac{\partial}{\partial x^i} + Y^i \frac{\partial}{\partial y^i}$$

where the functions X^i are projectable to $U \subset M$, the domain of the coordinates x^i, and where

$$X^i = \left.\frac{d\phi^i_t}{dt}\right|_{t=0}, \qquad Y^i = \left.\frac{d\psi^i_t}{dt}\right|_{t=0}$$

with $\phi^i_t = x^i \circ \psi_t$ and $\psi^\alpha_t = y^\alpha \circ \psi_t$.

Fix $x \in U$, and take $y \in E_x$ in the domain of the coordinates (x^i, y^i). From our earlier remarks about $y^i \circ \varphi$ we see that

$$y^i\left(\psi_{t;x}(y)\right) = \frac{c^i_j(t)y^j + c^i_0(t)}{c^0_k(t)y^k + c^0_0(t)}$$

for $t \in (-\varepsilon_y, \varepsilon_y)$ and for some functions $c^i_j, c^i_0, c^0_j, c^0_0 : (-\varepsilon_y, \varepsilon_y) \to \mathbb{R}$. Thus, putting $\boldsymbol{a}^i_j = dc^i_j/dt|_{t=0}$, etc. we see that

$$Y^i(y) = \boldsymbol{a}^i_0 + (\boldsymbol{a}^i_j - \delta^i_j \boldsymbol{a}^0_0)y^j(y) - \left(\boldsymbol{a}^0_j y^j(y)\right)y^i(y),$$

showing that ξ takes the required coordinate form, quadratic in the fibre coordinates.

For the converse, we observe that the fibres of $\mathcal{PWM} \to M$ are projective spaces and therefore compact, so that any vector field with the quadratic coordinate representation given above *and defined on a complete fibre of* $\mathcal{PWM}$ has vertically uniform flow boxes and so corresponds to an element of $A\mathcal{G}$. (The projective fibre coordinates y^i are not, of course, defined on complete fibres.) □

There are three remarks we should make about this result. The first is that there is a certain degree of indeterminacy in the coefficients of the quadratic expression for Y^i, which can be traced back to the fact that the coefficients in the formula for $y^i\left(\psi_{t;x}(y)\right)$ are determined only up to an overall scale factor. Since $\psi_{0;x}$ is the identity map of $\mathrm{P}\mathcal{W}_x M$, $c_0^0(t)$ is nonzero for t near 0, so one obvious way of fixing the scale is to choose $c_0^0(t) = 1$, so that $Y_0^0 = 0$. This is the choice we shall usually make.

The second remark is that in fact any vector field with the quadratic coordinate representation given above in any of the fibre coordinate systems specified for $\mathrm{P}\mathcal{W}M$ will be defined on a complete fibre of $\mathrm{P}\mathcal{W}M$. This is because such a vector field will be the image of a projectable vector field on the vector bundle $\mathcal{W}M \to M$ which is linear in the fibre coordinates, under the projectivization map $\mathrm{p} : \mathcal{W}M \to \mathrm{P}\mathcal{W}M$.

We can illustrate this (pretty obvious) fact in detail by considering the coordinates $y^i = y_0^i = w^i/w^0$ (where $w^0 \neq 0$). For any $v \in \mathcal{W}_x M$ we have

$$\mathrm{p}_*\left(\frac{\partial}{\partial w^0}\right)_v = -\left(\frac{w^i}{(w^0)^2}\right)\frac{\partial}{\partial y^i}\bigg|_{\mathrm{p}(v)}$$

$$\mathrm{p}_*\left(\frac{\partial}{\partial w^i}\right)_v = \left(\frac{1}{w^0}\right)\frac{\partial}{\partial y^i}\bigg|_{\mathrm{p}(v)},$$

and of course

$$\mathrm{p}_*\left(\frac{\partial}{\partial x^i}\bigg|_v\right) = \frac{\partial}{\partial x^i}\bigg|_{\mathrm{p}(v)}.$$

We note that also that

$$\mathrm{p}_*\left(w^0\frac{\partial}{\partial w^0} + w^i\frac{\partial}{\partial w^i}\right) = 0;$$

that is to say, the Liouville field Δ on $\mathcal{W}M$ projects to 0, as was to be expected, and indeed the kernel of p_* consists of scalar multiples of Δ. We therefore see that the projectable (to M) fibre-linear vector field

$$W = W^i\frac{\partial}{\partial x^i} + (W_0^0 w^0 + W_j^0 w^j)\frac{\partial}{\partial w} + (W_0^i w^0 + W_j^i w^j)\frac{\partial}{\partial w^i}$$

on $\mathcal{W}M$ is projectable to the vector field

$$\mathrm{p}_* W = W^i\frac{\partial}{\partial x^i} + \left(W_0^i + (W_j^i - W_0^0\delta_j^i)y^j - (W_j^0 y^j)y^i\right)\frac{\partial}{\partial y^i}$$

on $\mathrm{P}\mathcal{W}M$, so that we can obtain

$$\mathrm{p}_* W = X^i \frac{\partial}{\partial x^i} + \left(Y_0^i + Y_j^i y^j - (Y_j^0 y^j) y^i\right) \frac{\partial}{\partial y^i}$$

by taking

$$W = X^i \frac{\partial}{\partial x^i} + (Y_j^0 w^j) \frac{\partial}{\partial w^0} + (Y_0^i w^0 + Y_j^i w^j) \frac{\partial}{\partial u^i} ;$$

any other differs from this by a scalar multiple of Δ. There is therefore a 1-1 correspondence between equivalence classes modulo Δ of projectable fibre-linear vector fields on $\mathcal{W}M$ and projectable vector fields on $\mathrm{P}\mathcal{W}M$ of the required quadratic form with respect to the coordinates y^i. But W is defined on complete fibres of $\mathcal{W}M$, and therefore it, or its equivalence class modulo Δ, defines a unique vector field on complete fibres of $\mathrm{P}\mathcal{W}M$, whose representation with respect to any of the systems of coordinates y^i will be of the specified quadratic type.

When discussing the geometry of $\mathrm{P}\mathcal{W}M$ below, especially in Chaps. 8 and 9, we shall frequently take advantage of this result. We shall use the fibre coordinates y_0^i for preference, because they are valid in the neighbourhood of the image of the attachment section (which is the projection of $w^0 = 1$, $w^i = 0$, and is therefore given in those coordinates by $y_0^i = 0$). Our principle will be that in order to establish any result about elements of $\mathcal{A}^G$ it will be enough to establish it for the appropriate fibre-quadratic vector fields in those coordinates, because any such vector field will be the coordinate representation of one defined on a complete fibre of $\mathrm{P}\mathcal{W}M$.

We may also check that, for any $x \in M$, the fibre $K_x\mathcal{G}$ of the kernel may be identified with the Lie algebra $\mathfrak{sl}(n+1)$ by

$$\left(a_0^i + a_j^i y^j - (a_j^0 y^j) y^i\right) \left.\frac{\partial}{\partial y^i}\right|_{\mathrm{P}\mathcal{W}_x M} \mapsto \begin{pmatrix} -a_0^0 & a_j^0 \\ a_0^i & a_j^i - a_0^0 \delta_j^i \end{pmatrix}$$

where

$$a_0^0 = \frac{1}{n+1} a_i^i ,$$

but with the vector field bracket corresponding to the negative of the Lie bracket on $\mathfrak{sl}(n+1)$. In fact the fibre-linear vector fields

$$(b_0^0 w^0 + b_i^0 w^i) \left.\frac{\partial}{\partial w^0}\right|_{\mathcal{W}_x M} + (b_0^i w^0 + b_j^i w^j) \left.\frac{\partial}{\partial w^i}\right|_{\mathcal{W}_x M}$$

along $\mathcal{W}_x M$ form a Lie algebra anti-isomorphic to $\mathfrak{gl}(n+1)$ by

$$(b_0^0 w^0 + b_i^0 w^i) \left.\frac{\partial}{\partial w^0}\right|_{\mathcal{W}_x M} + (b_0^i w^0 + b_j^i w^j) \left.\frac{\partial}{\partial w^i}\right|_{\mathcal{W}_x M} \mapsto \begin{pmatrix} b_0^0 & b_j^0 \\ b_0^i & b_j^i \end{pmatrix} ,$$

and by adding a suitable multiple of Δ we can make the trace of the matrix zero, that is, obtain an element of $\mathfrak{sl}(n+1)$.

Our final remark about Proposition 4.4.2, perhaps the most significant, concerns the observation that the bundle $\mathrm{P}\mathcal{W}M \to M$ has compact fibres. We shall study this 'full' version of projective geometry on $\mathrm{P}\mathcal{W}M$ in Chap. 9; but there is also a restricted version of projective geometry defined on the tangent bundle of a manifold. Consider vector fields on TM taking the similar quadratic form

$$X^i \frac{\partial}{\partial x^i} + \left(Y^i_0 + Y^i_j u^j - (Y^0_j u^j) u^i\right) \frac{\partial}{\partial u^i}$$

in the tangent bundle coordinates (x^i, u^i), where the functions X^i, Y^i_0, Y^i_j and Y^0_j are all projectable to M. We may show directly that the set of all such vector fields along the fibres of TM also forms a Lie algebroid, noting that the linear coordinates u^i are defined on complete fibres; but there is now no reason why such a vector field should have vertically uniform flow boxes. For instance, taking $M = \mathbb{R}$ so that $TM = \mathbb{R} \times \mathbb{R}$, the vector field

$$\frac{\partial}{\partial x} - u^2 \frac{\partial}{\partial u}$$

is defined globally on TM but has flow

$$\big((x, u), t\big) \mapsto \left(x + t, \frac{u}{1 - tu}\right)$$

defined, at (x, u) with $u > 0$, only for $t < 1/u$. We therefore cannot assert that this Lie algebroid is derived from a Lie groupoid of fibre morphisms of TM.

We shall discuss this phenomenon in more detail in Chap. 9. For the moment, though, we shall shed some light on the matter by returning to our choice of $\mathrm{P}\mathcal{W}M \to M$ for the bundle of projective spaces over M, rather than the apparently more straightforward choice $\mathrm{P}(TM \times \mathbb{R}) \to M$. In the latter case there is a canonical identification of TM as an open-dense submanifold of $\mathrm{P}(TM \times \mathbb{R})$, where the elements of the projective space $\mathrm{P}(T_x M \oplus \mathbb{R})$ that cannot be identified with elements of $T_x M$ form a 'hyperplane at infinity'. But we require more flexibility, and we obtain this by using an appropriate family of inclusion maps $\bar{\vartheta} : TM \to \mathrm{P}\mathcal{W}M$, constructed using Ehresmann connection forms ϑ.

Recall that we started with the volume bundle $\nu : \mathcal{V}M \to M$; let ϑ be an Ehresmann connection form on $\mathcal{V}M$, so that it is an $\mathbb{R}$-valued 1-form (in other words, a scalar form) satisfying $\vartheta(\Upsilon) = 1$ and the invariance condition that $\mu_s^* \vartheta = \vartheta$ for any $s \in \mathbb{R}_+$. In coordinates, therefore,

$$\vartheta = (x^0)^{-1} dx^0 + \vartheta_j dx^j$$

where the functions ϑ_j are projectable to M as a consequence of equivariance. For example, any global section ς of $\mathcal{V}M \to M$ gives rise to an exact Ehresmann connection form

$$\vartheta = (x^0)^{-1}dx^0 - (\varsigma^0)^{-1}\frac{\partial\varsigma^0}{\partial x^j}dx^j$$

where $\varsigma^0 = x^0 \circ \varsigma$. In general, though, an Ehresmann connection form ϑ on $\mathcal{V}M$ need not be exact.

We have mentioned above that $\mathcal{W}M \to M$ may be viewed as the Atiyah algebroid of the principal bundle $\mathcal{V}M \to M$; any Ehresmann connection on the latter corresponds to an infinitesimal connection on the former.

Proposition 4.4.3 *Let X be any vector field on M, and let the horizontal lift of X by the Ehresmann connection ϑ be denoted by X^ϑ. The vector field X^ϑ, a section of $T\mathcal{V}M \to \mathcal{V}M$, then projects to a well-defined section of $\mathcal{W}M \to M$. The images of these sections determine a vector bundle morphism $\gamma^\vartheta : TM \to \mathcal{W}M$ which is a section of $\rho : \mathcal{W}M \to TM$, so that γ^ϑ is an infinitesmal connection whose image in $\mathcal{W}M$ is complementary to the line bundle $\langle e(M)\rangle$. This correspondence between Ehresmann connections and infinitesimal connections on $\mathcal{W}M$ is a bijection.*

Proof The projectability of X^ϑ to $\mathcal{W}M$, essentially its invariance under the derivative action μ_{s*}, is a consequence of the invariance of ϑ under μ_s. In coordinates, if $X = X^i\partial/\partial x^i$ then

$$X^\vartheta = X^i\left(\frac{\partial}{\partial x^i} - \vartheta_i\Upsilon\right)$$

so that for any $x \in M$ the map $T_xM \to \mathcal{W}_xM$ is given by

$$\left.\frac{\partial}{\partial x^i}\right|_x \mapsto \left.\frac{\partial}{\partial x^i}\right|_x - \vartheta_i(x)e(x)$$

and is clearly an infinitesimal connection on $\mathcal{W}M$ complementary to $\langle e(x)\rangle$. Two different Ehresmann connections must differ at some $x \in M$, so the corresponding infinitesimal connections will also differ at x.

Finally, suppose we are given an infinitesimal connection $\gamma : TM \to \mathcal{W}M$. At each $x \in M$ the image $\gamma(T_xM)$ is a hyperplane in $\mathcal{W}_xM$ not containing $e(x)$. As the map $\chi : T\mathcal{V}M \to \mathcal{W}M$ is a fibrewise isomorphism, we see that for each $z \in \mathcal{V}_xM$ the preimage $\chi^{-1}\big(\gamma(T_xM)\big) \subset T_z\mathcal{V}M$ is a hyperplane not containing Υ_z and so defines a cotangent vector ϑ_z which we may choose to satisfy $\langle\vartheta_z, \Upsilon_z\rangle = 1$. The resulting 1-form ϑ on $\mathcal{V}M$ is smooth because γ is smooth; it is the required Ehresmann connection form. □

The translated images $e(x) + \gamma^\vartheta(T_xM) \subset \mathcal{W}_xM$ determined by an Ehresmann connection form ϑ allow us to define a map $\bar{\vartheta} : TM \to \mathrm{P}\mathcal{W}M$ and hence identify TM as an open-dense submanifold $\bar{\vartheta}(TM) \subset \mathsf{P}\mathcal{W}M$. The elements of the projective space $\mathsf{P}\mathcal{W}_x M$ that cannot be identified in this way with elements of $T_x M$ form a 'hyperplane at infinity'. It is clear that not every projective transformation $\varphi : \mathsf{P}\mathcal{W}_x M \to \mathsf{P}\mathcal{W}_{\tilde{x}} M$ maps $\bar{\vartheta}(T_x M)$ to $\bar{\vartheta}(T_{\tilde{x}} M)$; for instance a nonsingular linear map $\mathcal{W}_x M \to \mathcal{W}_{\tilde{x}} M$ sending $e(x)$ to an element of $\bar{\vartheta}(T_{\tilde{x}} M)$ will give rise to a projective transformation mapping $[e(x)] \in \bar{\vartheta}(T_x M)$ to an 'infinite' element of $\mathsf{P}\mathcal{W}_{\tilde{x}} M$.

This problem does not, however, arise at the algebroid level, because $\bar{\vartheta}(TM)$ is open in $\mathsf{P}\mathcal{W}M$. If c is any curve in $\mathcal{G}$ with $c(0) = 1_x = \mathrm{id}_{\mathsf{P}\mathcal{W}_x M}$ and $\alpha(c(s)) = x$ then for each $y \in \bar{\vartheta}(T_x M)$ there is a value $\varepsilon_y > 0$ such that $c(s, y) \in \bar{\vartheta}(TM)$ for $|s| < \varepsilon_y$, giving a well-defined tangent vector at y, and hence a vector field along the fibre $\bar{\vartheta}(T_x M)$. Indeed, any vector field along the fibre $\mathsf{P}\mathcal{W}_x M$ that is an element of $\mathcal{A}^G$ restricts to give a vector field along the fibre $\bar{\vartheta}(T_x M)$, and as $\bar{\vartheta}(TM)$ is dense in $\mathsf{P}\mathcal{W}M$ we see that this correspondence is injective. We shall see the significance of this in Sect. 9.2.

4.5 Vector Bundles, Fibre Metrics and Euclidean Maps

We return to the consideration of a vector bundle $\pi : E \to M$ of rank k, and we now suppose that E carries a *fibre metric* g, a positive definite (or, more generally, nondegenerate) section of $S^2 E^* \to M$. In the positive definite case, if π is regarded as a $\mathsf{GL}(k)$-bundle then the fibre metric specifies a reduction of π to an $\mathsf{O}(k)$-bundle. We shall say that an affine map $\varphi : E_{\tilde{x}} \to E_x$, where $x, \tilde{x} \in M$, is an *orthogonal map* if φ satisfies

$$g_x\big(\varphi(v), \varphi(w)\big) = g_{\tilde{x}}(v, w), \quad v, w \in E_{\tilde{x}}.$$

We may also, though, regard π as an $\mathsf{A}(k)$-bundle as in Sect. 4.2, and then g specifies a reduction of π to an $\mathsf{E}(k)$ bundle where $\mathsf{E}(k)$, the k-dimensional Euclidean group, is the subgroup of $\mathsf{A}(k)$ containing matrices (A^a_b) with the additional property that the submatrix (A^α_β), $\alpha, \beta = 1, 2, \ldots, k$, is an element of the orthogonal group $\mathsf{O}(k)$. We shall say that an affine map $\varphi : E_{\tilde{x}} \to E_x$, where $x, \tilde{x} \in M$, is a *Euclidean map* if the associated linear map $\bar{\varphi} : E_{\tilde{x}} \to E_x$ given by $\bar{\varphi}(v) = \varphi(v) - \varphi(0)$ satisfies

$$g_x\big(\bar{\varphi}(v), \bar{\varphi}(w)\big) = g_{\tilde{x}}(v, w), \quad v, w \in E_u.$$

Proposition 4.5.1 *The set $\mathcal{G}_g$ of Euclidean maps $\varphi : E_{\tilde{x}} \to E_x$, for all $x, \tilde{x} \in M$, is a locally trivial Lie groupoid.*

Proof As before, this is immediate from Theorem 3.1.1; furthermore it is clear that $\mathcal{G}_g$ is a wide subgroupoid of the Lie groupoid of affine morphisms. □

In order to analyze fibre metrics and Euclidean maps, it is convenient to introduce the concept of a *generalized fibre metric*: this will be a positive definite (or nondegenerate) section of $S^2 V^*\pi \to E$, where $V\pi \to E$ is the vertical bundle of π and

$V^*\pi \to E$ is its dual. Any fibre metric g on π gives rise to a generalized fibre metric g^{V} by using, for $x \in M$ and $y \in E_x$, the vertical lift isomorphism $E_x \to T_y E_x = V_y\pi$ and setting $g^{\mathsf{V}}_y\big(v^{\mathsf{V}}, w^{\mathsf{V}}\big) = g_x(v, w)$ for $v, w \in E_x$ and $v^{\mathsf{V}}, w^{\mathsf{V}} \in V_y\pi$. Of course not every generalized fibre metric is a vertical lift, for the latter is constant on the fibres of E; in coordinates, if $g = g_{\alpha\beta}(x) y^\alpha \otimes y^\beta$ (where in that particular formula we think of the linear fibre coordinates y^α as sections of $E^* \to M$) then

$$g^{\mathsf{V}} = \pi^* g_{\alpha\beta}(x) dy^\alpha \otimes dy^\beta ,$$

whereas for an arbitrary generalized fibre metric h we would have $h = h_{\alpha\beta}(y) dy^\alpha \otimes dy^\beta$.

We must emphasize that a generalized fibre metric is not a tensor field on E in the usual sense, as its arguments are required to be vertical. We can, nevertheless, define the Lie derivative of h by a projectable (rather than more specifically a vertical) vector field Z on E: we set

$$(\mathcal{L}_Z h)(V, W) = \mathcal{L}_Z\big(h(V, W)\big) - h\big([Z, V], W\big) - h\big(V, [Z, W]\big)$$

where V, W are vertical vector fields on E, using the fact that the Lie bracket of a projectable vector field and a vertical vector field is again vertical.

Proposition 4.5.2 *If $\xi : M \to A\mathcal{G}_g$ is a Lie algebroid section regarded as a projectable vector field on E, so that the restriction of its flow to complete fibres of E consists of Euclidean maps, then $\mathcal{L}_\xi g^{\mathsf{V}} = 0$. If the Euclidean maps are, in addition, orthogonal maps then $\mathcal{L}_\xi \hat{g} = 0$ where $\hat{g} : E \to \mathbb{R}$ is the fibre-quadratic function $\hat{g}(y) = g(y, y)$.*

Conversely, if ξ is a projectable fibre-affine vector field on E satisfying the condition $\mathcal{L}_\xi g^{\mathsf{V}} = 0$ then the flow of ξ restricted to any fibre of π is a Euclidean map; if ξ satisfies the condition $\mathcal{L}_\xi \hat{g} = 0$ then the restriction of the flow to any fibre is an orthogonal map.

Proof We recall first that if $v, y \in E_{\tilde{x}}$ then $v^{\mathsf{V}} = \dot{c}(0) \in V_y\pi$ where $c : \mathbb{R} \to E_{\tilde{x}}$ is the curve $c(t) = y + tv$. If $\varphi : E_{\tilde{x}} \to E_x$ is an affine map with associated linear map $\bar{\varphi}$, and if $c_\varphi = \varphi \circ c$, then

$$c_\varphi(t) = \varphi(c(t)) = \varphi(y + tv) = \varphi(y) + t\bar{\varphi}(v)$$

so that

$$\varphi_*(v^{\mathsf{V}}) = \varphi_*(\dot{c}(0)) = \dot{c}_\varphi(0) = (\bar{\varphi}(v))^{\mathsf{V}} \in V_{\varphi(y)} .$$

We may therefore write the condition $g_x\big(\bar{\varphi}(v), \bar{\varphi}(w)\big) = g_{\tilde{x}}(v, w)$ for φ to be a Euclidean map in terms of the lifted fibre metric g^{V} and the tangent map φ_* as

$$g^{\mathsf{V}}_{\varphi(y)}\big(\varphi_*(v^{\mathsf{V}}), \varphi_*(w^{\mathsf{V}})\big) = g^{\mathsf{V}}_y(v^{\mathsf{V}}, w^{\mathsf{V}})$$

for every $v, w, y \in E_{\tilde{x}}$.

Let ψ_t be the flow of ξ on E. Let V and W be vertical vector fields on E so that, for any $y \in E$,

$$g^{\mathsf{V}}_{\psi_t(y)}\big(V_{\psi_t(y)}, W_{\psi_t(y)}\big) = g^{\mathsf{V}}_y\big(\psi_{-t*}V_{\psi_t(y)}, \psi_{-t*}W_{\psi_t(y)}\big)\,;$$

then

$$\begin{aligned}
\mathcal{L}_\xi\big(g^{\mathsf{V}}(V, W)\big)|_y &= \lim_{t\to 0}\frac{1}{t}\Big(g^{\mathsf{V}}_{\psi_t(y)}\big(V_{\psi_t(y)}, W_{\psi_t(y)}\big) - g^{\mathsf{V}}_y(V_y, W_y)\Big)\\
&= \lim_{t\to 0}\frac{1}{t}\Big(g^{\mathsf{V}}_y\big(\psi_{-t*}V_{\psi_t(y)}, \psi_{-t*}W_{\psi_t(y)}\big) - g^{\mathsf{V}}_y(V_y, W_y)\Big)\\
&= \lim_{t\to 0}\frac{1}{t}\Big(g^{\mathsf{V}}_y\big(\psi_{-t*}V_{\psi_t(y)}, \psi_{-t*}W_{\psi_t(y)}\big) - g^{\mathsf{V}}_y(V_y, \psi_{-t*}W_{\psi_t(y)})\Big)\\
&\quad + \lim_{t\to 0}\frac{1}{t}\Big(g^{\mathsf{V}}_y\big(V_y, \psi_{-t*}W_{\psi_t(y)}\big) - g^{\mathsf{V}}_y(V_y, W_y)\Big)\\
&= \lim_{t\to 0} g^{\mathsf{V}}_y\Big(\frac{1}{t}\big(\psi_{-t*}V_{\psi_t(y)} - V_y\big), \psi_{-t*}W_{\psi_t(y)}\Big)\\
&\quad + \lim_{t\to 0} g^{\mathsf{V}}_y\Big(V_y, \frac{1}{t}\big(\psi_{-t*}W_{\psi_t(y)} - W_y\big)\Big)\\
&= g^{\mathsf{V}}_y([\xi, V], W) + g^{\mathsf{V}}_y(V, [\xi, W])
\end{aligned}$$

so that $(\mathcal{L}_\xi g)(V, W)|_y = 0$.

If the Euclidean maps obtained from the flow ψ_t are also orthogonal maps, so that

$$g_{\overline{\psi}_t(x)}(\psi_t(v), \psi_t(w)) = g_x(v, w)$$

for any $v, w \in E_x$ where $\overline{\psi}_t$ is the projection to M of the flow ψ_t, then $\hat{g}(\psi_t(v)) = \hat{g}(v)$ for any $v \in E$, so that $\hat{g} = \hat{g} \circ \psi_t = \psi_t^*(\hat{g})$; therefore $\mathcal{L}_\xi \hat{g} = 0$.

For the converse, suppose that ξ is a fibre-affine projectable vector field on E (so that its flow ψ_t is defined on compete fibres) and suppose that $\mathcal{L}_\xi g^{\mathsf{V}} = 0$, so that

$$\mathcal{L}_\xi\big(g^{\mathsf{V}}(V, W)\big) = g^{\mathsf{V}}\big([\xi, V], W\big) + g^{\mathsf{V}}\big(V, [\xi, W]\big)$$

for any vertical vector fields V, W on E.

Fix $y, v, w \in E$, and let t_0 be such that $\psi_t(y)$ is defined; suppose without loss of generality that $t > 0$. Let V, W be vertical vector fields on E satisfying $V_{\psi_s(y)} = \psi_{s*}(v^{\mathsf{V}})$ and $W_{\psi_s(y)} = \psi_{s*}(w^{\mathsf{V}})$ for $s \in [0, t]$, where $v^{\mathsf{V}}, w^{\mathsf{V}} \in V_y\pi$; such vector fields always exist. With these choices $[\xi, V]_{\psi_s(y)} = [\xi, W]_{\psi_s(y)} = 0$ for all $s \in [0, t]$, so that

$$\mathcal{L}_\xi\big(g^{\mathsf{V}}(V, W)\big)|_{\psi_s(y)} = 0$$

for all $s \in [0, t]$. It follows that the function $s \mapsto g^{\mathsf{V}}(V, W)(\psi_s(y))$ is constant for $s \in [0, t]$ so that

$$g^{\mathsf{V}}_{\psi_t(y)}\big(\psi_{t*}(v^{\mathsf{V}}), \psi_{t*}(w^{\mathsf{V}})\big) = g^{\mathsf{V}}(V, W)(\psi_t(y)) = g^{\mathsf{V}}(V, W)(y) = g^{\mathsf{V}}_y(v^{\mathsf{V}}, w^{\mathsf{V}}).$$

Now suppose that ξ is a fibre-linear vector field on E, and suppose that $\mathcal{L}_\xi \hat{g} = 0$. In the same way $\hat{g}$ must be constant along the flow of ξ, so that $\hat{g}(\psi_t(v)) = \hat{g}(v)$ for any $v \in E$; thus by polarization

$$g_{\overline{\psi}_t(x)}(\psi_t(v), \psi_t(w)) = g_x(v, w)$$

for any $v, w \in E_x$. □

In that proposition we made the assumption that the vector field ξ was fibre-affine. In fact that assumption is superfluous.

Lemma 4.5.3 *If Z is any projectable vector field on E satisfying the condition $\mathcal{L}_Z g^{\mathsf{V}} = 0$ then it is fibre-affine.*

Proof We use vector bundle coordinates (x^i, y^α) on E. If

$$Z = Z^i \frac{\partial}{\partial x^i} + Z^\alpha \frac{\partial}{\partial y^\alpha}$$

then

$$(\mathcal{L}_Z g^{\mathsf{V}})_{\alpha\beta} = Z^i \frac{\partial g_{\alpha\beta}}{\partial x^i} + g_{\alpha\gamma} \frac{\partial Z^\gamma}{\partial y^\beta} + g_{\gamma\beta} \frac{\partial Z^\gamma}{\partial y^\alpha}.$$

Differentiating the condition $(\mathcal{L}_Z g^{\mathsf{V}})_{\alpha\beta} = 0$ with respect to y^γ, and using projectability, we obtain

$$g_{\alpha\delta} \frac{\partial^2 Z^\delta}{\partial y^\beta \partial y^\gamma} + g_{\beta\delta} \frac{\partial^2 Z^\delta}{\partial y^\alpha \partial y^\gamma} = 0,$$

so that

$$g_{\alpha\delta} \frac{\partial^2 Z^\delta}{\partial y^\beta \partial y^\gamma} = Z_{\alpha\beta\gamma}$$

is skew in its first pair of indices, symmetric in the second. But then

$$Z_{\alpha\beta\gamma} = -Z_{\beta\alpha\gamma} = -Z_{\beta\gamma\alpha} = Z_{\gamma\beta\alpha} = Z_{\gamma\alpha\beta} = -Z_{\alpha\gamma\beta} = -Z_{\alpha\beta\gamma},$$

so that

$$\frac{\partial^2 Z^\alpha}{\partial y^\beta \partial y^\gamma} = 0.$$

□

Corollary 4.5.4 *A projectable vector field Z on E corresponds to a section of $A\mathcal{G}_g$ if and only if with respect to vector bundle coordinates (x^i, y^α) it takes the form*

$$Z = X^i \frac{\partial}{\partial x^i} + (Y_0^\alpha + Y_\beta^\alpha y^\beta) \frac{\partial}{\partial y^\alpha}$$

where X^i, Y_0^α and Y_β^α are projectable to M and satisfy

$$X^i \frac{\partial g_{\alpha\beta}}{\partial x^i} + g_{\alpha\gamma} Y_\beta^\gamma + g_{\gamma\beta} Y_\alpha^\gamma = 0 . \qquad \square$$

We now take for $E \to M$ the tangent bundle $\tau_M : TM \to M$ and a Riemannian metric g on M, which we regard as a fibre metric on the tangent bundle. If we consider the Lie groupoid of orthogonal fibre morphisms $\mathcal{G}_g$, and consider path connections on $\mathcal{G}_g$ and infinitesimal connections on the Lie algebroid $A\mathcal{G}_g$, we may see that there is a canonical linear connection, the *Levi-Civita connection*, which is normally defined using covariant derivatives.

Proposition 4.5.5 *There is a unique symmetric linear connection*[3] *on $\mathcal{G}_g$ having the property that $\mathcal{L}_{X^{\mathsf{h}}} \hat{g} = 0$ for every vector field X on M, where $\hat{g} : TM \to \mathbb{R}$ is the fibre-quadratic function determined by g. There is consequently a canonical affine connection on $\mathcal{G}_g$.*

Proof Put

$$H_i = \frac{\partial}{\partial x^i} - \Gamma_{ij}^k y^j \frac{\partial}{\partial y^k} ;$$

then, using $\hat{g} = g_{hl} y^h y^l$,

$$\mathcal{L}_{H_i} \hat{g} = \left(\frac{\partial}{\partial x^i} - \Gamma_{ij}^k y^j \frac{\partial}{\partial y^k} \right) (g_{hl} y^h y^l) = \frac{\partial g_{hj}}{\partial x^i} y^h y^j - 2 g_{hk} \Gamma_{ij}^k y^j y^h$$

so that $\mathcal{L}_{X^{\mathsf{h}}} \hat{g} = 0$ implies

$$\frac{\partial g_{hj}}{\partial x^i} = g_{hk} \Gamma_{ij}^k + g_{jk} \Gamma_{ih}^k .$$

Thus, using $\Gamma_{ij}^k = \Gamma_{ji}^k$, we obtain

$$\Gamma_{ij}^k = \tfrac{1}{2} g^{hk} \left(\frac{\partial g_{hj}}{\partial x^i} + \frac{\partial g_{hi}}{\partial x^j} - \frac{\partial g_{ij}}{\partial x^h} \right)$$

[3] Previously, when considering an affine connection and its associated linear connection, we have written X^{h} for the horizontal lift and H_i for the horizontal vector fields given by the affine connection, and $X^{\bar{\mathsf{h}}}$ and $\bar{H}_i$ for those given by the linear connection. In general, though, we shall consider the horizontal lift and the horizontal vector fields associated only with a *linear* connection, so from now on we shall simplify the notation and use X^{h} and H_i (rather than $X^{\bar{\mathsf{h}}}$ and $\bar{H}_i$) for them.

where $g_{ij}g^{jk} = \delta^k_j$, and these are just the usual Christoffel symbols, the connection coefficients of the Levi-Civita connection given by g.

Using the bijection from Corollary 4.2.4 between linear and affine (rather than generalized affine) connections, we therefore see that there is a canonical affine connection on $\mathcal{G}_g$ given by

$$\gamma \circ X = X^i \left(\frac{\partial}{\partial x^i} - (\delta^k_i + \Gamma^k_{ij} y^j) \frac{\partial}{\partial y^k} \right). \qquad \square$$

Our final observation in this section is that each Riemannian metric g on M has an associated 'Riemannian volume form' which, despite its name, is not a genuine n-form as it incorporates a sign ambiguity; it is therefore a section ς_g of the volume bundle $\mathcal{V}M \to M$.

To construct ς_g we start with the index-lowering map $g^\flat : TM \to T^*M$, defined by

$$g^\flat(v)(w) = g_x(v, w), \qquad v, w \in T_x M,$$

and its inverse $g^\sharp = (g^\flat)^{-1} : T^*M \to TM$. Using the canonical identification of $\bigwedge^n(V^*)$ with the dual $(\bigwedge^n V)^*$ for any n-dimensional vector space V, we then define a map $g^\sharp : \bigwedge^n T^*M \to \bigwedge^n TM$ by

$$g^\sharp(\sigma^1 \wedge \sigma^2 \wedge \cdots \wedge \sigma^n) = g^\sharp(\sigma^1) \wedge g^\sharp(\sigma^2) \wedge \cdots \wedge g^\sharp(\sigma^n), \qquad \sigma^i \in T^*_x M.$$

If $\mathsf{v} \in \bigwedge^n T^*_x M$ is nonzero then $\langle \mathsf{v}, g^\sharp(\mathsf{v})\rangle = \lambda \neq 0$, so that $\pm|\lambda|^{-1/2}$ are the unique elements of $\mathbb{R}_0$ such that

$$\left\langle \pm |\lambda|^{-1/2}\mathsf{v},\ g^\sharp\left(\pm|\lambda|^{-1/2}\mathsf{v}\right)\right\rangle = \pm 1.$$

This procedure gives, to within sign, a 'section' of $\bigwedge^n T^*M$ and hence a section ς_g of $\mathcal{V}M \to M$. We shall regard $\mathcal{V}M$ as the positive half of the weighted line bundle $\mathcal{D}^p M$, and here we shall choose the weight $p = 1/n$.

Lemma 4.5.6 *The* Riemannian volume form $\varsigma_g : M \to \mathcal{D}^p M$ *is a smooth section of* $\mathcal{D}^p M \to M$ *given in weighted coordinates by*

$$x^0 \circ \varsigma_g = |\det g_{ij}|^{1/2n}.$$

Proof Let g^{-1} be the section of S^2TM defined by

$$g^{-1}_x(\sigma, \tau) = g_x\left(g^\sharp(\sigma), g^\sharp(\tau)\right), \qquad \sigma, \tau \in T^*_x M,$$

so that the functions g^{ij} used in the expression of the Christoffel symbols in the proof of Proposition 4.5.5 are just the coordinates of g^{-1}. We then note that $g^\sharp(dx^i) = g^{ij}\partial_j$, where $\partial_j = \partial/\partial x^j$, so that

$$\begin{aligned} g^\sharp(dx^1 \wedge dx^2 \wedge \cdots \wedge dx^n) &= g^\sharp(dx^1) \wedge g^\sharp(dx^2) \wedge \cdots \wedge g^\sharp(dx^n) \\ &= (g^{1j_1}\partial_{j^1}) \wedge (g^{2j_2}\partial_{j^2}) \wedge \cdots \wedge (g^{nj_n}\partial_{j^n}) \\ &= (\det g^{ij})\partial_1 \wedge \partial_2 \wedge \cdots \wedge \partial_n\,; \end{aligned}$$

thus

$$\varsigma_g = |\det g^{ij}|^{-1/2} dx^1 \wedge dx^2 \wedge \cdots \wedge dx^n = |\det g_{ij}|^{1/2} dx^1 \wedge dx^2 \wedge \cdots \wedge dx^n.$$

Using the weighted fibre coordinate x^0 we then obtain

$$x^0 \circ \varsigma_g = \varsigma_g^0 = |\det g_{ij}|^{1/2n}\,,$$

and we also see from this formula that the section ς_g is smooth. □

Chapter 5
Symmetries

A standard and powerful way to examine the properties of a geometric object is to study its symmetries. This applies in particular to connections, where both finite and infinitesimal symmetries may be considered, the former being diffeomorphisms with a property such as preserving geodesics, horizontal lifts or something similar, and the latter being vector fields whose flows have the same property.

We have described path connections on Lie groupoids and infinitesimal connections on transitive Lie algebroids, and so we shall describe symmetries and infinitesimal symmetries of these objects. The description we give in Sects. 5.2 and 5.3 applies to general locally trivial Lie groupoids and their Lie algebroids, although we shall of course be particularly interested in path connections on Lie groupoids of fibre morphisms and infinitesimal connections on the corresponding Lie algebroids of vector fields along fibres. To see how this approach relates to classical work we shall then consider, as a case study, the symmetries of affine connections on the tangent bundle.

First, though, we motivate the particular approach we have taken by expanding on the comment made at the end of Sect. 1.1, that a bisection of a Lie groupoid is a kind of generalized groupoid element.

5.1 Bisections and Automorphisms

Traditionally, an automorphism of a group is a bijection from the group to itself preserving the group operations; if the group is a Lie group then we would require both the automorphism and its inverse to be smooth. A particular class of automorphisms may be obtained by considering the inner automorphisms, given by conjugation with a group element.

We can take a similar approach with groupoids. An automorphism of a groupoid $\mathcal{G}$ is a groupoid morphism (P, p) from $\mathcal{G}$ to itself where both P and p are bijections and where (P^{-1}, p^{-1}) is also a groupoid morphism. In fact, just as with groups, the second part of the condition is satisfied automatically.

M. Crampin and D. Saunders, *Cartan Geometries and their Symmetries*,
Atlantis Studies in Variational Geometry 4, DOI 10.2991/978-94-6239-192-5_5

Lemma 5.1.1 *Let $\mathcal{G}$ and $\mathcal{H}$ be groupoids with sets M, N of identities. If $(\mathcal{P}, p)$ is a groupoid morphism from $\mathcal{G}$ to $\mathcal{H}$ where both $\mathcal{P}$ and p are bijections then $(\mathcal{P}^{-1}, p^{-1})$ is also a groupoid morphism.*

Proof First,

$$\alpha_{\mathcal{G}} \circ \mathcal{P}^{-1} = p^{-1} \circ p \circ \alpha_{\mathcal{G}} \circ \mathcal{P}^{-1} = p^{-1} \circ \alpha_{\mathcal{H}} \circ \mathcal{P} \circ \mathcal{P}^{-1} = p^{-1} \circ \alpha_{\mathcal{H}}$$

and similarly $\beta_{\mathcal{G}} \circ \mathcal{P}^{-1} = p^{-1} \circ \beta_{\mathcal{H}}$.

Now suppose that $\varphi_1, \varphi_2 \in \mathcal{H}$ and that (φ_1, φ_2) is an admissible pair, so that automatically $\big(\mathcal{P}^{-1}(\varphi_1), \mathcal{P}^{-1}(\varphi_2)\big)$ is an admissible pair. Then

$$\mathcal{P}\big(\mathcal{P}^{-1}(\varphi_1)\cdot\mathcal{P}^{-1}(\varphi_2)\big) = \big(\mathcal{P}\mathcal{P}^{-1}(\varphi_1)\big)\cdot\big(\mathcal{P}\mathcal{P}^{-1}(\varphi_2)\big) = \varphi_1\cdot\varphi_2 = \mathcal{P}\mathcal{P}^{-1}(\varphi_1\cdot\varphi_2);$$

but $\mathcal{P}$ is a bijection, so that $\mathcal{P}^{-1}(\varphi_1)\cdot\mathcal{P}^{-1}(\varphi_2) = \mathcal{P}^{-1}(\varphi_1\cdot\varphi_2)$. □

If $\mathcal{G}$ is a Lie groupoid then we shall also, of course, require an automorphism $(\mathcal{P}, p)$ to be such that both $\mathcal{P}$ and p are diffeomorphisms; the collection of all such automorphisms forms a subgroup $\mathsf{Aut}(\mathcal{G})$ of the diffeomorphism group $\mathsf{Diff}(\mathcal{G})$. But what could an 'inner automorphism' be? An inner automorphism of a group is given by conjugation with a fixed group element; but this cannot work for a groupoid as different elements will have different compatibility requirements and so no single element can be used for conjugation. Instead, given a Lie groupoid, we use a bisection.

Proposition 5.1.2 *Let $\Phi : M \to \mathcal{G}$ be a bisection of α, so that $\phi = \beta \circ \Phi$ is a diffeomorphism of M. Define the map $\tilde{\Phi} : \mathcal{G} \to \mathcal{G}$ by*

$$\tilde{\Phi}(\varphi) = \Phi\beta(\varphi) \cdot \varphi \cdot \Phi^{\iota}\alpha(\varphi)$$

where $\Phi^{\iota} = \iota \circ \Phi$ (that is, $\Phi^{\iota}(x) = \Phi(x)^{-1}$). The pair $(\tilde{\Phi}, \phi)$ is then a fibred diffeomorphism of $\mathcal{G}$ with respect to both the source map α and the target map β. In addition the map $\Phi \mapsto \tilde{\Phi}$ is a homomorphism from the group $\mathsf{B}(\mathcal{G})$ of bisections to the group $\mathsf{Aut}(\mathcal{G})$ of automorphisms.

If Φ is merely a local bisection of α, defined on an open neighbourhood U of $x \in M$, then $\tilde{\Phi}$ is defined on the open neighbourhood $V = \alpha^{-1}(U) \cap \beta^{-1}(U)$ of $1_x \in \mathcal{G}$.

Proof We note first that $\alpha(\Phi\beta(\varphi)) = \beta(\varphi)$ and that $\beta(\Phi^{\iota}\alpha(\varphi)) = \alpha(\Phi\alpha(\varphi)) = \alpha(\varphi)$, so that the products in the formula for $\tilde{\Phi}(\varphi)$ are defined. We then see that

$$\alpha\tilde{\Phi}(\varphi) = \alpha\big(\Phi\beta(\varphi) \cdot \varphi \cdot \Phi^{\iota}\alpha(\varphi)\big) = \alpha\Phi^{\iota}\alpha(\varphi) = \phi\alpha(\varphi)$$

and

$$\beta\tilde{\Phi}(\varphi) = \beta\big(\Phi\beta(\varphi) \cdot \varphi \cdot \Phi^{\iota}\alpha(\varphi)\big) = \beta\Phi\beta(\varphi) = \phi\beta(\varphi),$$

showing that $(\tilde{\Phi}, \phi)$ is a fibred map for both the source and target projections $\alpha, \beta : \mathcal{G} \to M$ and is therefore a morphism of groupoids. It is evidently smooth. To see that

$\tilde{\Phi}$ is a diffeomorphism, and hence that $(\tilde{\Phi}, \phi)$ is an automorphism of Lie groupoids, note that $\tilde{\Phi}^{-1} : \mathcal{G} \to \mathcal{G}$ defined by

$$\tilde{\Phi}^{-1}(\varphi) = \Phi^{\iota}\phi^{-1}\beta(\varphi) \cdot \varphi \cdot \Phi\phi^{-1}\alpha(\varphi)$$

satisfies

$$\begin{aligned}
\tilde{\Phi}^{-1}\tilde{\Phi}(\varphi) &= \Phi^{\iota}\phi^{-1}\beta\tilde{\Phi}(\varphi) \cdot \tilde{\Phi}(\varphi) \cdot \Phi\phi^{-1}\alpha\tilde{\Phi}(\varphi) \\
&= \Phi^{\iota}\beta(\varphi) \cdot \tilde{\Phi}(\varphi) \cdot \Phi\alpha(\varphi) \\
&= \Phi^{\iota}\beta(\varphi) \cdot \Phi\beta(\varphi) \cdot \varphi \cdot \Phi^{\iota}\alpha(\varphi) \cdot \Phi\alpha(\varphi) \\
&= 1_{\beta(\varphi)} \cdot \varphi \cdot 1_{\alpha(\varphi)} \\
&= \varphi
\end{aligned}$$

and similarly $\tilde{\Phi}\tilde{\Phi}^{-1}(\varphi) = \varphi$.

If Φ_1, Φ_2 are bisections then, using $\Phi_1 * \Phi_2 = (\Phi_1 \circ \phi_2) \cdot \Phi_2$,

$$\begin{aligned}
\widetilde{\Phi_1 * \Phi_2}(\varphi) &= (\Phi_1 * \Phi_2)\beta(\varphi) \cdot \varphi \cdot (\Phi_1 * \Phi_2)^{\iota}\alpha(\varphi) \\
&= \Phi_1\phi_2\beta(\varphi) \cdot \Phi_2\beta(\varphi) \cdot \varphi \cdot \Phi_2^{\iota}\alpha(\varphi) \cdot \Phi_1^{\iota}\phi_2\alpha(\varphi) \\
&= \Phi_1\beta\tilde{\Phi}_2(\varphi) \cdot \tilde{\Phi}_2(\varphi) \cdot \Phi_1^{\iota}\alpha\tilde{\Phi}_2(\varphi) \\
&= \tilde{\Phi}_1\tilde{\Phi}_2(\varphi)
\end{aligned}$$

making careful use of $(\Phi_1 * \Phi_2)^{\iota} = \iota \circ (\Phi_1 * \Phi_2)$; thus the map $\mathsf{B}(\mathcal{G}) \to \mathsf{Aut}(\mathcal{G})$ given by $\Phi \mapsto \tilde{\Phi}$ is a group homomorphism.

Finally, if Φ is a local bisection with domain U then the domain of Φ^{ι} is also U, so that the domain of $\tilde{\Phi}$ is the set V of all φ such that both $\alpha(\varphi) \in U$ and $\beta(\varphi) \in U$; if $x \in U$ then $1_x \in V$. □

The pair $(\tilde{\Phi}, \phi)$ will be called the *inner automorphism* of the Lie groupoid $\mathcal{G}$ generated by the bisection Φ; the group of all inner automorphisms will be denoted $\mathsf{Inn}(\mathcal{G})$. There is no reason in general why the surjective homomorphism $\mathsf{B}(\mathcal{G}) \to \mathsf{Inn}(\mathcal{G})$ should be an isomorphism. Take, for instance, $\mathcal{G}$ to be a Lie group G, so that $\mathsf{B}(\mathcal{G}) \cong G$ and $\mathsf{Inn}(\mathcal{G})$ is isomorphic to the ordinary group of inner automorphisms of G. If G is a nontrivial Abelian Lie group then $\mathsf{Inn}(\mathcal{G})$ is trivial, so that $\mathsf{B}(\mathcal{G})$ and $\mathsf{Inn}(\mathcal{G})$ cannot be isomorphic. More generally, we note that if the bisection Φ gives rise to the identity automorphism of $\mathcal{G}$ then $\phi = \mathrm{id}_M$ so that Φ is a vertical bisection and takes its values in the vertex groups $\mathcal{G}_x$.

We now specialize to the case where $\pi : E \to M$ is a fibre bundle with standard fibre F and structure group G, and where $\mathcal{G}$ is the Lie groupoid of fibre morphisms of π. We shall say that a fibred map (P, p) from π to itself is an *automorphism* of π if both P and p are diffeomorphisms, and if both (P, p) and $(\mathrm{P}^{-1}, p^{-1})$ are compatible fibre bundle maps as described in Sect. 3.3. As the image action of a compatible fibre bundle map is always effective by Proposition 3.3.1, we are necessarily considering the case where the action of G on F is effective.

We see first that an automorphism of π gives rise to a bisection of $\mathcal{G}$, and hence an inner automorphism of $\mathcal{G}$.

Proposition 5.1.3 *Given an automorphism* (P, p) *of the fibre bundle* $\pi : E \to M$, *define the map* $\Phi^{\mathrm{P}} : M \to \mathcal{G}$ *by*

$$\Phi^{\mathrm{P}}(x)(y) = \mathrm{P}(y), \qquad y \in E_x;$$

then Φ^{P} *is a bisection of* $\mathcal{G}$, *and the correspondence* $\mathsf{Aut}(\pi) \to \mathsf{B}(\mathcal{G})$ *given by* $\mathrm{P} \mapsto \Phi^{\mathrm{P}}$ *is a group homomorphism.*

Proof By construction $\alpha\Phi^{\mathrm{P}}(x) = \pi(y) = x$ so that $\alpha \circ \Phi^{\mathrm{P}} = \mathrm{id}_M$. Now put $\phi^{\mathrm{P}} = \beta \circ \Phi^{\mathrm{P}}$; then for any $y \in E_x$ we must have $\pi\big(\Phi^{\mathrm{P}}(x)(y)\big) = \beta\Phi^{\mathrm{P}}(x)$, but we also have $\pi\mathrm{P}(x) = p\pi(y) = p(x)$, so we see that

$$\phi^{\mathrm{P}}(x) = \beta\Phi^{\mathrm{P}}(x) = \pi\big(\Phi^{\mathrm{P}}(x)(y)\big) = \pi\mathrm{P}(x) = p(x),$$

showing that $\phi^{\mathrm{P}} = p$ and therefore that ϕ^{P} is a diffeomorphism of M. Thus Φ^{P} is a bisection.

Now let (P_1, p_1), (P_2, p_2) be two automorphisms of π. We see that, for any $x \in M$ and any $y \in E_x$,

$$\begin{aligned}\Phi^{\mathrm{P}_1} * \Phi^{\mathrm{P}_2}(x)(y) &= \big(\Phi^{\mathrm{P}_1} p_2(x) \cdot \Phi^{\mathrm{P}_2}(x)\big)(y) = \Phi^{\mathrm{P}_1} p_2(x)\big(\Phi^{\mathrm{P}_2}(x)(y)\big) \\ &= \Phi^{\mathrm{P}_1} p_2(x)\big(\mathrm{P}_2(y)\big) = \mathrm{P}_1\mathrm{P}_2(y) = \Phi^{\mathrm{P}_1\mathrm{P}_2}(x)(y)\end{aligned}$$

so that $\Phi^{\mathrm{P}_1} * \Phi^{\mathrm{P}_2} = \Phi^{\mathrm{P}_1\mathrm{P}_2}$. □

In fact we can say more: the groups $\mathsf{Aut}(\pi)$ and $\mathsf{B}(\mathcal{G})$ are isomorphic. This is the finite version of the infinitesimal result demonstrated in Theorem 3.2.1, that the Lie algebroids $\mathcal{A}^G$ and $A\mathcal{G}$ are isomorphic.

Theorem 5.1.4 *The correspondence* $\mathsf{Aut}(\pi) \to \mathsf{B}(\mathcal{G})$ *given by* $\mathrm{P} \mapsto \Phi^{\mathrm{P}}$ *is a group isomorphism.*

Proof We shall construct an inverse to the correspondence $\mathrm{P} \mapsto \Phi^{\mathrm{P}}$.

Given a bisection $\Phi : M \to \mathcal{G}$, define maps $\mathrm{P}^{\Phi} : E \to E$, $p^{\Phi} : M \to M$ by

$$\mathrm{P}^{\Phi}(y) = \big(\Phi\pi(y)\big)(y), \qquad p^{\Phi} = \phi = \beta \circ \Phi.$$

The pair $(\mathrm{P}^{\Phi}, p^{\Phi})$ is a fibred map because

$$\pi\mathrm{P}^{\Phi}(y) = \pi\Big(\big(\Phi\pi(y)\big)(y)\Big) = \beta\Phi\pi(y) = p^{\Phi}\pi(y),$$

and p^Φ is certainly a diffeomorphism. To see that P is also a diffeomorphism, define the smooth map $\bar{\mathrm{P}}^\Phi : E \to E$ by

$$\bar{\mathrm{P}}^\Phi(y) = \big(\Phi(p^\Phi)^{-1}\pi(y)\big)^{-1}(y);$$

this makes sense because

$$\alpha\big(\Phi(p^\Phi)^{-1}\pi(y)\big)^{-1} = \beta\Phi(p^\Phi)^{-1}\pi(y) = \phi(p^\Phi)^{-1}\pi(y) = \pi(y)$$

so that y is in the domain of $\big(\Phi(p^\Phi)^{-1}\pi(y)\big)^{-1}$, and also

$$\pi\bar{\mathrm{P}}^\Phi(y) = \beta\big(\Phi(p^\Phi)^{-1}\pi(y)\big)^{-1} = \alpha\Phi(p^\Phi)^{-1}\pi(y) = (p^\Phi)^{-1}\pi(y)$$

so that $\big(\bar{\mathrm{P}}^\Phi, (p^\Phi)^{-1}\big)$ is a fibred map. Then

$$\begin{aligned}\bar{\mathrm{P}}^\Phi\mathrm{P}^\Phi(y) &= \big(\Phi(p^\Phi)^{-1}\pi\mathrm{P}^\Phi(y)\big)^{-1}\mathrm{P}^\Phi(y) = \big(\Phi(p^\Phi)^{-1}\pi\big(\Phi\pi(y)(y)\big)\big)^{-1}\big(\Phi\pi(y)\big)(y)\\ &= \big(\Phi\pi(y)\big)^{-1}\big(\Phi\pi(y)\big)(y) = y\end{aligned}$$

because $\pi\big(\Phi\pi(y)(y)\big) = p^\Phi\pi(y)$; and

$$\begin{aligned}\mathrm{P}^\Phi\bar{\mathrm{P}}^\Phi(y) &= \big(\Phi\pi\bar{\mathrm{P}}^\Phi(y)\big)\bar{\mathrm{P}}^\Phi(y) = \big(\Phi\pi\big(\big(\Phi(p^\Phi)^{-1}\pi(y)\big)^{-1}(y)\big)\big)\big(\Phi(p^\Phi)^{-1}\pi(y)\big)^{-1}(y)\\ &= \big(\Phi(p^\Phi)^{-1}\pi(y)\big)\big(\Phi(p^\Phi)^{-1}\pi(y)\big)^{-1}(y) = y\end{aligned}$$

because $\pi\big(\big(\Phi(p^\Phi)^{-1}\pi(y)\big)^{-1}(y)\big) = (p^\Phi)^{-1}\pi(y)$. Thus P^Φ is indeed a diffeomorphism with inverse $\bar{\mathrm{P}}^\Phi$.

We must now check that $(\mathrm{P}^\Phi, p^\Phi)$ is indeed a compatible map of fibre bundles; it is sufficient to check that each local trivialization is projectable in a neighbourhood of each point. So let $\mathrm{T}_U : U \times F \to \pi^{-1}(U)$ be any local trivialization of π in a neighbourhood U of $x \in M$. The first projectability condition holds because p^Φ is a bijection, and the second holds because the projection between the standard fibres is the identity. The third condition is that the map $\mathrm{T}^{\mathrm{P}}_{p^\Phi(U)} : p^\Phi(U) \times F \to \pi^{-1}\big(p^\Phi(U)\big)$ given by $\mathrm{T}^{\mathrm{P}}_{p^\Phi(U)}\big(p^\Phi(x), q\big) = \mathrm{P}^\Phi\mathrm{T}_U(x, q)$ should also be a local trivialization; but $p^\Phi(x) = \phi(x)$ and

$$\mathrm{P}^\Phi\mathrm{T}_U(x, q) = \Phi\pi\big(\mathrm{T}_U(x, q)\big)\big(\mathrm{T}(x, q)\big) = \Phi(x)\big(\mathrm{T}_U(x, q)\big) = \mathrm{T}_{\hat{U}}\big(\phi(x), gq\big)$$

for some local trivialization $\mathrm{T}_{\hat{U}}$ around $\phi(x)$ and some $g \in G$ because $\Phi(x)$ is a fibre morphism from E_x to $E_{\phi(x)}$. Thus the third projectability condition holds by restricting T_U to $\phi^{-1}(\hat{U}) \times F$, so we see that $(\mathrm{P}^\Phi, p^\Phi)$ is indeed a compatible fibre bundle map. Of course the same argument holds for $\big((\mathrm{P}^\Phi)^{-1}, (p^\Phi)^{-1}\big)$, confirming that $(\mathrm{P}^\Phi, p^\Phi)$ is indeed a fibre bundle automorphism.

So now we have a group homomorphism $\mathsf{Aut}(\pi) \to \mathsf{B}(\mathcal{G})$ given by $\mathrm{P} \mapsto \Phi^{\mathrm{P}}$, and a map $\mathsf{B}(\mathcal{G}) \to \mathsf{Aut}(\pi)$ given by $\Phi \mapsto \mathrm{P}^{\Phi}$. But if $y \in E_x$ then

$$\Phi^{\mathrm{P}^{\Phi}}(x)(y) = \mathrm{P}^{\Phi}(y) = \big(\Phi\pi(y)\big)(y) = \Phi(x)(y)$$

showing that $\Phi^{\mathrm{P}^{\Phi}} = \Phi$, and

$$\mathrm{P}^{\Phi^{\mathrm{P}}}(y) = \big(\Phi^{\mathrm{P}}\pi(y)\big)(y) = \mathrm{P}(y)$$

showing that $\mathrm{P}^{\Phi^{\mathrm{P}}} = \mathrm{P}$. Thus the homomorphism $\mathsf{Aut}(\pi) \to \mathsf{B}(\mathcal{G})$ is a bijection, and hence it is an isomorphism. □

It follows from all this that, when studying Lie groupoids of fibre morphisms of a fibre bundle, it is of particular interest to consider the inner automorphisms of the groupoid as these arise, via bisections, from the automorphisms of the original bundle.

5.2 Symmetries of Connections

In this section and the next we consider symmetries and infinitesimal symmetries of path connections on a general locally trivial Lie groupoid $\mathcal{G}$, not necessarily a Lie groupoid of fibre morphisms, and of infinitesimal connections on the Lie algebroid $A\mathcal{G}$.

We start by considering symmetries of a path connection Γ on $\mathcal{G}$. In principle, we might expect a symmetry of Γ to be a diffeomorphism (or a local diffeomorphism) of $\mathcal{G}$ which maps lifted curves to lifted curves. We should, though, expect the diffeomorphism to be related to the groupoid structure, and so we shall insist from the start that a candidate for a symmetry must in fact be an automorphism of Lie groupoids. We shall say that the automorphism $(\mathcal{P}, p)$ is a *symmetry* of the path connection Γ if the projected path connection $\Gamma^{\mathcal{P}}$ is the same as Γ, so that for every curve $c : [0, 1] \to M$ we have

$$\mathcal{P} \circ c^{\Gamma} = (p \circ c)^{\Gamma^{\mathcal{P}}} = (p \circ c)^{\Gamma}$$

(Proposition 2.3.1). The same formula may be used if $(\mathcal{P}, p)$ is only a local automorphism of $\mathcal{G}$ defined on $\alpha^{-1}(U) \cap \beta^{-1}(U)$ where $U \subset M$ is some nonempty open set, and where we now consider curves $c : [0, 1] \to U$.

Our main interest will be in the case of symmetries which are inner automorphisms $\tilde{\Phi}$ arising from bisections Φ of $\mathcal{G}$, and we shall often refer to Φ as the symmetry, rather than $\tilde{\Phi}$. Indeed, our main concern will be with infinitesimal connections rather than path connections, so we need to investigate how a symmetry (as we have described it for a path connection Γ) relates to the corresponding infinitesimal connection γ.

Lemma 5.2.1 *If Φ is any bisection (or local bisection) of $\mathcal{G}$ and if $v \in A_x\mathcal{G} \subset T_{1_x}\mathcal{G}$ then $\tilde{\Phi}_*(v) \in A_{\phi(x)}\mathcal{G} \subset T_{1_{\phi(x)}}\mathcal{G}$.*

Proof As $\tilde{\Phi}(1_x) = 1_{\phi(x)}$ for any $x \in M$ (or for any $x \in U$ in the case of a local section) it follows that

$$\tilde{\Phi}_*(v) \in T_{1_{\phi(x)}}\mathcal{G}, \qquad \alpha_*\tilde{\Phi}_*(v) = \phi_*\alpha_*(v) = 0$$

so that $\tilde{\Phi}_*(v) \in A_{\phi(x)}\mathcal{G}$. □

Lemma 5.2.2 *If $\Phi : U \to \mathcal{G}$ is a symmetry of the path connection Γ, and if γ is the corresponding infinitesimal connection, then $\tilde{\Phi}_* \circ \gamma = \gamma \circ \phi_*$.*

Proof Take $v \in TM$. If $c : [0, 1] \to U$ is a curve with $v = dc/dt|_{t=0}$ then $\gamma(v) = dc^\Gamma/dt|_{t=0}$ so that

$$\begin{aligned}
\tilde{\Phi}_*\gamma(v) &= \tilde{\Phi}_* \left.\frac{dc^\Gamma}{dt}\right|_{t=0} \\
&= \left.\frac{d(\tilde{\Phi} \circ c^\Gamma)}{dt}\right|_{t=0} \\
&= \left.\frac{d(\phi \circ c)^\Gamma}{dt}\right|_{t=0} \\
&= \gamma\phi_*(v),
\end{aligned}$$

showing that $\tilde{\Phi}_* \circ \gamma = \gamma \circ \phi_*$. □

We can also see how a symmetry interacts with the kernel projection ω of γ.

Lemma 5.2.3 *If Φ satisfies the symmetry condition $\tilde{\Phi}_* \circ \gamma = \gamma \circ \phi_*$ then in addition*

$$\tilde{\Phi}_* \circ \omega = \omega \circ \tilde{\Phi}_*$$

as maps $A\mathcal{G} \to K\mathcal{G}$.

Proof Note first that $\tilde{\Phi}_* : A\mathcal{G} \to A\mathcal{G}$ restricts to a map $K\mathcal{G} \to K\mathcal{G}$, because $\rho \circ \tilde{\Phi}_* = \phi_* \circ \rho$. Now

$$\tilde{\Phi}_* = \tilde{\Phi}_* \circ j \circ \omega + \tilde{\Phi}_* \circ \gamma \circ \rho = \tilde{\Phi}_* \circ \omega + \gamma \circ \phi_* \circ \rho$$

from the symmetry condition, whereas it is always the case that

$$\tilde{\Phi}_* = j \circ \omega \circ \tilde{\Phi}_* + \gamma \circ \rho \circ \tilde{\Phi}_* = \omega \circ \tilde{\Phi}_* + \gamma \circ \phi_* \circ \rho,$$

from which the result follows. □

5.3 Infinitesimal Symmetries of Infinitesimal Connections

Our main concern will be with infinitesimal symmetries of the infinitesimal connection γ, so that we can use the Lie derivative construction from Chap. 1. If $\xi : M \to A\mathcal{G}$ is a section of the Lie algebroid, we would like to say that it was an infinitesimal symmetry of γ if $\mathcal{L}_\xi \gamma = 0$. One might imagine that this property would hold if the flow of the corresponding right-invariant vector field ξ^{R} on $\mathcal{G}$ consisted of symmetries of γ in the sense of Lemma 5.2.2, but this cannot be the case; we therefore have to use a slightly more involved construction.

Lemma 5.3.1 *Let $\xi : M \to A\mathcal{G}$ be a section of the Lie algebroid, with corresponding right-invariant vector field ξ^{R} on $\mathcal{G}$, and let Ψ_t be the flow of ξ^{R} with domain $V_t \subset \mathcal{G}$. Define $U_t \subset M$ by*

$$U_t = \{x \in M : 1_x \in V_t\},$$

define the maps $\Phi_t : U_t \to \mathcal{G}$ by $\Phi_t(x) = \Psi_t(1_x)$, and put $\phi_t(x) = \beta\Phi_t(x)$. Then each map Φ_t is a local bisection of $\mathcal{G}$, and the maps ϕ_t form the flow of the vector field $X = \rho \circ \xi$ on M with domain U_t. In addition $\Phi_t^\iota = \Phi_{-t} \circ \phi_t$.

Proof Note first that U_t is nonempty for t sufficiently small, so that the maps $\Phi_t : U_s \to \mathcal{G}$ may be defined. By right invariance, $\Psi_t(\varphi) = \Psi_t(1_{\beta(\varphi)}) \cdot \varphi$ for any $\varphi \in V_t$ satisfying $1_{\beta(\varphi)} \in V_t$. Thus

$$\alpha\Phi_t(x) = \alpha\Psi_t(1_x) = \alpha\Psi_t(1_{\beta(1_x)}) \cdot 1_x = \alpha(1_x) = x$$

so that Φ_t is a local section of α.

We also see that

$$\Psi_{s+t}(1_x) = \Psi_s(\Psi_t(1_x)) = \Psi_s(1_{\beta(\Psi_t(1_x))}) \cdot \Psi_t(1_x)$$

whenever both sides are defined. Putting $\phi_t(x) = \beta\Phi_(x)$, we then see that

$$\Phi_{s+t}(x) = \Phi_s\beta\Phi_t(x) \cdot \Phi_t(x) = \Phi_s\phi_t(x) \cdot \Phi_t(x).$$

Taking $s = -t$ then gives $1_x = \Phi_{-t}\phi_t(x) \cdot \Phi_t(x)$, so that $\Phi_t^\iota = \Phi_{-t} \circ \phi_t$. On the other hand, applying β gives

$$\phi_{s+t}(x) = \beta\Phi_{s+t}(x) = \beta\Phi_s\phi_t(x) = \phi_s\phi_t(x)$$

so that the maps ϕ_t form a pseudogroup; indeed

$$\left.\frac{d\phi_t(x)}{dt}\right|_{t=0} = \left.\frac{d(\beta(\Psi_t(1_x)))}{dt}\right|_{t=0} = \beta_*(\xi^{\mathsf{R}}_{1_x}) = \rho\xi(x)$$

so that ϕ_t is the flow of the vector field $X = \rho \circ \xi$ on M. In particular, therefore, each ϕ_t is a local diffeomorphism, so that each Φ_t is a bisection. □

We should emphasize that taking the bisections Φ_t and forming the corresponding (local) inner automorphisms $\tilde{\Phi}_t$ does *not* give the flow Ψ_t back again: these are not inverse operations. The flow of a right-invariant vector field on a Lie groupoid does not consist of groupoid automorphisms, just as it would not on a Lie group. Instead, the correspondence $\Psi_t \mapsto \Phi_t$ is like obtaining a group element from a right translation. This is the reason we have to use the more involved construction, obtaining the automorphisms from the flow using the bisections as intermediaries.

With this construction we are able to see the relationship between symmetries and infinitesimal symmetries of an infinitesimal connection. We shall say that a section $\xi : M \to A\mathcal{G}$ satisfying the condition $\mathcal{L}_\xi \gamma = 0$ is an *infinitesimal symmetry* of γ.

Proposition 5.3.2 *Let $\xi : M \to A\mathcal{G}$ be a section of the Lie algebroid. If the local bisections Φ_s constructed from the flow Ψ_s of ξ^{R} have the additional property that the symmetry equation for the path connection Γ holds wherever it makes sense, then ξ is an infinitesimal symmetry of γ.*

Proof In this proof we use pairs $(s, t) \in \mathbb{R}^2$ where s is the flow parameter for Ψ (or Φ) and t is the parameter of a curve c.

The symmetry condition on the local bisections Φ_s is that

$$\tilde{\Phi}_s c^\Gamma(t) = (\phi_s \circ c)^\Gamma(t)$$

for any curve $c : [0, 1] \to M$ whenever both sides are defined. As the images of c and c^Γ are compact, there will be some $\varepsilon > 0$ such that this equation holds for $(s, t) \in (-\varepsilon, \varepsilon) \times [0, 1]$.

Let Y be an arbitrary vector field on M; then for any $x \in M$ there is a nonempty interval $(-\varepsilon, \varepsilon)$ such that

$$\tilde{\Phi}_{-s*}\gamma(Y_{\phi_s(x)}) = \gamma\phi_{-s*}(Y_{\phi_s(x)}) \tag{1}$$

for $s \in (-\varepsilon, \varepsilon)$. As ϕ_s is the flow of $X = \rho \circ \xi$, it follows that

$$\left.\frac{d}{ds}\phi_{-s*}\big(Y_{\phi_s(x)}\big)\right|_{s=0} = (\mathcal{L}_X Y)_x = [X, Y]_x = [\rho \circ \xi, Y]_x$$

so that, as γ is linear and continuous on $T_x M$,

$$\left.\frac{d}{ds}\gamma\phi_{-s*}\big(Y_{\phi_s(x)}\big)\right|_{s=0} = \gamma\big([\rho \circ \xi, Y]_x\big) = \big(\gamma \circ [\rho \circ \xi, Y]\big)\big|_x .$$

If we can show that

$$\left.\frac{d}{ds}\tilde{\Phi}_{-s*}\gamma(Y_{\phi_s(x)})\right|_{s=0} = [\![\xi, \gamma \circ Y]\!]_x$$

it will follow from equation (1) above that the Lie algebroid section ξ satisfies the infinitesimal symmetry equation

$$\gamma \circ [\rho \circ \xi, Y] = [\![\xi, \gamma \circ Y]\!]$$

for any vector field Y on M, in other words that $\mathcal{L}_\xi \gamma \circ Y = 0$ for any Y, and hence that $\mathcal{L}_\xi \gamma = 0$.

Let χ_t be the flow of Y on M. Given $x \in M$ and putting $\chi_x(t) = \chi_t(x)$,

$$\begin{aligned}
\tilde{\Phi}_{-s*}\big(\gamma(Y_{\phi_s(x)})\big) &= \tilde{\Phi}_{-s*}\gamma\left(\left.\frac{d}{dt}\chi_{\phi_s(x)}(t)\right|_{t=0}\right) \\
&= \tilde{\Phi}_{-s*}\left.\frac{d}{dt}\chi^{\Gamma}_{\phi_s(x)}(t)\right|_{t=0} \\
&= \left.\frac{d}{dt}\tilde{\Phi}_{-s}\chi^{\Gamma}_{\phi_s(x)}(t)\right|_{t=0} \\
&= \left.\frac{d}{dt}\big(\Phi_{-s}(y) \cdot \chi^{\Gamma}_{\phi_s(x)}(t) \cdot \Phi^{\iota}_{-s}\phi_s(x)\big)\right|_{t=0}
\end{aligned}$$

where $y = \beta(\chi^{\Gamma}_{\phi_s(x)}(t))$ and $\phi_s(x) = \alpha(\chi^{\Gamma}_{\phi_s(x)}(t))$

$$\begin{aligned}
&= \left.\frac{d}{dt}\big(\Phi_{-s}(y) \cdot \chi^{\Gamma}_{\phi_s(x)}(t) \cdot \Phi_s(x)\big)\right|_{t=0} \\
&= \left.\frac{d}{dt}\big(\Psi_{-s}\big(\chi^{\Gamma}_{\phi_s(x)}(t) \cdot \Psi_s(1_x)\big)\big)\right|_{t=0} \\
&= \Psi_{-s*}\left(\left.\frac{d}{dt}\big(\chi^{\Gamma}_{\phi_s(x)}(t) \cdot \Psi_s(1_x)\big)\right|_{t=0}\right) \\
&= \Psi_{-s*}\Big((\gamma \circ Y)^{\mathsf{R}}_{\Psi_s(1_x)}\Big)
\end{aligned}$$

so that

$$\begin{aligned}
\left.\frac{d}{ds}\tilde{\Phi}_{-s*}\big(\gamma(Y_{\phi_s(x)})\big)\right|_{s=0} &= \left.\frac{d}{ds}\Psi_{-s*}\Big((\gamma \circ Y)^{\mathsf{R}}_{\Psi_s(1_x)}\Big)\right|_{s=0} \\
&= \big(\mathcal{L}_{\bar{\xi}}(\gamma \circ Y)^{\mathsf{R}}\big)_{1_x} \\
&= \big[\bar{\xi}, (\gamma \circ Y)^{\mathsf{R}}\big]_{1_x} \\
&= [\![\xi, \gamma \circ Y]\!]_x
\end{aligned}$$

as required. □

We may also see how symmetries and infinitesimal symmetries of γ may be expressed in terms of the kernel derivative ∇^{γ}.

Proposition 5.3.3 *If Φ is a symmetry satisfying $\tilde{\Phi}_* \circ \gamma = \gamma \circ \phi_*$ then*

$$\tilde{\Phi}_* \circ (\nabla^\gamma_Y \kappa) = \nabla^\gamma_{\phi_* \circ Y}(\tilde{\Phi}_* \circ \kappa)$$

for any vector field Y and any section κ of $K\mathcal{G}$.

Proof We expand the right-hand side to obtain

$$\begin{aligned}\nabla_{\phi_* \circ Y}(\tilde{\Phi}_* \circ \kappa) &= [\![\gamma \circ \phi_* \circ Y, \tilde{\Phi}_* \circ \kappa]\!] = [\![\tilde{\Phi}_* \circ \gamma \circ Y, \tilde{\Phi}_* \circ \kappa]\!] \\ &= \tilde{\Phi}_* \circ [\![\gamma \circ Y, \kappa]\!] = \tilde{\Phi}_* \circ (\nabla^\gamma_Y \kappa).\end{aligned}$$

This suggests that the version of this formula for an infinitesimal symmetry ξ, rather than a finite symmetry Φ, should be

$$[\![\xi, \nabla^\gamma_Y \kappa]\!] = \nabla^\gamma_Y\big([\![\xi, \kappa]\!]\big) + \nabla^\gamma_{[\rho \circ \xi, Y]}\kappa.$$

This is indeed the case, as we shall shortly show. But we first observe that, though until now we have dealt entirely with Lie algebroids which are derived from Lie groupoids, the components of this formula do not require us to be working in such a restricted context. In fact the formula holds for an infinitesimal connection on any transitive Lie algebroid, provided we have a definition of infinitesimal symmetry to hand. But it is obvious how to frame such a definition: we can take over the definition given earlier with no essential change. So for completeness we state the slightly more general definition as follows. Let $\mathcal{A}$ be a transitive Lie algebroid equipped with an infinitesimal connection γ. An *infinitesimal symmetry* of γ is a section $\xi : M \to \mathcal{A}$ such that $\mathcal{L}_\xi \gamma = 0$.

Proposition 5.3.4 *If ξ is an infinitesimal symmetry of γ and Y is any vector field on M then, as operators on sections of $\mathcal{K}$,*

$$\mathcal{L}_\xi \circ \nabla^\gamma_Y - \nabla^\gamma_Y \circ \mathcal{L}_\xi = \nabla^\gamma_{[\rho \circ \xi, Y]}.$$

Proof For any sections ξ of $\mathcal{A}$ and κ of $\mathcal{K}$ and for any vector field Y on M

$$\begin{aligned}&\mathcal{L}_\xi(\nabla^\gamma_Y \kappa) - \nabla^\gamma_Y(\mathcal{L}_\xi \kappa) - \nabla^\gamma_{[\rho \circ \xi, Y]}\kappa \\ &\quad = [\![\xi, \nabla^\gamma_Y \kappa]\!] - \nabla^\gamma_Y\big([\![\xi, \kappa]\!]\big) + \nabla^\gamma_{[\rho \circ \xi, Y]}\kappa \\ &\quad = [\![\xi, [\![\gamma \circ Y, \kappa]\!]]\!] - [\![\gamma \circ Y, [\![\xi, \kappa]\!]]\!] - [\![\gamma \circ [\rho \circ \xi, Y], \kappa]\!] \\ &\quad = [\![[\![\xi, \gamma \circ Y]\!], \kappa]\!] - [\![\gamma \circ [\rho \circ \xi, Y], \kappa]\!] = [\![\mathcal{L}_\xi \gamma \circ Y, \kappa]\!].\end{aligned}$$

Thus if $\mathcal{L}_\xi \gamma = 0$ then

$$\mathcal{L}_\xi \nabla^\gamma_Y \kappa - \nabla^\gamma_Y \mathcal{L}_\xi \kappa = \nabla^\gamma_{[\rho \circ \xi, Y]}\kappa.$$

□

For what it's worth, it is clear that under certain circumstances the result has a converse. Recall that $\mathcal{L}_\xi\gamma \circ Y$ is a section of $\mathcal{K}$. Thus if the Lie algebroid $\mathcal{A}$ has the property that the centre of its kernel $\mathcal{K}$ is trivial—that is to say that if $[\![\lambda, \kappa]\!] = 0$ for all κ implies that λ is the zero section of $\mathcal{K}$—then a section ξ of $\mathcal{A}$ such that $\mathcal{L}_\xi \circ \nabla^\gamma_Y - \nabla^\gamma_Y \circ \mathcal{L}_\xi = \nabla^\gamma_{[\rho\circ\xi, Y]}$ for all Y is an infinitesimal symmetry of γ.

The reason this result is interesting is that it is formally similar to a standard result about infinitesimal symmetries of linear connections in ordinary connection theory: indeed the formula displayed in the proposition may be used, mutatis mutandis, as the definition of such an infinitesimal symmetry. We have written $[\![\xi, \kappa]\!]$ as $\mathcal{L}_\xi\kappa$, and so on, to make the parallel more obvious. (Infinitesimal symmetries of linear connections are often called infinitesimal affine transformations: it is necessary to exercise caution in the use of this term in the present context, for reasons that will become apparent when we take up the topic again shortly.)

5.4 Properties of Infinitesimal Symmetries

We now investigate some properties of infinitesimal symmetries of an infinitesimal connection $\gamma : TM \to \mathcal{A}$ on a general transitive Lie algebroid $\mathcal{A}$, again not necessarily arising from a Lie groupoid.

Proposition 5.4.1 *The set of all infinitesimal symmetries of γ is a Lie algebra under the Lie algebroid bracket.*

Proof Certainly sums and constant multiples of infinitesimal symmetries are again infinitesimal symmetries; and if $\mathcal{L}_\xi\gamma = \mathcal{L}_\eta\gamma = 0$ then $\mathcal{L}_{[\![\xi,\eta]\!]}\gamma = 0$ as a consequence of Proposition 1.6.2. □

We denote by $\mathfrak{a}$ the Lie algebra of infinitesimal symmetries. It induces a Lie algebra of vector fields on M via the anchor, namely $\{\rho \circ \xi : \xi \in \mathfrak{a}\}$.

We may also express the condition for an infinitesimal symmetry in terms of the kernel projection ω of γ. As in Sect. 1.6 we write $\mathcal{L}_\xi\omega$ for the vector bundle morphism $\mathcal{A} \to \mathcal{K}$ given by

$$\mathcal{L}_\xi\omega \circ \eta = [\![\xi, \omega \circ \eta]\!] - \omega \circ [\![\xi, \eta]\!]$$

for any sections ξ, η of $\mathcal{A} \to M$.

Proposition 5.4.2 *The section ξ is an infinitesimal symmetry of γ if, and only if, $\mathcal{L}_\xi\omega = 0$.*

Proof Let η be a section of $\mathcal{A} \to M$; then $\omega \circ \eta = \eta - \gamma \circ \rho \circ \eta$, and so if ξ is an infinitesimal symmetry then

$$[\![\xi, \omega \circ \eta]\!] = [\![\xi, \eta]\!] - [\![\xi, \gamma \circ \rho \circ \eta]\!] = [\![\xi, \eta]\!] - \gamma \circ [\rho \circ \xi, \rho \circ \eta] = \omega \circ [\![\xi, \eta]\!],$$

so that $\mathcal{L}_\xi \omega = 0$. Conversely, suppose that $\mathcal{L}_\xi \omega = 0$, and for any vector field Y on M let η be a section of $\mathcal{A}$ such that $\rho \circ \eta = Y$. Then $\gamma \circ Y = \eta - \omega \circ \eta$, and so

$$[\![\xi, \gamma \circ Y]\!] = [\![\xi, \eta]\!] - [\![\xi, \omega \circ \eta]\!] = [\![\xi, \eta]\!] - \omega \circ [\![\xi, \eta]\!] = \gamma \circ \rho \circ [\![\xi, \eta]\!] = \gamma \circ [\rho \circ \xi, Y]. \quad \square$$

The following further alternative way of specifying when a section ξ of $\mathcal{A} \to M$ is an infinitesimal symmetry of γ can be useful in calculations. It depends on the use of γ and ω to decompose ξ into two summands.

Proposition 5.4.3 *Let ξ be a section of $M \to \mathcal{A}$, and set $X = \rho \circ \xi$ (a vector field on M) and $\kappa = \omega \circ \xi$ (a section of $\mathcal{K} \to M$). Then ξ is an infinitesimal symmetry of γ if, and only if,*

$$\nabla^\gamma \kappa + R^\gamma(X, \cdot) = 0.$$

Proof We have

$$\xi = \gamma \circ (\rho \circ \xi) + \omega \circ \xi = \gamma \circ X + \kappa.$$

For any vector field Y on M

$$\begin{aligned} \mathcal{L}_\xi \gamma \circ Y &= [\![\xi, \gamma \circ Y]\!] - \gamma \circ [\rho \circ \xi, Y] \\ &= [\![\gamma \circ X, \gamma \circ Y]\!] - \gamma \circ [X, Y] + [\![\kappa, \gamma \circ Y]\!] = -R^\gamma(X, Y) - \nabla^\gamma_Y \kappa, \end{aligned}$$

where R^γ is the curvature of γ, from which the result follows. $\square$

Proposition 5.4.4 *If ξ is an infinitesimal symmetry of an infinitesimal connection γ then $\mathcal{L}_\xi R^\gamma = 0$.*

Proof We know that, that for any section ξ of $\mathcal{A}$ and vector fields X and Y on M,

$$\mathcal{L}_\xi R^\gamma(X, Y) = \mathcal{L}_\xi \gamma \circ [X, Y] - [\![\mathcal{L}_\xi \gamma \circ X, \gamma \circ Y]\!] - [\![\gamma \circ X, \mathcal{L}_\xi \gamma \circ Y]\!],$$

(Corollary 2.6.6), from which the result follows directly. $\square$

Proposition 5.4.5 *If ξ and η are infinitesimal symmetries then*

$$[\![\omega \circ \xi, \omega \circ \eta]\!] - \omega \circ [\![\xi, \eta]\!] = -R^\gamma(\rho \circ \xi, \rho \circ \eta).$$

Proof From Lemma 2.2.5 we have

$$[\![\omega \circ \xi, \omega \circ \eta]\!] - \omega \circ [\![\xi, \eta]\!] = -\mathcal{L}_\xi \gamma \circ \rho \circ \eta + \mathcal{L}_\eta \gamma \circ \rho \circ \xi - R^\gamma(\rho \circ \xi, \rho \circ \eta)$$

for any sections ξ, η. $\square$

We therefore see that the kernel projection ω does not preserve the bracket of infinitesimal symmetries if the curvature R^γ is nonzero. Nevertheless, infinitesimal symmetries ξ for which $R^\gamma(\rho\circ\xi,\cdot)=0$ are of interest; we shall call such infinitesimal symmetries *flat*. Clearly any infinitesimal symmetry ξ for which $\rho\circ\xi=0$ is flat.

Proposition 5.4.6 *If ξ is a flat infinitesimal symmetry so are both $\omega\circ\xi$ and $\gamma\circ(\rho\circ\xi)$. If ξ and η are infinitesimal symmetries and η is flat then $[\![\xi,\eta]\!]$ is a flat infinitesimal symmetry. If ξ and η are both flat infinitesimal symmetries then $[\![\omega\circ\xi,\omega\circ\eta]\!]=\omega\circ[\![\xi,\eta]\!]$.*

Proof We have, for any ξ,

$$\mathcal{L}_{\omega\circ\xi}\gamma\circ X=[\![\omega\circ\xi,\gamma\circ X]\!]=[\![\xi,\gamma\circ X]\!]-[\![\gamma\circ(\rho\circ\xi),\gamma\circ X]\!]=\mathcal{L}_\xi\gamma\circ X+R^\gamma(\rho\circ\xi,X).$$

Thus if ξ is a flat infinitesimal symmetry so is $\omega\circ\xi$. Moreover, $\xi-\omega\circ\xi=\gamma\circ(\rho\circ\xi)$ is a symmetry, and since $\rho\circ\gamma\circ(\rho\circ\xi)=\rho\circ\xi$ it is flat.

If ξ and η are infinitesimal symmetries so is $[\![\xi,\eta]\!]$. Now for any ξ,η,

$$\begin{aligned}&\mathcal{L}_\xi R^\gamma(\rho\circ\eta,X)\\&\quad=[\![\xi,R^\gamma(\rho\circ\eta,X)]\!]-R^\gamma([\rho\circ\xi,\rho\circ\eta],X)-R^\gamma(\rho\circ\eta,[\rho\circ\xi,X])\\&\quad=[\![\xi,R^\gamma(\rho\circ\eta,X])]\!]-R^\gamma(\rho\circ[\![\xi,\eta]\!],X)-R^\gamma(\rho\circ\eta,[\rho\circ\xi,X]).\end{aligned}$$

If ξ is an infinitesimal symmetry then $\mathcal{L}_\xi R^\gamma=0$, and if in addition $R^\gamma(\rho\circ\eta,\cdot)=0$ then $R^\gamma(\rho\circ[\![\xi,\eta]\!],\cdot)=0$. Thus if ξ and η are infinitesimal symmetries with η flat then $[\![\xi,\eta]\!]$ is a flat infinitesimal symmetry.

The third assertion is clear from the formula for $[\![\omega\circ\xi,\omega\circ\eta]\!]-\omega\circ[\![\xi,\eta]\!]$. □

Corollary 5.4.7 *The set of flat infinitesimal symmetries of an infinitesimal connection γ is an ideal $\mathfrak{f}$ in the Lie algebra $\mathfrak{a}$ of all infinitesimal symmetries of γ. The kernel projection ω induces a homomorphism of $\mathfrak{f}$ into the Lie algebra of sections of $\mathcal{K}$, the kernel of $\mathcal{A}$; and ρ induces a homomorphism of $\mathfrak{f}$ into the Lie algebra of vector fields on M. Furthermore, $\omega|_\mathfrak{f}$ acts as the identity on its image, which coincides with the kernel of $\rho|_\mathfrak{f}$.* □

Proposition 5.4.8 *Let ξ be a section of $\mathcal{A}\to M$ such that $R^\gamma(\rho\circ\xi,\cdot)=0$. Then ξ is an infinitesimal symmetry of γ (necessarily flat) if, and only if, $\nabla^\gamma(\omega\circ\xi)=0$.*

Proof This follows from Proposition 5.4.3. □

We shall need the following results later.

Lemma 5.4.9 *For any section ξ of $\mathcal{A}$ and any $f\in C^\infty(M)$,*

$$\mathcal{L}_{f\xi}\gamma\circ X=f(\mathcal{L}_\xi\gamma\circ X)-X(f)\omega\circ\xi.$$

Proof

$$
\begin{aligned}
\mathcal{L}_{f\xi}\gamma \circ X &= [\![f\xi, \gamma \circ X]\!] - \gamma \circ [\rho \circ (f\xi), X] \\
&= f[\![\xi, \gamma \circ X]\!] - X(f)\xi - f(\gamma \circ [\rho \circ \xi, X]) + X(f)\gamma \circ (\rho \circ \xi) \\
&= f(\mathcal{L}_\xi \gamma \circ X) - X(f)\omega \circ \xi.
\end{aligned}
$$

□

Lemma 5.4.10 *If ξ is a flat symmetry of γ and $f \in C^\infty(M)$ then $\mathcal{L}_{f\xi} R^\gamma = 0$.*

Proof We have, in general,

$$
\begin{aligned}
\mathcal{L}_{f\xi} R^\gamma(X, Y) &= [\![f\xi, R^\gamma(X, Y)]\!] - R^\gamma([\rho \circ (f\xi), X], Y) - R^\gamma(X, [\rho \circ (f\xi), Y]) \\
&= f\mathcal{L}_\xi R^\gamma(X, Y) + X(f)R^\gamma(\rho \circ \xi, Y) + Y(f)R^\gamma(X, \rho \circ \xi),
\end{aligned}
$$

from which the result follows. □

We shall also consider infinitesimal symmetries which are projectable in the sense of Sects. 1.4 and 2.3. Suppose that $\mathcal{A} \to M$ and $\mathcal{B} \to N$ are transitive Lie algebroids, not necessarily derived from Lie groupoids, and suppose that (P, p) is a full Lie algebroid morphism from $\mathcal{A}$ to $\mathcal{B}$. Let γ be a projectable infinitesimal connection on $\mathcal{A}$ with projection γ^P on $\mathcal{B}$.

Lemma 5.4.11 *If the projectable section ξ of A is an infinitesimal symmetry of γ then its projection ξ^P is an infinitesimal symmetry of γ^P.*

Proof By definition $\mathcal{L}_\xi \gamma \circ X = [\![\xi, \gamma \circ X]\!] - \gamma \circ [\rho_A \circ \xi, X]$ for any vector field X on $\mathcal{M}$. We therefore see that $\mathcal{L}_\xi \gamma = 0$ implies

$$
\begin{aligned}
[\![\xi^P, \gamma^P \circ X^P]\!] &= [\![\xi^P, (\gamma \circ X)^P]\!] = [\![\xi, \gamma \circ X]\!]^P = (\gamma \circ [\rho_A \circ \xi, X])^P \\
&= \gamma^P \circ [\rho_A \circ \xi, X]^P = \gamma^P \circ [(\rho_A \circ \xi)^P, X^P] = \gamma^P \circ [\rho_B \circ \xi^P, X^P]
\end{aligned}
$$

so that $\mathcal{L}_{\xi^P} \gamma^P = 0$. □

5.5 Case Study: Affine Transformations

We now apply this theory to the case of 'affine transformations' of a manifold M with $\dim M = n$: these will be symmetries of a path connection on the Lie groupoid $\mathcal{G}$ of affine morphisms of TM introduced in Sect. 4.2, or infinitesimal symmetries of an infinitesimal connection on the Lie algebroid $\mathcal{A}^G = \Lambda(A\mathcal{G})$ of fibre-affine projectable vector fields introduced in the same section.

A word about nomenclature is needed here. We have defined an affine *morphism* as an affine map between two fibres of a vector bundle $E \to M$, and so this is independent of any connection on the bundle. A classical affine *transformation*, on the other hand, is a diffeomorphism of M which is, in addition, a symmetry of a given *linear* connection on M (that is to say, on the particular vector bundle TM); in the infinitesimal case a classical affine transformation is a vector field on M which is an infinitesimal symmetry of the linear connection. In this section we shall investigate the infinitesimal symmetries of an infinitesimal connection on the Lie algebroid $\mathcal{A}^G$ of fibre-affine projectable vector fields along fibres of TM, and examine the relationship between our theory and the classical version.

In preparation for the latter discussion we first provide a specific characterisation of bisections, those maps $\Phi : M \to \mathcal{G}$ that will be candidates to be symmetries of path connections. For each $x \in M$ the image $\Phi(x)$ is an affine map $T_x M \to T_{\phi(x)} M$ where $\phi = \beta \circ \Phi$. But there is also a linear map $\phi_* : T_x M \to T_{\phi(x)} M$, so that we may construct the affine map $\phi_*^{-1} \circ \Phi(x) : T_x M \to T_x M$. In this way the bisection Φ determines, and is determined by, the diffeomorphism $\phi : M \to M$, a vector field $X \in \mathfrak{X}(M)$ and a type $(1, 1)$ tensor field $\Theta \in \Omega(M) \otimes \mathfrak{X}(M)$ such that, for each $v \in T_x M$,

$$\phi_*^{-1}\bigl(\Phi(x)(v)\bigr) = X_x + \Theta_x(v).$$

We shall find a condition for the tensor field Θ to be the identity tensor, so that the linear part of the affine map $\Phi(x)$ will necessarily be the tangent map ϕ_{*x}. This condition involves the *vertical endomorphism* on the tangent bundle $\tau_M : TM \to M$: this is a type $(1, 1)$ tensor field Σ on TM defined as the composite $TTM \to \tau_M^* TM \to TTM$, where the first map is $w \mapsto \bigl(\tau_{TM}(w), \tau_{M*}(w)\bigr)$ and the second is the vertical lift mapping $\tau_{M*}(w)$ to $\bigl(\tau_{M*}(w)\bigr)^{\mathsf{V}} \in T_{\tau_{TM}(w)} TM$. In coordinates (x^i, y^i) on TM we have

$$\Sigma = \frac{\partial}{\partial y^i} \otimes dx^i .$$

Proposition 5.5.1 *Let Φ be a bisection with $\phi = \beta \circ \Phi$ such that $\phi_*^{-1}\bigl(\Phi(x)(v)\bigr) = X_x + \Theta_x(v)$ for all $x \in M$ and all $v \in T_x M$. Then $\Theta = I$, so that the tangent map ϕ_{*x} is the linear map associated to the affine map $\Phi(x)$, precisely when $\Sigma \circ \tilde{\Phi}_* = \tilde{\Phi}_* \circ \Sigma$, where $\Sigma : \mathcal{A}^G \to \mathcal{A}^G$ is the vertical endomorphism on the tangent bundle and we use the identification $\mathcal{A}^G \cong A\mathcal{G} \subset T\mathcal{G}$.*

Proof We use the coordinates on $A\mathcal{G}$ and $\mathcal{A}^G$ described in Sect. 4.2. Fix $x \in M$ and put

$$\Phi(\hat{x})(v)^i = C^i_j(\hat{x}) v^j + C^i(\hat{x})$$

for $\hat{x}$ sufficiently close to x, and for $v \in T_{\hat{x}} M$. For any $\varphi \in \mathcal{G}$ with $\alpha(\varphi) = x$ and with $\beta(\varphi)$ sufficiently close to x we will have $\varphi(v)^i = P^i_j v^j + P^i$ where $P^i_j = z^i_j(\varphi)$ and $P^i = z^i_0(\varphi)$; if $w \in T_{\phi(x)} M$ we will therefore also have

$$\begin{aligned}\tilde{\Phi}(\varphi)(w)^i &= \left(\Phi_{\beta(\varphi)}\varphi\Phi^{\iota}_x(w)\right)^i \\ &= C^i_j(\beta(\varphi))\Big(P^j_k\bar{C}^k_l(x)\big(w^l - C^l(x)\big) + P^j\Big) + C^i(\beta(\varphi))\end{aligned}$$

where $\bar{C}^k_l C^j_k = \delta^j_l$, so that

$$\begin{aligned}x^i(\tilde{\Phi}(\varphi)) &= x^i(\phi(\beta(\varphi))) \\ z^i_0(\tilde{\Phi}(\varphi)) &= -C^i_j(\beta(\varphi))z^j_k(\varphi)\bar{C}^k_l(x)C^l(x) + C^i_j(\beta(\varphi))z^j_0(\varphi) + C^i(\beta(\varphi)) \\ z^i_l(\tilde{\Phi}(\varphi)) &= C^i_j(\beta(\varphi))z^j_k(\varphi)\bar{C}^k_l(x).\end{aligned} \tag{2}$$

Now take $\boldsymbol{a} \in A_x\mathcal{G}$. As in the proof of Theorem 3.2.1, let ξ be a section of $A\mathcal{G} \to M$ satisfying $\xi_x = \boldsymbol{a}$. Let ξ^{R} be the corresponding right-invariant vector field on $\mathcal{G}$, and let Ψ_t be the flow of ξ^{R} in a neighbourhood of 1_x, so that $s \mapsto \Psi_t(1_x)$ is a curve in $\mathcal{G}$ and $d\Psi_t(1_x)/dt|_{t=0} = \boldsymbol{a}$; then from equation (2) we obtain

$$\begin{aligned}\dot{X}^i(\tilde{\Phi}_*(\boldsymbol{a})) &= \left.\frac{\partial\phi^i}{\partial x^j}\right|_x \dot{X}^j(\boldsymbol{a}) \\ \dot{Z}^i_0(\tilde{\Phi}_*(\boldsymbol{a})) &= \left.\frac{d}{dt}\right|_{t=0} z^i_0(\tilde{\Phi}\Psi_t(1_x)) \\ &= \left.\frac{d}{dt}\right|_{t=0} \Big(-C^i_j(\beta\Psi_t(1_x))z^j_k(\Psi_t(1_x))\bar{C}^k_l(x)C^l(x) \\ &\qquad + C^i_j(\beta\Psi_t(1_x))z^j_0(\Psi_t(1_x)) + C^i(\beta\Psi_t(1_x))\Big) \\ &= -\left.\frac{\partial C^i_j}{\partial x^k}\right|_x \dot{X}^k(\boldsymbol{a})\bar{C}^j_l(x)C^l(x) - C^i_j(x)\dot{Z}^j_k(\boldsymbol{a})\bar{C}^k_l(x)C^l(x) \\ &\qquad + C^i_j(x)\dot{Z}^j_0(\boldsymbol{a}) + \left.\frac{\partial C^i}{\partial x^k}\right|_x \dot{X}^k(\boldsymbol{a}) \\ \dot{Z}^i_l(\tilde{\Phi}_*(\boldsymbol{a})) &= \left.\frac{d}{dt}\right|_{t=0} z^i_l(\tilde{\Phi}\Psi_t(1_x)) \\ &= \left.\frac{d}{ts}\right|_{t=0} \Big(C^i_j(\beta\Psi_t(1_x))z^j_k(\Psi_t(1_x))\bar{C}^k_l(x)\Big) \\ &= \left.\frac{\partial C^i_j}{\partial x^k}\right|_x \dot{X}^k(\boldsymbol{a})\bar{C}^j_l(x) + C^i_j(x)\dot{Z}^j_k(\boldsymbol{a})\bar{C}^k_l(x)\end{aligned}$$

using $z^j_k(c(0)) = z^j_k(1_x) = \delta^j_k$ and $z^j_0(c(0)) = z^j_0(1_x) = 0$. Identifying $A\mathcal{G}$ with $A^{\mathcal{G}}$, so that an element of $A\mathcal{G}$ is regarded as a vector field along a fibre of TM, if

$$\boldsymbol{a} = \boldsymbol{a}^i \left.\frac{\partial}{\partial x^i}\right|_{T_xM} + (\boldsymbol{a}^i_0 + \boldsymbol{a}^i_j y^j) \left.\frac{\partial}{\partial y^i}\right|_{T_xM}$$

(so that $\boldsymbol{a}^i = \dot{X}^i(\boldsymbol{a})$, $\boldsymbol{a}^i_0 = \dot{Z}^j_0(\boldsymbol{a})$ and $\boldsymbol{a}^i_j = \dot{Z}^i_j(\boldsymbol{a})$) then

$$\tilde{\Phi}_*(\boldsymbol{a}) = \boldsymbol{b}^i \left.\frac{\partial}{\partial x^i}\right|_{T_{\phi(x)}M} + (\boldsymbol{b}^i_0 + \boldsymbol{b}^i_j y^j) \left.\frac{\partial}{\partial y^i}\right|_{T_{\phi(x)}M}$$

where

$$\boldsymbol{b}^i = \left.\frac{\partial \phi^i}{\partial x^j}\right|_x \boldsymbol{a}^j$$

$$\boldsymbol{b}^i_0 = \left(C^l(x)\bar{C}^j_l(x) \left.\frac{\partial C^i_j}{\partial x^k}\right|_x + \left.\frac{\partial C^i}{\partial x^k}\right|_x \right)\boldsymbol{a}^k + C^i_j(x)\boldsymbol{a}^j_k\bar{C}^k_l(x)C^l(x) + C^i_j(x)\boldsymbol{a}^j_0$$

$$\boldsymbol{b}^i_l = \left.\frac{\partial C^i_j}{\partial x^k}\right|_x \boldsymbol{a}^k\bar{C}^j_l(x) + C^i_j(x)\boldsymbol{a}^j_k\bar{C}^k_l(x).$$

We therefore see that

$$\Sigma(\tilde{\Phi}_*(\boldsymbol{a})) = \boldsymbol{a}^j \left.\frac{\partial \phi^i}{\partial x^j}\right|_x \left.\frac{\partial}{\partial y^i}\right|_{T_xM}, \qquad \tilde{\Phi}_*(\Sigma(\boldsymbol{a})) = \boldsymbol{a}^j C^i_j(x) \left.\frac{\partial}{\partial y^i}\right|_{T_xM}$$

so that $\Sigma \circ \tilde{\Phi}_* = \tilde{\Phi}_* \circ \Sigma$ exactly when $C^i_j = \partial\phi^i/\partial x^j$, in other words when, for each x, the linear map associated to the affine map $\Phi(x)$ is ϕ_{x*}. □

We may interpret this result as saying that a bisection 'respects the vertical lift' when it arises purely from a prolonged diffeomorphism ϕ_* and a 'translation' given by a vector field on M, with tensor field $\Theta = I$. The infinitesimal version, that a section of the Lie algebroid $\mathcal{A}^G$ respects the vertical lift precisely when it is itself the sum of a complete lift and a vertical lift, may be demonstrated as follows.

Proposition 5.5.2 *A section ξ of $\mathcal{A}^G$ satisfies the condition $\mathcal{L}_\xi\Sigma = 0$ if, and only if, it takes the form $\xi = Z_1^{\mathrm{C}} + Z_2^{\mathrm{V}}$ for vector fields Z_1 and Z_2 on M.*

Proof With

$$\xi = X^i\frac{\partial}{\partial x^i} + (Y^i + Y^i_j y^j)\frac{\partial}{\partial y^i}$$

we note that

$$\mathcal{L}_\xi\frac{\partial}{\partial y^k} = -Y^i_k\frac{\partial}{\partial y^i}, \qquad \mathcal{L}_\xi dx^k = \frac{\partial X^k}{\partial x^i}dx^i$$

so that, as $\Sigma = \partial/\partial y^k \otimes dx^k$,

$$\mathcal{L}_\xi\Sigma = \left(\frac{\partial X^k}{\partial x^i} - Y^k_i\right)\frac{\partial}{\partial y^k} \otimes dx^i.$$

Thus

$$Z_1 = X^i \frac{\partial}{\partial x^i} = \rho \circ \xi, \quad Z_2 = Y^i \frac{\partial}{\partial x^i}.$$

□

We now consider infinitesimal symmetries of an infinitesimal connection γ on $\mathcal{A}^G$, and begin by working out the symmetry conditions explicitly, using the method of Proposition 5.4.3. This involves expressing the section ξ of $\mathcal{A} \to M$ in the form $\xi = \gamma \circ X + \kappa$ where $X = \rho \circ \xi$ and $\kappa = \omega \circ \xi$; the condition for ξ to be an infinitesimal symmetry is that $\nabla^\gamma \kappa + R^\gamma(X, \cdot) = 0$. We shall work in terms of local coordinates (x^i, y^i) where the y^i are canonical linear fibre coordinates, and it will be helpful to write

$$H_i = \gamma \circ \frac{\partial}{\partial x^i} = \frac{\partial}{\partial x^i} - \left(\delta_i^j + \Gamma_{ik}^j y^k\right) \frac{\partial}{\partial y^j}.$$

We shall assume that the corresponding linear connection is symmetric, so that we have

$$R_{ij}^\gamma = R_{lij}^k y^l \frac{\partial}{\partial y^k}$$

by Corollary 4.2.8.

Proposition 5.5.3 *Let* $\xi = \gamma \circ X + \kappa$ *where*

$$X = X^i \frac{\partial}{\partial x^i}, \quad \kappa = \left(Y^i + Y_j^i y^j\right) \frac{\partial}{\partial y^i},$$

so that

$$\xi = X^i H_i + \left(Y^i + Y_j^i y^j\right) \frac{\partial}{\partial y^i}.$$

Then ξ *is an infinitesimal symmetry of* γ *if, and only if,*

$$Y_j^i = Y_{|j}^i, \quad Y_{j|k}^i = R_{jkl}^i X^l,$$

where the rule indicates covariant differentiation with respect to the linear connection.

Proof We have

$$\nabla_{\partial/\partial x^i}^\gamma \kappa = \left(Y_{|i}^j - Y_i^j + Y_{k|i}^j y^k\right) \frac{\partial}{\partial y^j}$$

from Lemma 4.2.5, so ξ will be an infinitesimal symmetry if and only if

$$Y_j^i = Y_{|j}^i, \quad Y_{j|k}^i = R_{jkl}^i X^l.$$

□

Thus infinitesimal symmetries take the form

$$X^i H_i + \left(Y^i + Y^i_{|j} y^j\right)\frac{\partial}{\partial y^i} \quad \text{where } Y^i_{|jk} = R^i_{jkl} X^l.$$

It is clear from this result that there can be absurdly many infinitesimal symmetries. For example, when the associated linear connection is flat there is no restriction on X, and all vector fields of the form

$$X^i H_i + \left(Y^i + Y^i_{|j} y^j\right)\frac{\partial}{\partial y^i} \quad \text{where } Y^i_{|jk} = 0$$

are infinitesimal symmetries (which of course are flat as symmetries), and in particular every vector field on M determines at least one infinitesimal symmetry, namely its horizontal lift. (We note in passing that in general the flat infinitesimal symmetries are those for which $R^i_{jkl} X^l = 0$ and $Y^i_{|jk} = 0$—an instance of Proposition 5.4.8.)

We shall therefore narrow down our consideration of infinitesimal symmetries to those which respect the vertical lift: that is, we shall consider the case where ξ may be regarded as the sum of a complete lift and a vertical lift, by virtue of Proposition 5.5.2. In such a case we find that the complete lift may be identified with a classical infinitesimal symmetry of the associated linear connection [46], and the vertical lift may be regarded as a 'translation' and arises because we are using affine connections so that additional symmetries may be obtained by considering translations of the fibres of $TM \to M$.

A word about classical infinitesimal symmetries of linear connections, or infinitesimal affine transformations, before proceeding. A vector field X on M is an infinitesimal affine transformation of a symmetric linear connection with covariant derivative operator ∇ if, and only if, for every vector field Y

$$\mathcal{L}_X \circ \nabla_Y - \nabla_Y \circ \mathcal{L}_X = \nabla_{[X,Y]};$$

this is the analogue of the formula in Proposition 5.3.4. The condition can be expressed as

$$X^i_{|jk} = R^i_{jkl} X^l,$$

or more explicitly as

$$\frac{\partial^2 X^i}{\partial x^j \partial x^k} - \Gamma^l_{jk}\frac{\partial X^i}{\partial x^l} + \Gamma^i_{lk}\frac{\partial X^l}{\partial x^j} + \Gamma^i_{jl}\frac{\partial X^l}{\partial x^k} + X^l \frac{\partial \Gamma^i_{jk}}{\partial x^l} = 0.$$

In preparation for our discussion it is worth noting that the complete lift of a vector field X on M may be expressed in terms of the horizontal vector fields H_i of an infinitesimal connection on $\mathcal{A}^G$ as follows:

$$X^{\mathrm{c}} = X^i \frac{\partial}{\partial x^i} + \frac{\partial X^i}{\partial x^j} y^j \frac{\partial}{\partial y^i} = X^i H_i + \left(X^i + X^i_{|j} y^j\right)\frac{\partial}{\partial y^i}.$$

Thus an infinitesimal symmetry may be expressed as

$$X^i H_i + \left(Y^i + Y^i_{|j} y^j\right)\frac{\partial}{\partial y^i} = X^{\mathsf{C}} + \left((Y^i - X^i) + (Y^i_{|j} - X^i_{|j})y^j\right)\frac{\partial}{\partial y^i}$$
$$= X^{\mathsf{C}} + \left(Z^i + Z^i_{|j} y^j\right)\frac{\partial}{\partial y^i}$$

say, and the symmetry condition becomes

$$Z^i_{|jk} = -X^i_{|jk} + R^i_{jkl} X^l.$$

Theorem 5.5.4 *Let ξ be an infinitesimal symmetry of γ. If ξ respects the vertical lift, so that it can be written as a sum of a complete lift and a vertical lift,*

$$\xi = Z_1^{\mathsf{C}} + Z_2^{\mathsf{V}},$$

where Z_1 and Z_2 are vector fields on M, then Z_1 is a classical infinitesimal affine transformation of the infinitesimal linear connection $\bar{\gamma}$ associated to γ, and Z_2 is covariant constant. If more particularly ξ is a fibre-linear projectable vector field then it is itself a complete lift of a classical infinitesimal affine transformation of $\bar{\gamma}$.

Proof If ξ is a general infinitesimal symmetry of γ we may write

$$\xi = X^{\mathsf{C}} + \left(Z^i + Z^i_{|j} y^j\right)\frac{\partial}{\partial y^i}$$

where $X = \rho \circ \xi$, and $Z^i_{|jk} = -X^i_{|jk} + R^i_{jkl} X^l$. We now consider the additional conditions imposed in the theorem. We have $Z_1^i = X^i$, $Z_2^i = Z^i$, and the conditions further require that $Z^i_{|j} = 0$: thus $X^i_{|jk} = R^i_{jkl} X^l$ and $X = Z_1$ is a classical infinitesimal affine transformation of $\bar{\gamma}$; and Z_2 is covariant constant. In particular, if $\xi = X^{\mathsf{C}}$ then X is a classical infinitesimal affine transformation of $\bar{\gamma}$. □

Denote by $\mathfrak{a}_0$ the subset of $\mathfrak{a}$ consisting of the infinitesimal symmetries of the form $X^{\mathsf{C}} + Y^{\mathsf{V}}$. From Theorem 5.5.4 a vector field of this form is an infinitesimal symmetry if and only if X is a classical infinitesimal affine transformation and Y is covariant constant, which we write $\nabla Y = 0$; denote by $\mathfrak{c}$ the classical symmetries, and $\mathfrak{v}_0$ the vertical lifts. Then as vector spaces $\mathfrak{a}_0 = \mathfrak{c} \oplus \mathfrak{v}_0$. In fact:

Proposition 5.5.5 *$\mathfrak{a}_0$ is a Lie subalgebra of $\mathfrak{a}$, and is the semidirect sum of $\mathfrak{c}$ and $\mathfrak{v}_0$.*

Proof The crucial observation is that for any vector fields X, Y,

$$[X^{\mathsf{C}}, Y^{\mathsf{V}}] = [X, Y]^{\mathsf{V}} = (\nabla_X Y)^{\mathsf{V}} - (\nabla_Y X)^{\mathsf{V}};$$

if Y is covariant constant the first term in the final expression vanishes, and we can write the second as the vertical lift of $(\nabla X)(Y)$, considering the covariant differential ∇X as a type $(1, 1)$ tensor field. Thus for the bracket of two elements of $\mathfrak{a}_0$ we have

$$[X^{\mathrm{c}} + Y^{\mathrm{v}}, \hat{X}^{\mathrm{c}} + \hat{Y}^{\mathrm{v}}] = [X, \hat{X}]^{\mathrm{c}} + \left(-(\nabla X)(\hat{Y}) + (\nabla \hat{X})(Y)\right)^{\mathrm{v}}.$$

This must of course belong to $\mathfrak{a}$; evidently it actually belongs to $\mathfrak{a}_0$, which is therefore a subalgebra of $\mathfrak{a}$.

To show that $\mathfrak{a}_0$ is the semidirect sum of $\mathfrak{c}$ and $\mathfrak{v}_0$ it is enough to show that $\mathfrak{c}$ is a Lie subalgebra of $\mathfrak{a}_0$ and $\mathfrak{v}_0$ is an ideal of $\mathfrak{a}_0$. If we set $Y = \hat{Y} = 0$ we see that $\mathfrak{c}$ is a subalgebra of $\mathfrak{a}_0$. If we set $\hat{X} = 0$ we find that

$$[X^{\mathrm{c}} + Y^{\mathrm{v}}, \hat{Y}^{\mathrm{v}}] = -\left((\nabla X)(\hat{Y})\right)^{\mathrm{v}},$$

which shows that $\mathfrak{v}_0$ is an ideal of $\mathfrak{a}_0$. □

Notice also that $\mathfrak{v}_0$ is abelian (we expect 'translations' to commute). There is an additional point of interest: in such circumstances we know that $\mathrm{ad} : \mathfrak{c} \to \mathfrak{gl}(\mathfrak{v}_0)$ is a representation of $\mathfrak{c}$; it is given explicitly by $X \mapsto -\nabla X$.

We shall continue the study of the infinitesimal symmetries of an infinitesimal affine connection in Sect. 8.1, where we impose some additional conditions, to be explained in the next chapter, which fix a symmetry of the more substantial structure which we call an infinitesimal Cartan affine geometry.

Chapter 6
Cartan Geometries

We now move on to the main topic of this volume, the study of the structures on differentiable manifolds knows as Cartan geometries. We shall define a (finite) Cartan geometry as a special kind of fibre-morphism groupoid with a path connection, and use this to motivate a detailed investigation of infinitesimal Cartan geometries given in terms of Lie algebroids. In fact our main concern in subsequent chapters will be with the infinitesimal geometries, and so we shall spend some time defining them in an intrinsic way using Lie algebra actions; as we shall see, there are infinitesimal Cartan geometries which are 'incomplete' and which cannot be obtained as infinitesimal versions of finite Cartan geometries. We conclude the chapter with an introduction to the concept of a symmetry for a Cartan geometry.

6.1 Cartan Geometries as Fibre Morphism Groupoids

We start with a fibre bundle $\pi : E \to M$ with standard fibre F and group G, where G acts transitively on F with isotropy subgroup G_o, so that F can be identified with the homogeneous space G/G_o. Let $\Pi : P \to M$ be a principal bundle with group G, with which $E \to M$ is associated by the left action of G on $F = G/G_o$. Then P is reducible to G_o, a closed subgroup of G, if and only if the bundle $E \to M$ admits a section: see for example [29]. We assume that $E \to M$ has this property: we may say that E itself is reducible to the group G_o. Then there is a global section $\varsigma : M \to E$ of π. We may identify $\varsigma(M)$, the image of M in E by ς, with M itself, and we may thereby regard E_x as a copy of F attached to M at $\varsigma(x)$: so we call ς the *attachment section*. Assume finally that $\dim F = \dim M$.

One can think of such a bundle $\pi : E \to M$ as obtained by attaching to each point of M a homogeneous space $F = G/G_o$ of the same dimension. That is to say, this construction realises Cartan's concept of 'un espace généralizé'. We call such a bundle a generalized space modelled on F, or on G and G_o; or a *generalized F-space*.

M. Crampin and D. Saunders, *Cartan Geometries and their Symmetries*, Atlantis Studies in Variational Geometry 4, DOI 10.2991/978-94-6239-192-5_6

Given a generalized F-space, let $\mathcal{G}$ be the Lie groupoid of fibre morphisms of the G-bundle π, namely of those maps $\varphi : E_{\tilde{x}} \to E_x$, for all $x, \tilde{x} \in M$, which respect the action of G. We also have a Lie groupoid $\mathcal{G}_o$ corresponding to the G_o action, which we may regard as a wide Lie subgroupoid of $\mathcal{G}$.

Finally, assume that we have a path connection on $\mathcal{G}$, which satisfies the following *nondegeneracy condition*. Let c be any curve in M which is regular at zero, and c^Γ its groupoid lift to $\mathcal{G}$. We require c^Γ to intersect the subgroupoid $\mathcal{G}_o$ transversally at $c^\Gamma(0) = 1_{c(0)}$.

We call a structure with these features a *Cartan geometry*. Thus a Cartan geometry consists of

1. a fibre bundle $\pi : E \to M$ with standard fibre F and group G, such that $\dim F = \dim M$ and G acts transitively on F;
2. a reduction of π to the group G_o where G_o is the isotropy subgroup of G of some point $o \in F$ (so that $F = G/G_o$), together with the corresponding attachment section ς of π;
3. the Lie groupoid $\mathcal{G}$ of fibre morphisms of E which when expressed in terms of local trivializations belong to G, and the corresponding Lie subgroupoid $\mathcal{G}_o$;
4. a path connection Γ on $\mathcal{G}$ which satisfies the nondegeneracy condition with respect to $\mathcal{G}_o$.

We illustrate the definition with a trivial but fundamental example.

Take $M = F = G/G_o$, so that G_o is the isotropy group of $o \in M$, and let $E = M \times M$. We may identify $\mathcal{G}$ with the trivial groupoid $M \times G \times M$. There is a natural section ς of the fibre bundle $\mathrm{pr}_1 : M \times M \to M$ given by $\varsigma(x) = (x, x)$ (perhaps not the most obvious section, but the most natural one). With reference to item (3) above, the subgroupoid $\mathcal{G}_o$ consists of those fibre morphisms $\varphi : E_{\tilde{x}} \to E_x$ such that $\varphi(\varsigma(\tilde{x})) = \varsigma(x)$. In the present context, with $\varphi \sim (x, g, \tilde{x}) \in M \times G \times M$, this condition is

$$\varphi(\tilde{x}, \tilde{x}) = (x, g\tilde{x}) = (x, x),$$

so $\mathcal{G}_o$ consists of the elements of $M \times G \times M$ of the form (gx, g, x); that is, it is the action groupoid corresponding to the action of G on M, regarded as a subgroupoid of $M \times G \times M$ as specified in Corollary 1.3.4. Finally, we take the trivial path connection Γ given by

$$c^\Gamma(t) = (c(t), 1_G, c(0))$$

for any curve c in M. This structure is thus a Cartan geometry. We call it the *model Cartan geometry* for the homogeneous space G/G_o; it stands in the same relation to our theory as the Klein geometry does to the conventional one.

We shall attempt to clarify our choice of section. First, we go back to the construction of the section when P is reducible in the general case. The associated bundle $E \to M$ with standard fibre F is defined by $E = (P \times F)/G$ where G acts to the right on $P \times F$ by $r_g(p, y) = (pg, g^{-1}y)$, $p \in P$, $y \in F$. When $F = G/G_o$, E may be identified with P/G_o as follows. We define a map $\mu : P \to E$ by $\mu(p) = [(p, o)]$, the orbit of (p, o) under the right action. Clearly $\mu(p) = \mu(p')$ if and only if $p' = pg$

for some $g \in G_o$: that is, μ is constant on G_o orbits, and induces a diffeomorphism $P/G_o \to E$. The bundle $P \to M$ is reducible to $G_o \subset G$ if there is a sub-bundle $Q \subset P$ invariant under the right action of G_o, so that $Q \to M$ is itself a principal bundle with group G_o. For $q \in Q$ and $g \in G_o$, $\mu(qg) = \mu(q)$. We define a map $\varsigma : M \to E$ by $\varsigma(x) = \mu(q)$ for any $q \in Q$ with $\Pi(q) = x$. Then ς is well defined, and is the required section.

To apply this construction in our particular case, we need a principal bundle with group G, and we take the trivial bundle $M \times G = P$. The associated bundle is $P/G_o = M \times M$, and $\mu : M \times G \to M \times M$ is given by $\mu(x, g) = [(x, g, o)] = (x, go)$. Evidently $M \times G_o$ is a reduction of $M \times G$ to G_o. In this case $\varsigma(x) = (x, o)$. But in this special situation there is another reduction of P to G_o. Recall that G is itself a principal G_o-bundle over M. Moreover we may inject G into $M \times G$ by $g \mapsto (go, g)$. The image Q is a sub-bundle of $M \times G$ invariant under the right action of G_o, and so defines a reduction of $M \times G$ to G_o. In this case ς is defined as follows. For $x \in M$, take any g such that $x = go$. Then $(x, g) \in Q$, and $\varsigma(x) = \mu(x, g) = (x, go) = (x, x)$.

The slight sense of oddity associated with this section will be resolved if one recognises that one isn't using the most appropriate trivialization. Let us choose a local section of $G \to M$, say $x \mapsto g(x)$, $x \in U$: then $g(x)o = x$. The map $\mathrm{T} : U \times M \to M \times M$ by $\mathrm{T}(x, y) = (x, g(x)y)$ is a local trivialization (and we must include the word 'local' because we may not assume that G is trivial over M). It satisfies $\mathrm{T}(x, o) = (x, x)$. Moreover, if $\varphi : E_{\tilde{x}} \to E_x$ satisfies $\varphi(\tilde{x}, \tilde{x}) = (x, x)$, then

$$\mathrm{T}_x^{-1} \circ \varphi \circ \mathrm{T}_{\tilde{x}}(o) = \mathrm{T}_x^{-1}(\varphi(\tilde{x}, \tilde{x})) = \mathrm{T}_x^{-1}(x, x) = o\,,$$

so that relative to this local trivialization φ respects the action of G_o.

6.2 The Lie Algebroid of a Cartan Geometry

Suppose given a Cartan geometry on a generalized F-space $\pi : E \to M$. As we know from Chap. 3 we may realize $A\mathcal{G}$, the Lie algebroid of $\mathcal{G}$, as a Lie algebroid of projectable vector fields along fibres of $E \to M$, with anchor the projection π_* and bracket of sections the ordinary bracket of vector fields on E. (It must be borne in mind, however, that the anchor acts on vector fields along fibres while in normal usage π_* acts on individual vectors at points of E.) We denoted this second algebroid by $\mathcal{A}^G$, and it is this that we shall almost invariably use when discussing the Lie algebroid of a Cartan geometry.

The Lie algebroid $\mathcal{A}^G$ is transitive. Any element of its kernel $\mathcal{K}^G$ is a vertical vector field along a fibre of E, and can be identified via a local trivialization of E with a fundamental vector field of the action of G on F. Thus for each $x \in M$, $\mathcal{K}_x^G$ is isomorphic to $\mathfrak{g}_{\mathrm{R}}$, the opposite Lie algebra of G. The fibre dimension (rank) of the vector bundle $\mathcal{A}^G \to M$ is thus $\dim M + \dim \mathfrak{g}$.

A section of $\mathcal{A}^G \to M$ is a projectable vector field on E, belonging to a certain $C^\infty(M)$-module of projectable vector fields, closed under bracket. In terms of a local trivialization of $\pi : E \to M$ an element of $\mathcal{A}^G_x$ can be expressed as a pair $(v, a^\sharp)$ with $v \in T_x M$, $a \in \mathfrak{g}_\mathsf{R}$, and $a^\sharp$ the fundamental vector field on F corresponding to a. A local section of $\mathcal{A}^G$ takes the form $(X, \xi^\sharp)$ where X is a local vector field on M, ξ a local $\mathfrak{g}_\mathsf{R}$-valued function on M, and for $x \in M$, $\xi^\sharp|_x = \xi(x)^\sharp$. The Lie algebroid bracket is given by

$$[\![(X, \xi^\sharp), (Y, \eta^\sharp)]\!] = \left([X, Y], (X(\eta) - Y(\xi) + \{\xi, \eta\}_\mathsf{R})^\sharp\right),$$

as described in Proposition 3.2.3.

Let $\{e_\alpha\}$ be a basis for $\mathfrak{g}_\mathsf{R}$. Let $e^\sharp_\alpha$ be the corresponding fundamental vector fields of the action of G on F. Take a local trivialization T of $E \to M$ over $U \subset M$, with $e^{\sharp\mathsf{T}}_\alpha$ vertical vector fields on $\pi^{-1}(U)$ forming a local basis for the kernel $\mathcal{K}^G$, corresponding to the vector fields $e^\sharp_\alpha$ on F. An element of $\mathcal{A}^G_x$ may be written as

$$\left(v^i \frac{\partial}{\partial x^i} + a^\alpha e^{\sharp\mathsf{T}}_\alpha\right)_{E_x}$$

where $v^i, a^\alpha \in \mathbb{R}$. A local section of $\mathcal{A}^G$ takes the form

$$X^i \frac{\partial}{\partial x^i} + \xi^\alpha e^{\sharp\mathsf{T}}_\alpha$$

with $X^i, \xi^\alpha \in C^\infty(U)$; in terms of the previous representation we have $\xi = \xi^\alpha e_\alpha$.

We regard the Lie algebroid $A\mathcal{G}_o$ of $\mathcal{G}_o$ as a transitive Lie subalgebroid $\mathcal{A}^G_o \subset \mathcal{A}^G$ with kernel $\mathcal{K}_o$, the latter being a Lie algebra sub-bundle of $\mathcal{K}^G$. For $x \in M$, $\mathcal{A}^G_{o,x}$ consists of those elements of $\mathcal{A}^G_x$ which, as projectable vector fields along E_x, are tangent to $\varsigma(M)$ at $\varsigma(x)$ (recall that if vector fields on a manifold are tangent to a submanifold so is their bracket). In terms of a local trivialization, $(v, a^\sharp)$ represents an element of $\mathcal{A}^G_{o,x}$ when $a \in \mathfrak{g}_{o\mathsf{R}}$. The fibre dimension of $\mathcal{A}^G_o$ is $\dim M + \dim \mathfrak{g}_o = \dim F + \dim \mathfrak{g}_o = \dim \mathfrak{g}$, which is of course the fibre dimension of $\mathcal{K}^G$.

The infinitesimal version of the nondegeneracy condition for the path connection Γ may be formulated as follows. Let γ be the corresponding infinitesimal connection on $\mathcal{A}^G$. Then for each $x \in M$, $\gamma_x(T_x M)$ is a subspace of $\mathcal{A}^G_x$, complementary to $\mathcal{K}^G_x$; we require that $\mathcal{A}^G_{o,x} \cap \gamma_x(T_x M) = \{0\}$.

The Lie algebroid of the model Cartan geometry for G/G_o consists of the following items:

- the homogeneous space $F = M = G/G_o$;
- the generalized F-space $\mathrm{pr}_1 : M \times M \to M$ with the attachment section $\varsigma : M \to M \times M$ where $\varsigma(x) = (x, x)$;
- the Lie algebroid $A\mathcal{G}$ of the trivial Lie groupoid $M \times G \times M$;

- its Lie subalgebroid $A\mathcal{G}_o$, which is the Lie algebroid of the action groupoid corresponding to the action of G on M;
- the infinitesimal connection γ corresponding to the path connection Γ defined in the previous section.

The Lie algebroids $A\mathcal{G}$ and $A\mathcal{G}_o$ are specified in Proposition 1.3.2 and Corollary 1.3.4 respectively: in particular, $A\mathcal{G}$ can be identified as a vector bundle with $TM \oplus_M (M \times \mathfrak{g})$ where $\mathfrak{g}$ is the Lie algebra of G. But we prefer to work with their representations $\mathcal{A}^G$ and $\mathcal{A}^{G_o}$ by projectable vector fields along fibres of $M \times M \to M$, in the manner given by Proposition 3.2.3. Then $\mathcal{A}^G_x$ consists of vector fields along $\{x\} \times M$ of the form $(v, a^\sharp)$ for $v \in T_x M$, $a \in \mathfrak{g}$, where

$$a^\sharp_y = \frac{d}{dt}(l_{\exp(-ta)}(y))_{t=0} ,$$

the fundamental vector field of the *right* action of G on M. A section ξ of $\mathcal{A}^G \to M$ is determined by a vector field X on M and a $\mathfrak{g}$-valued function $\bar{\xi}$ on M by

$$\xi(x) = \big(X_x, \bar{\xi}(x)^\sharp\big);$$

we shall write $\xi = (X, \bar{\xi}^\sharp)$. The bracket of such vector fields is given by

$$\big[(X, \bar{\xi}^\sharp), (Y, \bar{\eta}^\sharp)\big] = \big([X, Y], X(\bar{\eta})^\sharp - Y(\bar{\xi})^\sharp + \{\bar{\xi}, \bar{\eta}\}^\sharp\big) .$$

Here $\{\cdot, \cdot\}$ is the ordinary Lie algebra bracket on $\mathfrak{g}$, that is, the one determined by left-invariant vector fields on G.

The subalgebroid $\mathcal{A}^G_o$ consists of those elements of $\mathcal{A}^G$ which are tangent to the attachment section, that is, it consists of projectable vector fields along fibres of the form $(a^\sharp_x, a^\sharp)$, $a \in \mathfrak{g}$. So $\xi = (X, \bar{\xi}^\sharp)$ is a section of $\mathcal{A}^G_o$ if and only if $X_x = \bar{\xi}(x)^\sharp_x$ for all $x \in M$. It follows that the Lie algebroid bracket of sections ξ and η of $\mathcal{A}^G_o \to M$ is given by ζ with

$$\bar{\zeta}(x) = \bar{\xi}^\sharp_x(\bar{\eta}) - \bar{\eta}^\sharp_x(\bar{\xi}) + \{\bar{\xi}, \bar{\eta}\} .$$

As expected, the Lie algebroid $\mathcal{A}^G_o$ is isomorphic to the action algebroid of the action of G on M.

The infinitesimal connection γ is given simply by

$$\gamma_x(v) = (v, 0) .$$

It is evidently flat. Its kernel projection is

$$\omega_x(v, a^\sharp) = (0, a^\sharp) .$$

6.3 Lie Algebra Actions

It is usually more convenient in practice to deal with a Cartan geometry via its infinitesimal version, that is, via the Lie algebroid and its infinitesimal connection, rather than the Lie groupoid and its path connection. It is therefore of interest to attempt to specify an infinitesimal Cartan geometry intrinsically, by listing its defining properties. Now basic to the definition of a Cartan geometry is the action of a Lie group G on a manifold F. For the definition of an infinitesimal Cartan geometry we must replace this by the action of a Lie algebra $\mathfrak{g}$ on a manifold F. In this section we discuss Lie algebra actions, in preparation for defining infinitesimal Cartan geometries in the following one. A detailed analysis of Lie algebra actions can be found in [1].

Let $\mathfrak{g}$ be a (real) Lie algebra. By a (smooth) *action* of $\mathfrak{g}$ on a manifold F we mean a Lie algebra homomorphism of $\mathfrak{g}$ into the Lie algebra $\mathfrak{X}(F)$ of smooth vector fields on F, denoted by $a \mapsto a^\sharp$. Alternatively, and somewhat less formally, one may regard a (finite-dimensional) Lie algebra of vector fields on F as defining an action. Of course when a Lie group G acts on a manifold F on the left, its opposite Lie algebra $\mathfrak{g}_R$ acts on F in this sense. On the other hand, it is by no means necessary that a Lie algebra action derives from a Lie group action. If the action of a Lie algebra $\mathfrak{h}$ is *complete*, in the sense that for each $a \in \mathfrak{h}$, $a^\sharp$ is a complete vector field, then by a theorem of Palais [33] there is a Lie group G with Lie algebra $\mathfrak{g}$ satisfying $\mathfrak{h} = \mathfrak{g}_R$ and an action of G on F giving rise to the action of $\mathfrak{h}$. We do not wish to restrict our attention to complete Lie algebra actions, however.

We say that a Lie algebra action is *transitive* if for all $p \in F$, the linear map $\mathfrak{g} \to T_pF$ by $a \mapsto a_p^\sharp$ is surjective. For any fixed $o \in F$, the kernel of the linear map $a \mapsto a_o^\sharp$ is a subalgebra $\mathfrak{g}_o$ of $\mathfrak{g}$ (since the bracket of two vector fields which vanish at a point also vanishes at that point). We call $\mathfrak{g}_o$ the *isotropy algebra* of o. We shall be mainly interested in transitive Lie algebra actions and their isotropy algebras.

We shall need to address the definition of the effectiveness of a Lie algebra action. Unfortunately the terminology for effectiveness of Lie algebra actions in the literature is not uniform. We can distinguish several related but differing concepts, as follows. We say that a transitive action of $\mathfrak{g}$ on F is

- *effective* if the homomorphism $a \to a^\sharp$ of $\mathfrak{g} \to \mathfrak{X}(F)$ is injective, so that if $a^\sharp = 0$ then $a = 0$;
- *locally effective at* $p \in F$ if there is no nonzero element a of $\mathfrak{g}$ such that $a^\sharp$ vanishes on a neighbourhood of p;
- *locally effective* if it is locally effective at all $p \in F$, so that if $a^\sharp$ vanishes on an open subset of F then $a = 0$;
- *algebraically effective* if there is no nontrivial ideal of $\mathfrak{g}$ contained in the isotropy algebra $\mathfrak{g}_o$ of $o \in F$.

We shall show below that in fact the middle two are equivalent: if an action is locally effective at any single point of F then it is locally effective. An action which is algebraically effective is effective in all of the other senses. This is because, in the first place, if X is a vector field that vanishes on a neighbourhood U of o and Y is any vector field then $[X, Y]$ also vanishes on U, so for an arbitrary action the set of elements a of $\mathfrak{g}$ for which $a^\sharp$ vanishes on a neighbourhood of o is an ideal of $\mathfrak{g}$ contained in $\mathfrak{g}_o$. In the other case, for an arbitrary action the kernel of the homomorphism $a \to a^\sharp$ is again an ideal of $\mathfrak{g}$ contained in $\mathfrak{g}_o$. Finally, if an action is locally effective then it is effective: for if $a \in \mathfrak{g}$ is such that $a^\sharp = 0$, then certainly $a^\sharp$ vanishes on every neighbourhood of every point.

The definition of algebraic effectiveness comes from Sharpe [36]. He calls a pair $(\mathfrak{g}, \mathfrak{g}_o)$, where $\mathfrak{g}$ is a Lie algebra and $\mathfrak{g}_o$ a subalgebra of $\mathfrak{g}$, an infinitesimal Klein geometry or a Klein pair. He calls a Klein pair for which there is no nontrivial ideal of $\mathfrak{g}$ contained in $\mathfrak{g}_o$ effective (without qualification). Alekseevsky and Michor [1], on the other hand, who deal only with Lie algebra actions, call an action effective if the map $\mathfrak{g} \to \mathfrak{X}(F)$ is injective. They do use the property that we have called local effectiveness as a condition in several important results, but do not give it a specific name.

If the data are analytic, as for example in the case of the transitive Lie algebra action derived from a transitive Lie group action, and F is connected, then effectiveness implies local effectiveness, since an analytic vector field on a connected analytic manifold cannot vanish on an open set without vanishing everywhere. On the other hand, it is easy to construct smooth, but not transitive, Lie algebra actions which are effective as actions but not locally effective, by the use of bump functions. We shall show below that this phenomenon cannot arise if the action is transitive and F is connected: an effective transitive action of a Lie algebra on a connected manifold is locally effective. This will turn out to have a significant consequence for the infinitesimal symmetries of an infinitesimal Cartan geometry.

We first recall some relevant facts. For a vector field X on a manifold F and a diffeomorphism $\varphi : F \to F$ the vector field $\varphi_* X$ is defined as follows:

$$(\varphi_* X)_p = \varphi_{*\varphi^{-1}(p)}\big(X_{\varphi^{-1}(p)}\big), \quad p \in F$$

(where of course $\varphi_{*q} : T_q F \to T_{\varphi(q)} F$ is the induced linear map of tangent spaces); and similarly for a local diffeomorphism, with the obvious provisos. Let X and Y be vector fields on F; let φ_t be the flow of X. Then the following formula holds, where it makes sense:

$$\left(\frac{d}{ds}(\varphi_{s*} Y)\right)_{s=t} = -[X, \varphi_{t*} Y].$$

Next, consider a Lie algebra $\mathfrak{g}$. Any $b \in \mathfrak{g}$ defines a linear map ad_b of the vector space $\mathfrak{g}$ to itself, where $\mathrm{ad}_b\, a = \{b, a\}$. It is of course a derivation, indeed an inner derivation, of the Lie algebra bracket of $\mathfrak{g}$. Now $\exp(t\ \mathrm{ad}_b) : \mathfrak{g} \to \mathfrak{g}$ is defined for all t, and is a Lie algebra isomorphism, that is, an element of $\mathsf{Aut}(\mathfrak{g})$. It is clear, by expanding $\exp((s + t)\ \mathrm{ad}_b)$, differentiating with respect to s and then replacing s by $s - t$, that

$$\left(\frac{d}{ds}(\exp(s\ \mathrm{ad}_b)a)\right)_{s=t} = \big\{b, \exp(t\ \mathrm{ad}_b)a\big\}.$$

Now suppose that $\mathfrak{g}$ acts on F. Then from the last formula

$$\mathcal{L}_{\partial/\partial t}\big(\exp(t\ \mathrm{ad}_b)a\big)^\sharp = \big\{b, \exp(t\ \mathrm{ad}_b)a\big\}^\sharp = \big[b^\sharp, (\exp(t\ \mathrm{ad}_b)a)^\sharp\big]$$

using the homomorphism property of the action in the last step. Denote by φ^b_t the flow of $b^\sharp$ on F, and consider the time-dependent vector field

$$Z(t) = \varphi^b_{t*}a^\sharp - (\exp(-t\ \mathrm{ad}_b)a)^\sharp\,.$$

At each $p \in F$ this defines a curve Z_p in T_pF, which we may differentiate to give another curve Z'_p, and therefore another time-dependent vector field Z' on F. From the computations above we see that

$$\begin{aligned} Z'(t) &= -\big[b^\sharp, \varphi^b_{t*}a^\sharp\big] + \big[b^\sharp, (\exp(-t\ \mathrm{ad}_b)a)^\sharp\big] \\ &= -\big[b^\sharp, Z(t)\big]\,, \end{aligned}$$

which at each $p \in F$ is a first-order homogeneous linear ordinary differential equation for the curve Z_p. But the initial condition is given by $Z(0) = a^\sharp - a^\sharp = 0$ so that we must have $Z(t) = 0$, that is,

$$\varphi^b_{t*}a^\sharp = (\exp(-t\ \mathrm{ad}_b)a)^\sharp$$

whenever the left-hand side is defined.

This formula is useful because it relates something going on in F (the left-hand side) with something going on in $\mathfrak{g}$ (the right-hand side, if one ignores the $^\sharp$).The first point to note is that it shows that $\varphi^b_{t*}a^\sharp$, a vector field on F, is obtained, via the action, from an element of $\mathfrak{g}$. As a further example, it allows one to relate the isotropy algebras at different points of F. Suppose for convenience that $\mathfrak{g}$ acts transitively. If one unpacks the definition of the left-hand side, and evaluates at $o \in F$, one finds that for $p = \varphi^b_t(o)$

$$a^\sharp_p = \varphi^b_{t*o}\big(\exp(t\ \mathrm{ad}_b)a\big)^\sharp_o\,.$$

The isotropy algebra $\mathfrak{g}_p$ at p is simply the kernel of the map $a \mapsto a^\sharp_p$. But φ^b_{t*o} is an isomorphism of vector spaces $T_oF \to T_pF$, so $a \in \mathfrak{g}_p$ if and only if $\exp(t\ \mathrm{ad}_b)a \in \mathfrak{g}_o$, the isotropy algebra of o. Thus $\mathfrak{g}_p = \exp(-t\ \mathrm{ad}_b)(\mathfrak{g}_o)$ (recall that $\exp(-t\ \mathrm{ad}_b)$ is a Lie algebra isomorphism for the usual Lie bracket, so $\exp(-t\ \mathrm{ad}_b)(\mathfrak{g}_o)$ is indeed a Lie subalgebra of $\mathfrak{g}$, given that $\mathfrak{g}_o$ is). Since $\exp(-t\ \mathrm{ad}_b) = \mathrm{Ad}(\exp(-tb))$ this generalizes the usual formula in the case of a group action. Suppose further that F is connected. Then for a transitive action of $\mathfrak{g}$, every point of F can be reached from o by a succession of local diffeomorphisms of the form φ^b_t, and the basic formulæ such as $\varphi^b_{t*}a^\sharp = (\exp(-t\ \mathrm{ad}_b)a)^\sharp$ can clearly be iterated.

Proposition 6.3.1 *If a transitive action of a Lie algebra $\mathfrak{g}$ on a connected manifold F is locally effective at $o \in F$ then it is locally effective.*

Proof Suppose that $a \in \mathfrak{g}$ is such that $a^\sharp$ vanishes on a neighbourhood V of p, where $p = \varphi_t^b(o)$ for some $t \in \mathbb{R}$, $b \in \mathfrak{g}$. Now φ_t^b maps sufficiently small neighbourhoods of o diffeomorphically onto neighbourhoods of p. So we may assume (by shrinking it if necessary) that $V = \varphi_t^b(U)$ where U is a neighbourhood of o. Then for all $p' \in V$, $0 = a^\sharp_{p'} = \varphi^b_{t*o'}(\exp(t\ \mathrm{ad}_b)a)^\sharp_{o'}$ where $p' = \varphi_t^b(o')$, $o' \in U$. It follows that $(\exp(t\ \mathrm{ad}_b)a)^\sharp$ vanishes on the neighbourhood U of o. So $\exp(t\ \mathrm{ad}_b)a = 0$, and therefore $a = 0$. We may repeat the argument step by step to cover the whole of F. □

Proposition 6.3.2 *Suppose that $\mathfrak{g}$ acts effectively and transitively on F, which is connected. Then the action is locally effective.*

Proof For $p \in F$ let

$$\tilde{\mathfrak{g}}_p = \{a \in \mathfrak{g} : a^\sharp \text{ vanishes on a neighbourhood of } p\}.$$

Then $\tilde{\mathfrak{g}}_p$ is an ideal of $\mathfrak{g}$ contained in $\mathfrak{g}_p$, the isotropy algebra of p.

If $\mathfrak{h}$ is any ideal of a Lie algebra $\mathfrak{g}$, then for any $a \in \mathfrak{h}$ and $b \in \mathfrak{g}$, $\mathrm{ad}_b\, a \in \mathfrak{h}$, and so $\exp(t\ \mathrm{ad}_b)a \in \mathfrak{h}$ also: that is, $\exp(t\ \mathrm{ad}_b)$ maps $\mathfrak{h}$ onto itself.

Let $p, q \in F$: we relate $\tilde{\mathfrak{g}}_p$ and $\tilde{\mathfrak{g}}_q$. Suppose first that $q = \varphi_t^b(p)$ for some $t \in \mathbb{R}$, $b \in \mathfrak{g}$. Let $a \in \tilde{\mathfrak{g}}_p$. Then since φ_t^b maps any sufficiently small open set containing p diffeomorphically onto an open set containing q, $\varphi^b_{t*}a^\sharp$ (which is a vector field defined on a neighbourhood of q) vanishes on a neighbourhood of q: in fact

$$\left(\varphi^b_{t*}a^\sharp\right)_{q'} = \varphi^b_{t*p'}a^\sharp_{p'} \quad \text{where } q' = \varphi_t^b(p'),$$

so if $a^\sharp_{p'} = 0$ then $(\varphi^b_{t*}a^\sharp)_{q'} = 0$. But

$$\varphi^b_{t*}a^\sharp = (\exp(-t\ \mathrm{ad}_b)a)^\sharp.$$

Thus if $a \in \tilde{\mathfrak{g}}_p$ then $\exp(-t\ \mathrm{ad}_b)a \in \tilde{\mathfrak{g}}_q$. Now $\tilde{\mathfrak{g}}_q$ is an ideal of $\mathfrak{g}$, so $\exp(t\ \mathrm{ad}_b)$ maps $\tilde{\mathfrak{g}}_q$ onto itself, and therefore

$$\exp(t\ \mathrm{ad}_b)\big(\exp(-t\ \mathrm{ad}_b)a\big) = a \in \tilde{\mathfrak{g}}_q.$$

That is to say, $\tilde{\mathfrak{g}}_p \subset \tilde{\mathfrak{g}}_q$; similarly $\tilde{\mathfrak{g}}_q \subset \tilde{\mathfrak{g}}_p$; so $\tilde{\mathfrak{g}}_p = \tilde{\mathfrak{g}}_q$. Now if F is connected any two of its points may be mapped one to the other by a finite product of terms φ_t^b: so for all $p, q \in F$, $\tilde{\mathfrak{g}}_p = \tilde{\mathfrak{g}}_q$. So there is a single ideal $\tilde{\mathfrak{g}}$ of $\mathfrak{g}$ such that $\tilde{\mathfrak{g}}_p = \tilde{\mathfrak{g}}$ for all p. Moreover, $\tilde{\mathfrak{g}}$ is contained in the isotropy algebra $\mathfrak{g}_p$ for every p. That is, if $a \in \tilde{\mathfrak{g}}$ then $a^\sharp_p = 0$ for every $p \in F$: but the action is effective, so $a = 0$. □

6.4 Infinitesimal Cartan Geometries

In this section we shall define an infinitesimal Cartan geometry.

We should like to define such a geometry as, in the first place, a transitive Lie algebroid $\mathcal{A}$ of projectable vector fields along fibres of a bundle $\pi : E \to M$, with standard fibre F. The kernel $\mathcal{K}$ should be such that at each $x \in M$, $\mathcal{K}_x$, which is a Lie algebra of vector fields on (and tangent to) E_x, is isomorphic to a fixed Lie algebra $\mathfrak{g}$ acting transitively on F: that is to say, the isomorphism $E_x \cong F$ given by any trivialization of the fibre bundle $E \to M$ should give rise to the isomorphism between $\mathcal{K}_x$ and the Lie algebra of fundamental vector fields on F arising from the action of $\mathfrak{g}$, in a way which is similar to the restriction of the isomorphism described in Proposition 3.2.2.

The question then arises, will an infinitesimal Cartan geometry, so defined, necessarily be derivable from a Cartan geometry defined as a groupoid? We have in effect already pointed out, in Sect. 4.4, that the answer to this question is no. Defining an infinitesimal Cartan geometry, therefore, involves generalizing to some degree the original concept.

In particular, to construct the bundle $\pi : E \to M$ we need a Lie group acting on F, and there is now no group ready to hand. In order to get over this difficulty we consider those diffeomorphisms of F which preserve the action, in the sense we now explain.

Suppose we have two actions, $\mathfrak{g}$ on F and $\mathfrak{g}'$ on F'. We say that they are *isomorphic* if there is a diffeomorphism $\phi : F \to F'$ intertwining them, so that there is a Lie algebra isomorphism $\Phi : \mathfrak{g} \to \mathfrak{g}'$ such that $\phi_*(a^\sharp) = \Phi(a)^\sharp$.

With this in mind, we say that a diffeomorphism $\phi : F \to F$ *preserves the action* of $\mathfrak{g}$ on F if there is a map $\Phi : \mathfrak{g} \to \mathfrak{g}$ such that for all $a \in \mathfrak{g}$, $\phi_*(a^\sharp) = \Phi(a)^\sharp$. The set of all such ϕ evidently forms a group under composition; each Φ is a Lie algebra isomorphism when the action is effective, and $\phi \mapsto \Phi$ is a group homomorphism from the action-preserving diffeomorphisms of F to the Lie algebra isomorphisms of $\mathfrak{g}$. When $\mathfrak{g}$ derives from a Lie group action, say to the left, the left action of $g \in G$ is action-preserving, and the corresponding Lie algebra isomorphism is $\operatorname{Ad} g$.

We therefore propose to begin the definition of an infinitesimal Cartan geometry as follows. Let $\mathfrak{g}$ be a Lie algebra which acts effectively and transitively on a manifold F. Let $\widetilde{G}$ be a Lie group of diffeomorphisms of F which preserve the action. Let $\pi : E \to M$ be a fibre bundle with standard fibre F and group $\widetilde{G}$. Let $\mathcal{A}$ be a transitive Lie algebroid of projectable vector fields on E whose anchor is π_* and whose bracket is the ordinary bracket of vector fields. Then for any $x \in M$, $\mathcal{K}_x$ is a Lie algebra of vector fields on (tangent to) E_x. We require that in any local trivialisation of E this Lie algebra is isomorphic to $\mathfrak{g}$ acting on F in the manner explained above. By construction, if this requirement is satisfied in one trivialisation it is satisfied in all.

In addition we choose a point $o \in F$, with isotropy algebra $\mathfrak{g}_o$. We require that $\widetilde{G}$ fixes o. Then π admits a global section $\varsigma : M \to E$. Finally we require that $\dim F = \dim M$.

To summarize: an *infinitesimal Cartan geometry* consists of

1. a manifold F on which a Lie algebra $\mathfrak{g}$ acts effectively and transitively, a point $o \in F$ with isotropy algebra $\mathfrak{g}_o$, and a Lie group $\widetilde{G}$ of diffeomorphisms of F which preserve the action of $\mathfrak{g}$ and fix o;
2. a fibre bundle $\pi : E \to M$ with standard fibre F and group $\widetilde{G}$, such that $\dim F = \dim M$, with the associated global section ς of π;
3. a transitive Lie algebroid $\mathcal{A}$ of projectable vector fields along fibres of E, whose bracket is the ordinary bracket of vector fields on E and whose anchor is π_*, such that the kernel $\mathcal{K}$ of $\mathcal{A}$ is a Lie algebra bundle where each fibre is isomorphic to $\mathfrak{g}$.

For each $x \in M$ let $\mathcal{A}_{o,x}$ be the subspace of $\mathcal{A}_x$ consisting of those elements which are tangent to $\varsigma(M)$ at $\varsigma(x)$, and let $\mathcal{A}_o$ be the vector sub-bundle of $\mathcal{A}$ whose fibre over x is $\mathcal{A}_{o,x}$. Now if vector fields on a manifold are tangent to a submanifold, so is their bracket: so $\mathcal{A}_o$ is in fact a Lie subalgebroid of $\mathcal{A}$. Moreover, $\mathcal{K}_{o,x} = \mathcal{A}_{o,x} \cap \mathcal{K}_x$ is a Lie subalgebra of $\mathcal{K}_x$. We require that the following further conditions are satisfied:

4. the Lie algebroid $\mathcal{A}_o$ is also transitive;
5. $\mathcal{K}_o = \mathcal{A}_o \cap \mathcal{K}$ is a Lie algebra bundle where each fibre is isomorphic to $\mathfrak{g}_o$;
6. at each point p of E_x, $(\mathcal{K}_x/\mathcal{K}_{o,x})_p$ is isomorphic as a vector space to T_pE_x;
7. $\mathcal{A}$ is equipped with an infinitesimal connection γ such that $\mathcal{A}_{o,x} \cap \gamma_x(T_xM) = \{0\}$ for each $x \in M$.

Proposition 6.4.1 *In an infinitesimal Cartan geometry the fibre dimensions of $\mathcal{A}_o$ and $\mathcal{K}$ are equal.*

Proof When the conditions above hold

$$\dim \mathcal{K}_x - \dim \mathcal{K}_{o,x} = \dim \mathfrak{g} - \dim \mathfrak{g}_o = \dim F = \dim M,$$

and so $\dim \mathcal{A}_{o,x} = \dim \mathcal{K}_{o,x} + \dim M = \dim \mathcal{K}_x$. □

As before, an element of $\mathcal{A}$ may be expressed locally as

$$\left(v^i \frac{\partial}{\partial x^i} + a^\alpha e_\alpha^{\sharp\mathrm{T}} \right)_{E_x}$$

where the $e_\alpha^{\sharp\mathrm{T}}$ are vector fields tangent to E_x corresponding in a trivialization T to a basis $\{e_\alpha\}$ of $\mathfrak{g}$, and where $v^i, a^\alpha \in \mathbb{R}$. A local section of $\mathcal{A}$ over $U \subset M$ takes the form

$$X^i \frac{\partial}{\partial x^i} + \xi^\alpha e_\alpha^{\sharp\mathrm{T}}$$

with $X^i, \xi^\alpha \in C^\infty(U)$.

A Cartan geometry gives rise to an infinitesimal Cartan geometry according to this definition, with $\mathcal{A} = A\mathcal{G}$ and $\mathcal{A}_o = A\mathcal{G}_o$; but it is not the case that every infinitesimal

Cartan geometry arises in this way. Sharpe, in his definition of a Cartan geometry, actually dispenses with the group G. He requires only the following ingredients for a model geometry: a Lie algebra $\mathfrak{g}$ and subalgebra $\mathfrak{h}$; a Lie group H whose Lie algebra is $\mathfrak{h}$; and a representation of H on $\mathfrak{g}$ which reduces to the adjoint representation on $\mathfrak{h}$. If $\mathfrak{g}$ and H act on F, and H preserves the action of $\mathfrak{g}$, fixes o, and produces the action of $\mathfrak{g}_o$, then there is indeed a representation of H on $\mathfrak{g}$ which reduces to the adjoint representation on $\mathfrak{g}_o$. We may then use H instead of $\widetilde{G}$ in the definition of an infinitesimal Cartan geometry. But the definition above is still more general than this.

As we shall discuss in more detail later, vector fields of the form

$$\left(Y^i + Z^i_j u^j + (Y_j u^j) u^i\right) \frac{\partial}{\partial u^i}$$

on $\mathbb{R}^n$ (with coordinates u^i) form a Lie algebra $\mathfrak{g}$; those with $\eta^i = 0$ form a subalgebra $\mathfrak{g}_o$ which fixes the origin. Linear transformations of $\mathbb{R}^n$ preserve $\mathfrak{g}$ (or, if we think in terms of vector fields arising from a Lie algebra action, preserve the action), and also fix the origin. The tangent bundle TM of any n-dimensional manifold M has standard fibre $\mathbb{R}^n$ and group $\mathsf{GL}(n)$, so we can construct an infinitesimal Cartan algebra from these data. As we pointed out in Sect. 4.4, such an infinitesimal Cartan geometry cannot be derived from a Cartan geometry defined as a groupoid.

We end this section by discussing an infinitesimal version of the model Cartan geometry introduced in Sect. 6.1. We begin with some fairly general observations.

Let $\mathfrak{g}$ be a Lie algebra which acts on a manifold M, which we take to be connected. Consider the trivial bundle $\mathrm{pr}_1 : M \times M \to M$; we may of course consider it as a bundle with standard fibre M. For $v \in T_x M$, $a \in \mathfrak{g}$, $(v, a^\sharp)$ is a pr_1-projectable vector field on the fibre $\{x\} \times M$ (the second component defines a vector field along—tangent to—$\{x\} \times M$). The collection of all such projectable vector fields on fibres is a vector bundle $\mathcal{A}$ over M.

As before, a section ξ of $\mathcal{A} \to M$ is determined by a vector field X on M and a $\mathfrak{g}$-valued function $\bar{\xi}$ on M by

$$\xi(x) = \left(X_x, \bar{\xi}(x)^\sharp\right).$$

The bracket of such vector fields is given by

$$\left[(X, \bar{\xi}^\sharp), (Y, \bar{\eta}^\sharp)\right] = \left([X, Y], X(\bar{\eta})^\sharp - Y(\bar{\xi})^\sharp + \{\bar{\xi}, \bar{\eta}\}^\sharp\right).$$

With this bracket (and the obvious anchor) $\mathcal{A}$ is a transitive Lie algebroid. Notice that $\mathcal{K}_x$ consists effectively of fundamental vector fields $a^\sharp$, and the Lie algebra bracket of sections of $\mathcal{K}$ is effectively the bracket of $\mathfrak{g}$.

There is a natural flat infinitesimal connection on $\mathcal{A}$ given by $\gamma(X) = (X, 0)$. Moreover, there is a section of pr_1 with respect to which this connection is nondegenerate, namely the diagonal section $\varsigma : x \mapsto (x, x)$: as before, we take ς as the attachment section. The standard fibre is evidently of the same dimension as M. So provided that $\mathfrak{g}$ acts transitively and effectively on M, $\mathcal{A}$ with the infinitesimal connection γ is an infinitesimal Cartan geometry. We might call it the *model infinitesimal Cartan geometry defined by the action*. Evidently the infinitesimal Cartan geometry derived from a model Cartan geometry, as described in Sect. 6.2, is an example. As in that case, the subalgebroid $\mathcal{A}_o$ consisting of those elements of $\mathcal{A}$ which are tangent to the image of the attachment section is isomorphic to the action algebroid of the action of $\mathfrak{g}$ on M.

6.5 Nondegeneracy and Soldering

The nondegeneracy condition on the infinitesimal connection of an infinitesimal Cartan geometry, that $\mathcal{A}_{o,x} \cap \gamma_x(T_x M) = \{0\}$, has the important consequence that because of it the bundle E is soldered to M along the image of the attachment section, as we now explain. We first define soldering, a concept which is independent of the existence of an infinitesimal connection; as we mentioned in the introduction, it is supposed to capture the idea that for each $x \in M$, the fibre E_x is not just attached to M at $\varsigma(x)$, but is tangent to it. An early recognition of the significance of soldering in the context of Cartan connection theory may be found in Kobayashi [28].

Let $E \to M$ be a fibre bundle, with standard fibre F, where $\dim F = \dim M$, which has a global cross-section $\varsigma : M \to E$. Then E is *soldered* to M along the image of ς if there is a map $\mathfrak{s} : TM \to TE$, the *soldering map*, satisfying the conditions that if $v \in T_x M$ then $\mathfrak{s}(v) \in T_{\varsigma(x)} E$, and that the restriction $\mathfrak{s}|_{T_x M}$ is a linear isomorphism $T_x M \to T_{\varsigma(x)} E_x$.

The standard example of a soldered bundle is the tangent bundle $TM \to M$, where the global section ς is the zero section, and the soldering map is the vertical lift operator, or more exactly the identification of the vector space $T_x M$ with its own tangent space at the origin.

For another example we consider the Cartan projective bundle $\mathrm{P}\mathcal{W}M \to M$, which we discussed in Sects. 4.3 and 4.4. Recall that, for each choice of weight p, $\mathrm{P}\mathcal{W}M \to M$ has a canonical section $x \mapsto [e(x)]$; we write $[e(x)]$ for the ray generated by $e(x)$ when thinking of it as an element of the projective space $\mathrm{P}\mathcal{W}_x M$, and $\langle e(x) \rangle$ when thinking of it as a one-dimensional subspace of $\mathcal{W}_x M$. There is also a canonical soldering map $\mathfrak{s} : TM \to T\mathrm{P}\mathcal{W}M$ which we construct as follows. Given $x \in M$ let $\mathrm{p}_x : \mathcal{W}_x M \to \mathrm{P}\mathcal{W}_x M$ be the projection; then

$$\mathrm{p}_{x*} : T_{e(x)} \mathcal{W}_x M \to T_{[e(x)]}(\mathrm{P}\mathcal{W}_x M)$$

is a surjective linear map whose kernel is the 1-dimensional subspace $T_{e(x)} \langle e(x) \rangle \subset T_{e(x)} \mathcal{W}_x M$. Now $T_{e(x)} \mathcal{W}_x M$ is canonically isomorphic to $\mathcal{W}_x M$ using the inverse of the vertical lift, with $T_{e(x)} \langle e(x) \rangle$ corresponding to $\langle e(x) \rangle \subset \mathcal{W}_x M$ itself. Thus $T_{[e(x)]}(\mathrm{P}\mathcal{W}_x M)$ is isomorphic to the quotient space $\mathcal{W}_x M / \langle e(x) \rangle \cong T_x M$, and we may let

$$\mathfrak{s}_x : T_x M \to T_{[e(x)]}(\mathrm{P}\mathcal{W}_x M) \subset T_{[e(x)]}(\mathrm{P}\mathcal{W}M)\big|_{\mathrm{P}\mathcal{W}_x M}$$

be the inverse map. (The construction works in fact for any weight p, but we shall need it only for the case $p = 1/(n+1)$.)

As a final example we take the bundle $\mathrm{pr}_1 : M \times M \to M$ which underlies the model Cartan geometries. Clearly the map $T_x M \to T_{(x,x)}(M \times M)$ by $v \mapsto (0, v)$ defines a soldering of $M \times M$ to M along the diagonal, the image of the attachment section.

We now turn to infinitesimal Cartan geometries. An infinitesimal connection γ on a Lie algebroid can be represented by its kernel projection ω, which, it will be recalled, is defined as follows: for any $x \in M$, ω_x is the linear map $\mathcal{A}_x \to \mathcal{K}_x$ where, for $\boldsymbol{a} \in \mathcal{A}_x$,

$$\omega_x(\boldsymbol{a}) = \boldsymbol{a} - \gamma(\rho(\boldsymbol{a}))\,.$$

Thus ω_x is projection of $\mathcal{A}_x$ to $\mathcal{K}_x$ along $\gamma(T_x M)$. Now under the assumption that $\mathcal{A}_{o,x} \cap \gamma(T_x M) = \{0\}$, the restriction of ω_x to $\mathcal{A}_{o,x}$ is injective; and then since $\dim \mathcal{A}_{o,x} = \dim \mathcal{K}_x$ it is an isomorphism $\mathcal{A}_{o,x} \to \mathcal{K}_x$. That is to say, the nondegeneracy condition on the infinitesimal connection in the definition of an infinitesimal Cartan geometry can be stated equivalently as the requirement that the kernel projection ω is such that for each $x \in M$ the linear map $\omega_x : \mathcal{A}_x \to \mathcal{K}_x$ restricts to an isomorphism $\mathcal{A}_{o,x} \to \mathcal{K}_x$.

Let us interpret this result in terms of the representation of elements of $\mathcal{A}$ as projectable vector fields along fibres of $\pi : E \to M$. For an infinitesimal Cartan geometry we have a global section $\varsigma : M \to E$, and $\mathcal{A}_{o,x}$ consists of those elements of $\mathcal{A}_x$ which, considered as vector fields along E_x, are tangent to the submanifold $\varsigma(M)$. Moreover, the elements of $\mathcal{K}_{o,x}$, considered as vector fields tangent to E_x, vanish at $\varsigma(x)$. But at each point p of E_x, $T_p E_x$ (the tangent space to the fibre, that is, the vertical subspace of $T_p E$) is isomorphic to $\mathcal{K}_x/\mathcal{K}_{o,x}$; so in particular the vector space spanned by the values of the elements of $\mathcal{K}_x$ at $\varsigma(x)$ is the whole of $T_{\varsigma(x)} E_x$. Now ω_x induces an isomorphism of $\mathcal{A}_{o,x}$ with $\mathcal{K}_x$; on restriction to $\varsigma(x)$ this induces an isomorphism of $T_{\varsigma(x)}\varsigma(M)$ with $T_{\varsigma(x)} E_x$. But $T_{\varsigma(x)}\varsigma(M)$ is just an isomorphic copy of $T_x M$. We have shown that in any Cartan geometry the kernel projection induces for each $x \in M$ an isomorphism $T_x M \to T_{\varsigma(x)} E_x$. That is to say, for any Cartan geometry the bundle E is soldered to M along the image of the section ς, the soldering being defined by the kernel projection of the infinitesimal connection.

We examine this construction in more detail.

Let $\{e_\alpha\} = \{e_i, e_\mu\}$ be a basis for $\mathfrak{g}$, with $\{e_\mu\}$ a basis for $\mathfrak{g}_o$. Take a local trivialization T of $E \to M$ over $U \subset M$, with $\{e_\alpha^{\sharp\mathrm{T}}\} = \{e_i^{\sharp\mathrm{T}}, e_\mu^{\sharp\mathrm{T}}\}$ corresponding vertical vector fields on $\pi^{-1}(U)$ forming a basis for $\mathcal{K}$. Then $e_\mu^{\sharp\mathrm{T}}$ vanishes on $\varsigma(U)$.

Any element $\boldsymbol{a} \in \mathcal{A}_x$ may be written as

$$\boldsymbol{a} = \left(v^i \frac{\partial}{\partial x^i} + a^\alpha e_\alpha^{\sharp\mathrm{T}} \right)_{E_x}$$

where $v^i, a^\alpha \in \mathbb{R}$ and $(a^\alpha) = (a^i, a^\mu)$; then $\boldsymbol{a} \in \mathcal{A}_{o,x}$ if $a^i = 0$, so that $\mathcal{A}_{o,x}$ is spanned over $\mathbb{R}$ by the vector fields $\partial/\partial x^i$, $e_\mu^{\sharp\mathrm{T}}$ along E_x. Of course any element of $\mathcal{K}_x$ takes the form $a^\alpha e_\alpha^{\sharp\mathrm{T}}|_{E_x}$. The tangent space to $\varsigma(U)$ at $\varsigma(x)$ is spanned by the $\partial/\partial x^i$ (considered as elements of $T_{\varsigma(x)}E$).

An infinitesimal connection γ on $\mathcal{A}$ is given over U by

$$\gamma \circ \frac{\partial}{\partial x^i} = \frac{\partial}{\partial x^i} - \Gamma_i^\alpha e_\alpha^{\sharp\mathrm{T}}$$

where $\Gamma_i^\alpha \in C^\infty(U)$; this is to be regarded as a projectable vector field on $\pi^{-1}(U)$. Now for $v \in T_x M, \gamma(v) \in \mathcal{A}_{o,x}$ if and only if $v^j \Gamma_j^i(x) = 0$. Thus $\mathcal{A}_{o,x} \cap \gamma_x(T_x M) = \{0\}$ if and only if the matrix $\big(\Gamma_j^i(x)\big)$ is nonsingular. That is, the infinitesimal connection γ satisfies the nondegeneracy condition if and only if the matrix $\big(\Gamma_j^i\big)$ is everywhere nonsingular.

For the kernel projection ω we have, for $\boldsymbol{a} \in \mathcal{A}_x$,

$$\omega_x(\boldsymbol{a}) = \omega_x\left(v^i \frac{\partial}{\partial x^i} + a^\alpha e_\alpha^{\sharp\mathrm{T}}\right) = (a^\alpha + v^i \Gamma_i^\alpha) e_\alpha^{\sharp\mathrm{T}} .$$

Evaluation at $\varsigma(x) \in E_x$ gives

$$\omega_x\left(v^i \frac{\partial}{\partial x^i} + a^\alpha e_\alpha^{\sharp\mathrm{T}}\right)_{\varsigma(x)} = (a^i + v^j \Gamma_j^i) e_i^{\sharp\mathrm{T}}\Big|_{\varsigma(x)}$$

since $e_\mu^{\sharp\mathrm{T}}\big|_{\varsigma(x)} = 0$. If $\boldsymbol{a} \in \mathcal{A}_{o,x}$ then $a^i = 0$ so that

$$\boldsymbol{a}_{\varsigma(x)} = v^i \left.\frac{\partial}{\partial x^i}\right|_{\varsigma(x)} \in \varsigma_*(T_x M), \qquad \omega(\boldsymbol{a})_{\varsigma(x)} = v^j \Gamma_j^i(x) e_i^{\sharp\mathrm{T}}(\varsigma(x)) \in T_{\varsigma(x)} E_x .$$

We therefore obtain a linear map $\mathfrak{s} : T_x M \to T_{\varsigma(x)} E_x$ by starting with a tangent vector $v \in T_x M$, extending $\varsigma_*(v)$ in an arbitrary way to an algebroid element $\boldsymbol{a} \in \mathcal{A}_{o,x}$, applying the kernel projection ω to give an element of $\mathcal{K}_x$, and evaluating the result at $\varsigma(x)$. This evidently does not depend on the particular extension from $\varsigma_*(v)$ to $\boldsymbol{a}$. When γ satisfies the nondegeneracy condition, so that the matrix $\big(\Gamma_j^i\big)$ is everywhere nonsingular, $\mathfrak{s}$ is an isomorphism, and is a soldering map; and conversely.

We have established the following important results.

Proposition 6.5.1 *Consider an infinitesimal Cartan geometry on a generalized F-space $\pi : E \to M$, with Lie algebroid $\mathcal{A}$, kernel $\mathcal{K}$, and subalgebroid $\mathcal{A}_o$. Let ω be the kernel projection corresponding to the infinitesimal connection γ of the geometry. Then*

- *the restriction of ω to $\mathcal{A}_o$ is a vector bundle isomorphism $\mathcal{A}_o \to \mathcal{K}$ over the identity of M;*

- *E is soldered to M along the attachment section* ς, *where the soldering map* $\mathfrak{s} : TM \to TE$ *is defined by* $\mathfrak{s}(v) = \omega_x(\boldsymbol{a})_{\varsigma(x)}$, *where* $v \in T_x M$ *and* $\boldsymbol{a} \in \mathcal{A}_{o,x}$ *is such that* $\boldsymbol{a}_{\varsigma(x)} = \varsigma_*(v)$. □

We pointed out at the beginning of this section that the concept of soldering is independent of the existence of a connection, and we gave three examples of bundles with canonical soldering maps. When dealing with infinitesimal Cartan geometries on a generalized F-space which is canonically soldered to its base it is natural to prefer those whose infinitesimal connections define solderings that match the canonical one, and this we shall always do. The two cases of interest are the affine and projective ones. In particular, with reference to the terminology of Sect. 4.2, we shall deal only with affine, never with generalized affine, connections.

6.6 Curvature and Torsion of an Infinitesimal Cartan Geometry

The curvature of an infinitesimal Cartan geometry is of course the curvature R^γ of its infinitesimal connection γ. It has all of the properties of curvature set out in previous chapters: in particular it is a $\mathcal{K}$-valued 2-form on M (Lemma 2.2.4).

For each pair of vector fields X, Y on M, we may interpret $R^\gamma(X, Y)$ as a vector field on E tangent to its fibres. We may restrict this vector field to the image of the attachment section: then

$$x \mapsto \mathfrak{s}^{-1}\big(R^\gamma(X_x, Y_x)|_{\varsigma(x)}\big)\,,$$

where $\mathfrak{s}$ is the soldering map, defines a vector field on M. If we set

$$T(X_x, Y_x) = \mathfrak{s}^{-1}\big(R^\gamma(X_x, Y_x)|_{\varsigma(x)}\big)$$

then T is a type (2, 1) tensor field on M which is skew-symmetric in X and Y: we call it the *torsion* of the infinitesimal Cartan geometry.

A more conventional way of defining torsion would be as the projection of $R^\gamma(X, Y)$ into $\mathcal{K}/\mathcal{K}_o$. But sections of $\mathcal{K}_o$, considered as vertical vector fields on E, are precisely those sections of $\mathcal{K}$ which vanish on the image of ς, so the two definitions are essentially equivalent; the first, however, brings out more clearly the tensorial nature of the torsion. In particular it makes it clear that we may consistently require that an infinitesimal Cartan geometry be torsion free, something we shall normally do without further comment.

As before, we take a local trivialization T of E, with $\{e_\alpha^{\sharp\mathrm{T}}\}$ a local basis of vertical vector fields corresponding to a basis $\{e_\alpha\} = \{e_i, e_\mu\}$ of $\mathfrak{g}$ with $\{e_\mu\}$ a basis for $\mathfrak{g}_o$. With

$$\gamma \circ \frac{\partial}{\partial x^i} = \frac{\partial}{\partial x^i} - \Gamma_i^\alpha e_\alpha^{\sharp\mathrm{T}}\,,$$

where the coefficients are local functions on M, we find that

$$R^\gamma\left(\frac{\partial}{\partial x^i}, \frac{\partial}{\partial x^j}\right) = \left(\frac{\partial\Gamma^\alpha_j}{\partial x^i} - \frac{\partial\Gamma^\alpha_i}{\partial x^j} - \Gamma^\beta_i\Gamma^\gamma_j C^\alpha_{\beta\gamma}\right) e^{\sharp\mathrm{T}}_\alpha$$

where the $C^\alpha_{\beta\gamma}$ are the structure constants of $\mathfrak{g}$ in the given basis. Then

$$T^k_{ij} = \bar{\Gamma}^k_l \left(\frac{\partial\Gamma^l_j}{\partial x^i} - \frac{\partial\Gamma^l_i}{\partial x^j} - \Gamma^\beta_i\Gamma^\gamma_j C^l_{\beta\gamma}\right),$$

where the $\bar{\Gamma}^k_l$ are the elements of the matrix inverse to Γ^k_l, since

$$\mathfrak{s}_x\left(\frac{\partial}{\partial x^i}\right) = \Gamma^j_i(x) e^{\sharp\mathrm{T}}_j\Big|_{\varsigma(x)}.$$

In many cases of interest $\Gamma^i_j(x) = \delta^i_j$, and in those cases

$$T^k_{ij} = -\Gamma^\beta_i\Gamma^\gamma_j C^k_{\beta\gamma}.$$

A more revealing expression is obtained as follows. Take fibre coordinates y^i on E, such that the attachment section is given by $y^i = 0$, and such that on the attachment section

$$e^{\sharp\mathrm{T}}_i = \frac{\partial}{\partial y^i};$$

and suppose that $\Gamma^i_j(x) = \delta^i_j$. In many such cases of interest we may write

$$\gamma\left(\frac{\partial}{\partial x^i}\right) = \frac{\partial}{\partial x^i} - (\delta^j_i + \Gamma^j_{ik} y^k + \cdots)\frac{\partial}{\partial y^j}$$

where the coefficients Γ^j_{ik} are local functions on M. It is then easy to see that

$$T^k_{ij} = \Gamma^k_{ij} - \Gamma^k_{ji}.$$

Thus the requirement for such an infinitesimal Cartan geometry to be torsion free is that the coefficients Γ^i_{jk} be symmetric in their lower indices: this is of course formally the same as the condition satisfied by the connection coefficients of a torsion-free linear connection.

6.7 Symmetries of Cartan Geometries

In Chap. 5 we defined a symmetry of a path connection to be a diffeomorphism of the Lie groupoid $\mathcal{G}$ on which the path connection was defined; the diffeomorphism was obtained from a bisection of $\mathcal{G}$. We intend now to define a symmetry of a Cartan geometry, and we could do so in terms of such a diffeomorphism. We shall, however, use instead a diffeomorphism of the fibre bundle $\pi : E \to M$.

Let $\Theta : E \to E$ be a fibre-preserving diffeomorphism, and $\theta : M \to M$ the induced diffeomorphism of M, so that $\pi \circ \Theta = \theta \circ \pi$; the pair (Θ, θ) is a bundle automorphism of π. If, for every $x \in M$, the map $\Theta|_{E_x} : E_x \to E_{\theta(x)}$ belongs to the groupoid $\mathcal{G}$, we say that (Θ, θ) is an automorphism of π^G; if in addition $\Theta|_{E_x} \in \mathcal{G}_o$ then (Θ, θ) is an automorphism of π^{G_o}. The requirement that (Θ, θ) be an automorphism of π considered as a G_o-bundle (rather than a G-bundle) will encapsulate the idea that a symmetry should preserve the attachment section. We shall therefore say that the G_o-bundle automorphism (Θ, θ) is a *symmetry* of the Cartan geometry with path connection Γ if the bisection $\Phi : M \to \mathcal{G}_o$, $\Phi(x) = \Theta|_{E_x}$, is a symmetry of Γ in the sense of Chap. 5: that is, writing $\phi = \theta : M \to M$ for the projection of Φ, if $\tilde{\Phi} \circ c^\Gamma = (\phi \circ c)^\Gamma$ for every curve $c : [0, 1] \to M$.

We shall say that a symmetry (Θ, θ) of a Cartan geometry is a *vertical symmetry* if $\theta = \mathrm{id}_M$. One feature of our definition is that there are no nontrivial vertical symmetries when the standard fibre F is connected, so that in such a case if $\theta = \mathrm{id}_M$ then also $\Theta = \mathrm{id}_E$. To show this, we start with an observation about the effect of any symmetry on developed curves.

Lemma 6.7.1 *Let (Θ, θ) be a symmetry of a Cartan geometry on $\pi : E \to M$ with connected standard fibre F. If $x \in M$ and $c : [0, 1] \to M$ is any curve with $c(0) = x$ then, for each $t \in [0, 1]$,*

$$\Theta\big(c_\Gamma(t)\big) = (\theta \circ c)_\Gamma(t) \in E_{\theta(x)} .$$

In particular if $\theta = \mathrm{id}_M$ then $\Theta\big(c_\Gamma(t)\big) = c_\Gamma(t)$ for any $t \in I$.

Proof Let (Φ, ϕ) be the bisection of $\mathcal{G}_o$ corresponding to (Θ, θ). Then Φ must preserve the path connection, so that for any curve c in M we must have $\tilde{\Phi}\big(c^\Gamma(t)\big) = (\phi \circ c)^\Gamma(t)$ for each $t \in [0, 1]$; and Φ must also preserve the attachment section, so that $\varsigma\phi(x) = \Phi(x)\big(\varsigma(x)\big)$ for any $x \in M$.

Now the development of c is given by

$$(\phi \circ c)_\Gamma(t) = \big((\phi \circ c)^\Gamma(t)\big)^{-1}\big(\varsigma\phi c(t)\big)$$

where

$$\big((\phi \circ c)^\Gamma(t)\big)^{-1} = \big(\tilde{\Phi}\big(c^\Gamma(t)\big)\big)^{-1} = \big(\Phi(c(t)) \cdot c^\Gamma(t) \cdot \Phi^\iota(c(0))\big)^{-1} = \Phi(c(0)) \cdot \big(c^\Gamma(t)\big)^{-1} \cdot \Phi^\iota(c(t))$$

showing that

$$(\phi \circ c)_\Gamma(t) = \Phi(c(0))\Big(\big(c^\Gamma(t)\big)^{-1}\big(\varsigma(c(t))\big)\Big) = \Phi(c(0))\big(c_\Gamma(t)\big),$$

from which the result follows. □

Thus if $\theta = \mathrm{id}_M$ then Θ is the identity when restricted to those points in E_x that can be reached by a developed curve from $\varsigma(x)$. But, by the nondegeneracy condition for a Cartan geometry, every point in a neighbourhood of $\varsigma(x)$ can be reached by a developed curve.

Proposition 6.7.2 *A vertical symmetry Θ is the identity in a neighbourhood of $\varsigma(M)$.*

Proof We know from Lemma 3.4.1 that, for injective curves c,

$$\dot{c}_\Gamma(0) = \upsilon(\varsigma_* \dot{c}(0))$$

where $\upsilon : T_y E \to V_y E$ is the vertical projection determined by the path connection, and by the nondegeneracy condition $\upsilon \circ \varsigma_* : T_x M \to V_{\varsigma(x)} E$ is an isomorphism.

Now let ∇ be any symmetric linear connection on M (this need have no relation to the path connection of the Cartan geometry) and let $\exp : U_x \to U$ be an exponential map where $U \subset M$ is a neighbourhood of x and $U_x \subset T_x M$ is a neighbourhood of 0_x, so that if $u \in U_x$ and if $c : [0, 1] \to M$ is the ∇-geodesic with $c(0) = x$ and $\dot{c}(0) = u$ then $\exp(u) = c(1)$. We may suppose U_x to be sufficiently small that c is injective.

We now define a map $\varepsilon : U \to E_x$ by taking, for any $\hat{x} \in U$, the ∇-geodesic c satisfying $c(0) = x$ and $\dot{c}(0) = \exp^{-1}(\hat{x})$, and setting $\varepsilon(\hat{x}) = c_\Gamma(1)$. Then ε is smooth, $\varepsilon(c(t)) = c_\Gamma(t)$, and $\varepsilon_* : T_x M \to V_{\varsigma(x)} E$ satisfies

$$\varepsilon_* \dot{c}(0) = \dot{c}_\Gamma(0) = \upsilon(\varsigma_* \dot{c}(0)).$$

As $T_x M$ is spanned by tangent vectors of the form $\dot{c}(0) \in U_x$ where c is injective, it follows that ε_* is nonsingular so that ε is smoothly invertible in a neighbourhood $N_x \subset E_x$ of $\varsigma(x)$. Thus every point $y \in N_x$ lies on the development of a curve in M, and so $\Theta_x(y) = y$ for all such y. □

In order to complete our task we need a lemma about manifolds and maps which are real analytic rather than merely C^∞.

Lemma 6.7.3 *Let F be a connected real analytic manifold, and let $\phi : F \to F$ be a real analytic map. If there is a nonempty subset $U \subset F$ such that $\phi(x) = x$ whenever $\phi \in U$, then $\phi = \mathrm{id}_F$.*

Proof Let S be the set of points $x \in F$ with the property that there is an open set U_x containing x such that, for all $y \in U_x$, $\phi(y) = y$. Then $S = \bigcup_{x \in S} U_x$ is clearly clearly open; we now show that it is also closed.

Let z_k be a sequence of points of S converging to a point z; then $\phi(z) = z$ as ϕ is continuous. We need to show that there is a neighbourhood of z on which $\phi(y) = y$.

Take a coordinate neighbourhood W of z from the analytic atlas with z as origin, and let x^i be the coordinates on W. Since $\phi(z) = z$, the open subset $V = W \cap \phi^{-1}(W)$ of W contains z and satisfies $\phi(V) \subset W$. At any point of V, ϕ is represented by functions ϕ^i which are analytic functions of the x^i, and in addition $\phi^i(0) = 0$. Furthermore, for all k sufficiently large, $z_k \in V$.

Now ϕ^i is analytic, and $\phi(y) = y$ on a neighbourhood of z_k, so the Taylor expansion of ϕ^i about z_k is just $z^i_k + (x^i - z^i_k) = x^i$ on that neighbourhood.

We now consider the Taylor expansion of ϕ^i about the limit point z.

The partial derivatives of each ϕ^i are continuous, so the value of each partial derivative of ϕ^i at z is the limit of the sequence of the values of the same partial derivative at z_k. So the Taylor series of ϕ^i about z is also x^i. Since ϕ^i is analytic this means that $\phi^i = x^i$ on some neighbourhood of z, so that $\phi(y) = y$ on that neighbourhood, as required. That is, S is both open and closed; but S contains the nonempty set U, so it must be the whole of F. □

Theorem 6.7.4 *A vertical symmetry of a Cartan geometry with connected standard fibre must be trivial.*

Proof Take $x \in M$, and let $\mathrm{T} : U \times F \to E_U$ be a trivialization of π around x, so that $\mathrm{T}^{-1} \circ \Theta|_{E_x} \circ \mathrm{T}$ is represented by the action of an element $g \in G_o$ on F where $o \in F$ is the point corresponding to $\varsigma(x)$, so that there is some neighbourhood $N \subset F$ of o such that $gq = q$ for $q \in N$.

We now use a result of Chevalley [14] that if G is a Lie group and G_o a closed subgroup then $F = G/G_o$ admits the structure of a real analytic manifold in such a way that the action of G on F is real analytic. If follows from this, the connectedness of F and the previous lemma that $gq = q$ for all $q \in F$, so that $g = 1_{G_o}$. □

Our reason for regarding Θ (rather than Φ) as the symmetry of a Cartan geometry is related to our preference for considering infinitesimal symmetries of an infinitesimal connection γ, sections of the Lie algebroid, as projectable vector fields on the fibre bundle. Given an infinitesimal Cartan geometry, we consider sections ξ of $\mathcal{A}_o \to M$ rather than of $\mathcal{A} \to M$, and we say that such a section is an *infinitesimal symmetry* of the infinitesimal Cartan geometry based on γ if it is an infinitesimal symmetry in the sense of Chap. 5: that is, if $\mathcal{L}_\xi \gamma = 0$. That is, an infinitesimal symmetry of an infinitesimal Cartan geometry is a projectable vector field ξ on E which is a section of $\mathcal{A}_o \to M$ such that for every vector field X on M,

$$[\xi, \gamma \circ X] = \gamma \circ [\pi_* \xi, X] .$$

We shall say that an infinitesimal symmetry ξ of an infinitesimal Cartan geometry is *vertical* if $\pi_* \xi = 0$ (or, in other words, that it is a section of $\mathcal{K}_o \to M$). This is the infinitesimal analogue of the statement that a finite symmetry (Θ, θ) is vertical when $\theta = \mathrm{id}_M$. As with finite symmetries, it is again the case that a vertical infinitesimal symmetry of an infinitesimal Cartan geometry must be trivial. If the infinitesimal geometry is derived from a finite Cartan geometry then we could consider the flow of the corresponding right invariant vector field on the Lie groupoid and appeal to the theorem about finite vertical symmetries. But if the infinitesimal geometry is not derived from a finite geometry then this approach is no longer open to us. Nevertheless the result still holds, and we give a proof in Chap. 11.

6.8 Projectability of Cartan Geometries and Their Symmetries

In subsequent chapters we shall need to consider two fibre bundles related by a projection, and we shall need to see whether an infinitesimal Cartan geometry on the first bundle, together with its Lie algebra of infinitesimal symmetries, projects to a similar geometry and Lie algebra on the second. The flat infinitesimal symmetries introduced in Sect. 5.4 will help us investigate this question, and so we first establish some more of their properties.

Suppose first that we have an infinitesimal Cartan geometry on a fibre bundle $\pi : E \to M$ with standard fibre F where $\dim F = \dim M$, with attachment section $\varsigma : M \to E$, and with infinitesimal connection γ. Denote by $\mathcal{A}$ the Lie algbroid of the geometry, regarded as consisting of vector fields on E projectable to M, and $\mathcal{A}_o$ the subalgebroid of $\mathcal{A}$ consisting of vector fields tangent to the image of ς. Let $\mathcal{K}$ be the kernel of $\mathcal{A}$, $\mathcal{K}_o$ that of $\mathcal{A}_o$. We know (Proposition 6.5.1) that ω defines a vector bundle isomorphism $\mathcal{A}_o \to \mathcal{K}$.

Consider the Lie algebra $\mathfrak{a}$ of infinitesimal symmetries of the geometry: its elements are sections of $\mathcal{A}_o$ which are infinitesimal symmetries of the connection γ. Let $\mathfrak{f}$ be the subalgebra of $\mathfrak{a}$ consisting of those symmetries of γ which are flat. Then $\mathfrak{f}$ is an ideal in $\mathfrak{a}$, by Proposition 5.4.6, together with the fact that $\mathfrak{a}$ is closed under the algebroid bracket. Moreover, $\omega(\mathfrak{f})$ consists of symmetries of γ, though not necessarily of symmetries of the geometry: that is, for $\xi \in \mathfrak{f}$, $\omega \circ \xi$ is a section of $\mathcal{K}$, but not in general one of $\mathcal{K}_o$. In fact for $\xi \in \mathfrak{f} \neq 0$, $\omega \circ \xi$ cannot be a symmetry of the geometry because, using the result mentioned at the end of the previous section, there are no nonzero vertical infinitesimal symmetries of the geometry: and if $\omega \circ \xi = 0$ then $\xi = 0$ because ω is an isomorphism of vector bundles $\mathcal{A}_o \to \mathcal{K}$. Now $\omega(\mathfrak{f})$ is a Lie algebra isomorphic to $\mathfrak{f}$. Consider $\mathfrak{f} \oplus \omega(\mathfrak{f})$. For any $\xi, \xi', \eta, \eta' \in \mathfrak{f}$,

$$\begin{aligned} [\![\xi + \omega \circ \xi', \eta + \omega \circ \eta']\!] &= [\![\xi, \eta]\!] + [\![\xi, \omega \circ \eta']\!] + [\![\omega \circ \xi', \eta]\!] + [\![\omega \circ \xi', \omega \circ \eta']\!] \\ &= [\![\xi, \eta]\!] + \omega \circ ([\![\xi, \eta']\!] + [\![\xi', \eta]\!] + [\![\xi', \eta']\!]), \end{aligned}$$

using the fact that $\mathcal{L}_\xi \omega = \mathcal{L}_\eta \omega = 0$ and the homomorphism property of ω in this case. Thus $\mathfrak{f} \oplus \omega(\mathfrak{f})$ is a Lie algebra isomorphic to the semidirect sum of $\mathfrak{f}$ with itself induced by the adjoint action of $\mathfrak{f}$ on itself. Set $\mathfrak{f} \oplus \omega(\mathfrak{f}) = \mathfrak{F}$. Then $\mathfrak{F}$ is a Lie algebra of symmetries of γ, $\rho(\mathfrak{F})$ is a Lie algebra of vector fields on M isomorphic to $\mathfrak{f}$, and the kernel of $\rho|_{\mathfrak{F}}$ is a Lie algebra of vertical vector fields on E also isomorphic to $\mathfrak{f}$; moreover, $\dim \mathfrak{F} = 2 \dim \mathfrak{f}$.

Now suppose we have two bundles $\pi : E \to M$, $\pi' : E' \to M'$ and a fibred map (P, p), specified by this diagram.

$$\begin{array}{ccc} E & \xrightarrow{\mathrm{P}} & E' \\ {\scriptstyle\pi}\downarrow & & \downarrow{\scriptstyle\pi'} \\ M & \xrightarrow[p]{} & M' \end{array}$$

We assume that

- (P, p) is a full map of fibred manifolds;
- E carries an infinitesimal Cartan geometry, with Lie algebra $\mathfrak{f}$ of flat infinitesimal symmetries;
- for each $u \in E$, the kernel of $\mathrm{P}_{*u} : T_u E \to T_{\mathrm{P}(u)} E'$ is $\mathfrak{F}_u$, the subspace of $T_u E$ spanned by the elements of $\mathfrak{F}$, considered as vector fields on E, evaluated at the point u.

To put the last requirement another way. The Lie algebra $\mathfrak{F}$, considered as a Lie algebra of vector fields on E, is finite dimensional; but it defines a distribution (in the sense of Frobenius) $D_{\mathfrak{F}}$ by $D_{\mathfrak{F}}|_u = \mathfrak{F}_u$. Again: we may think of $\mathfrak{F}$ as an abstract Lie algebra, with an effective action on E; then $D_{\mathfrak{F}}$ is a realization of the action algebroid generated by the action. Since $\mathfrak{F}$ is a Lie algebra, that is, closed under bracket, $D_{\mathfrak{F}}$ is integrable in the sense of Frobenius. We require that the fibres of the surjective submersion $\mathrm{P} : E \to E'$ are the integral submanifolds of $D_{\mathfrak{F}}$. Similarly, the fibres of $p : M \to M'$ are the integral submanifolds of the integrable distribution on M generated in the same way by the action of $\mathfrak{f}$.

We should like to see that E' also carries an infinitesimal Cartan geometry. Now the fibre dimension of E' is the dimension of M', since each is the corresponding dimension on the left-hand side of the diagram reduced by $\dim \mathfrak{f}$. This is a good start, but clearly more is required, namely an infinitesimal version of compatibility as defined in Sect. 3.3.

So consider the standard fibre F of $E \to M$, on which the Lie algebra $\mathfrak{g}$ acts effectively and transitively. We may regard $\mathfrak{f}$ as a subalgebra of $\mathfrak{g}$, obtained by representing $\omega(\mathfrak{f})$ on F via a local trivialisation. We construct the integrable distribution $D_{\mathfrak{f}}$ as before, and assume that its integral submanifolds form the fibres of a surjective submersion $\varpi : F \to K$. But we also want a Lie algebra action on K, so we need to consider the conditions for a vector field X on F to be projectable to K. The necessary and sufficient condition for projectability is that $[X, D_{\mathfrak{f}}] \subset D_{\mathfrak{f}}$. Therefore the projectable elements a of $\mathfrak{g}$ are those for which $[a^\sharp, D_{\mathfrak{f}}] \subset D_{\mathfrak{f}}$, which amounts to $[a, \mathfrak{f}] \subset \mathfrak{f}$.

Now recall that the normalizer $\mathfrak{n}$ of $\mathfrak{f}$ in $\mathfrak{g}$ is defined by

$$\mathfrak{n} = \left\{ a \in \mathfrak{g} : \{a, b\} \in \mathfrak{f} \quad \text{for all } b \in \mathfrak{f} \right\};$$

$\mathfrak{n}$ is a subalgebra of $\mathfrak{g}$ containing $\mathfrak{f}$, and $\mathfrak{f}$ is an ideal of $\mathfrak{n}$. So the projectable elements of $\mathfrak{g}$ are those in the normalizer of $\mathfrak{f}$ in $\mathfrak{g}$. Since ϖ_* preserves brackets of projectable vector fields, and maps $\mathfrak{f}$ to zero, we obtain an action of the Lie algebra $\mathfrak{n}/\mathfrak{f}$ on K by projection, which is clearly effective. In order for (P, p) to define an infinitesimal Cartan geometry on E' we require that this action is transitive.

Proposition 6.8.1 *Let $\pi : E \to M$, $\pi' : E' \to M'$ be two bundles, and suppose that the three conditions itemised above are satisfied. If the foliation given by the integral manifolds of $D_{\mathfrak{f}}$ defines a fibred manifold $\varpi : F \to K$, and if the action of $\mathfrak{n}/\mathfrak{f}$ on K is transitive, then the infinitesimal connection γ and attachment section ς on $E \to M$ project to $E' \to M'$ and define an infinitesimal Cartan geometry. If $\mathfrak{a}$ denotes the algebra of infinitesimal symmetries of the geometry on $E \to M$ then $\mathfrak{a}/\mathfrak{f}$ is a Lie algebra of infinitesimal symmetries of the geometry on $E' \to M'$.*

Proof Consider first the infinitesimal connection γ. The conditions for it to be projectable are that

- for any vector field X on M projectable to M', $\gamma \circ X$ must be projectable to E';
- assuming this to be so, for any vector fields X_1 and X_2 on M such that $p_* X_1 = p_* X_2$, $\mathrm{P}_*(\gamma \circ X_1) = \mathrm{P}_*(\gamma \circ X_2)$.

Now the condition for any vector field X on M to be projectable to M' is that, for any $a \in \mathfrak{f}$ with fundamental vector field $a^\sharp$ on M, $[a^\sharp, X] \in D_{\mathfrak{f}}^M$, the distribution on M generated by the action of $\mathfrak{f}$. Similarly, the condition for $\gamma \circ X$ to be projectable to E' is that for any $b \in \mathfrak{F}$ with fundamental vector field $b^\sharp$ on E, $[b^\sharp, \gamma \circ X] \in D_{\mathfrak{F}}$. Since b is a symmetry of γ,

$$[b^\sharp, \gamma \circ X] = \mathcal{L}_{b^\sharp}\gamma \circ X + \gamma \circ ([\pi_* b^\sharp, X]) = \gamma \circ ([\pi_* b^\sharp, X]) .$$

But $\pi_* b^\sharp$ is a fundamental vector field of the action of $\mathfrak{f}$ on M, and every such fundamental vector field takes this form for some $b \in \mathfrak{f}$. Thus if X is projectable to M', $[\pi_* b^\sharp, X] \in D_{\mathfrak{f}}^M$. It follows from Proposition 5.4.6, by a simple extension, that if $[\pi_* b^\sharp, X] \in D_{\mathfrak{f}}^M$ then $\gamma \circ ([\pi_* b^\sharp, X]) \in D_{\mathfrak{F}}$, and therefore $[b^\sharp, \gamma \circ X] \in D_{\mathfrak{F}}$. That is to say, whenever X is projectable to M', $\gamma \circ X$ is projectable to E', so that the first condition is satisfied.

For the second, it is enough to show that if $p_* X = 0$ then $\mathrm{P}_*(\gamma \circ X) = 0$. Indeed, it is enough to show that for any $a \in \mathfrak{f}$, $\mathrm{P}_*(\gamma \circ \pi_* a^\sharp) = 0$. But from Proposition 5.4.6 we know that $\gamma \circ \pi_* a^\sharp \in \mathfrak{F}$, so certainly $\mathrm{P}_*(\gamma \circ \pi_* a^\sharp) = 0$.

It follows that γ projects to $E' \to M'$ and defines an infinitesimal connection γ' there.

The condition for projectability of the attachment section is that $\mathrm{P}(\varsigma(x_1)) = P(\varsigma(x_2))$ whenever $p(x_1) = p(x_2)$. This may be translated into to the following equivalent differential condition: for any $v \in T_x M$ such that $p_{*x}(v) = 0$, $P_{*\varsigma(x)}(\varsigma_{*x}(v)) = 0$. But if $p_{*x}(v) = 0$ then $\varsigma_{*x}(v) \in D_{\mathfrak{f}}|_{\varsigma(x)}$, so certainly $P_{*\varsigma(x)}(\varsigma_{*x}(v)) = 0$; thus $\varsigma : M \to E$ projects to an attachment section $\varsigma' : M' \to E'$.

The infinitesimal symmetries of the Cartan geometry on E form a Lie algebra $\mathfrak{a}$, of which $\mathfrak{f}$ is an ideal. This means that the infinitesimal symmetries are projectable to E', and by Lemma 5.4.11 they project to infinitesimal symmetries of γ'; since they are tangent to the attachment section of E, which projects to the attachment section of E', they are infinitesimal symmetries of the infinitesimal geometry on E'. That is to say, $\mathfrak{a}/\mathfrak{f}$ is an algebra of infinitesimal symmetries of the projected geometry. □

We explore, finally, whether $\mathfrak{a}/\mathfrak{f}$ contains all the infinitesimal symmetries of the projected geometry.

Suppose that ξ', a vector field on E', is an infinitesimal symmetry of the Cartan geometry on E'. It is, at least locally, the projection of some vector field ξ on E. Let $\{\Upsilon_\alpha\}$ be a basis for $\mathfrak{f}$ on E, and set $\omega(\Upsilon_\alpha) = \Delta_\alpha$, so that $\{\Delta_\alpha\}$ is a basis for $\omega(\mathfrak{f})$. Then any other vector field which projects onto ξ' and is a section of $\mathcal{A}_o$ takes the form $\eta = \xi + f^\alpha \Upsilon_\alpha$ for some functions f^α on M. The task is to choose the coefficients f^α so as to obtain an element of $\mathfrak{a}$. From Lemma 5.4.9

$$\mathcal{L}_\eta \gamma \circ X = \mathcal{L}_\xi \gamma \circ X - (X f^\alpha)\Delta_\alpha .$$

Now ξ projects onto ξ', which is an infinitesimal symmetry of γ'. That is to say, for every projectable X, $\mathcal{L}_\xi \gamma \circ X$ (which is necessarily projectable) projects to zero. Now $\mathcal{L}_\xi \gamma \circ X$ is always vertical, so must be a linear combination over $C^\infty(M)$ of the Δ_α. Indeed, since $\mathcal{L}_\xi \gamma$ is tensorial, and $p : M \to M'$ is surjective, we may write, for any vector field X on M,

$$\mathcal{L}_\xi \gamma \circ X = \theta_\xi^\alpha(X)\Delta_\alpha$$

where the θ_ξ^α are 1-forms on M. We have

$$\theta_\eta^\alpha = \theta_\xi^\alpha - d f^\alpha,$$

and see therefore that $d\theta_\xi^\alpha$ depends only on ξ', not on which particular representative ξ we choose amongst those vector fields that project onto ξ'. We obtain another interpretation of this as follows. With $\mathcal{L}_\xi \gamma \circ X = \theta_\xi^\alpha(X)\Delta_\alpha$,

$$\begin{aligned}
\mathcal{L}_\xi R^\gamma(X, Y) &= \mathcal{L}_\xi \gamma \circ [X, Y] - [\mathcal{L}_\xi \gamma \circ X, \gamma \circ Y] - [\gamma \circ X, \mathcal{L}_\xi \gamma \circ Y] \\
&= \theta_\xi^\alpha([X, Y])\Delta_\alpha - [\theta_\xi^\alpha(X)\Delta_\alpha, \gamma \circ Y] - [\gamma \circ X, \theta_\xi^\alpha(Y)\Delta_\alpha] \\
&= \big(\theta_\xi^\alpha([X, Y]) + Y(\theta_\xi^\alpha(X)) - X(\theta_\xi^\alpha(Y))\big)\Delta_\alpha \\
&= -d\theta_\xi^\alpha(X, Y)\Delta_\alpha ,
\end{aligned}$$

using the fact that $[\Delta_\alpha, \gamma \circ Y] = [\gamma \circ X, \Delta_\alpha] = 0$ because Δ_α is an infinitesimal symmetry of γ and is vertical. From Lemma 5.4.10 we see that

$$\mathcal{L}_\eta R^\gamma = \mathcal{L}_{\xi + f^\alpha \Upsilon_\alpha} R^\gamma = \mathcal{L}_\xi R^\gamma .$$

We know that $\mathcal{L}_\xi R^\gamma$ projects to zero, and indeed the necessary and (locally) sufficient condition for there to be an infinitesimal symmetry of the infinitesimal Cartan geometry on E which projects onto the infinitesimal symmetry ξ' of the infinitesimal Cartan geometry on E' is that for one, and therefore any, ξ which projects onto ξ', $\mathcal{L}_\xi R^\gamma = 0$.

Chapter 7
A Comparison with Alternative Approaches

The title of this volume is 'Cartan geometries and their symmetries: a Lie algebroid approach', and as we explain in the Introduction there are several other ways to approach the study of Cartan geometry. In this chapter we compare our theory with several of these alternatives.

The conventional modern account of Cartan geometry involves the use of Cartan connection forms on principal bundles: see for example [10, 17, 31, 36]. In this theory a Cartan connection is an algebra-valued 1-form on the bundle satisfying certain conditions. The Lie algebra is not, though the algebra of the bundle group (that would be an Ehresmann connection) but instead a larger algebra with dimension equal to the dimension of the total space of the bundle rather than the dimension of its fibre. We discuss this approach in the first two sections.

In Sect. 7.3 we introduce the 'tractor connection'. The approach using tractor bundles has been described in, for example, [2, 10, 8], and offers the advantage that these are vector bundles, generally easier to deal with than principal bundles. We have, indeed, introduced aspects of the tractor approach in Sect. 4.3, and we shall make use of these bundles again in Chaps. 9 and 10.

The following section relates our work to that of A.D. Blaom [4, 5]. Blaom also makes some use of Lie algebroids, although he does not relate them to fibre morphisms of a bundle in the explicit way that we have. We do, though, make use of a connection which we call the 'Blaom connection' to indicate that it is based on his construction; we explain the relation between this connection and the tractor connection.

The remaining sections of this chapter explore some extensions of our definition.

7.1 Cartan Gauge

In this section and the following one we examine the relation between our definition of a Cartan geometry and that given by Sharpe in [36]. Sharpe's definition comes in two forms. The first, which is called the base definition of a Cartan geometry, uses

M. Crampin and D. Saunders, *Cartan Geometries and their Symmetries*,
Atlantis Studies in Variational Geometry 4, DOI 10.2991/978-94-6239-192-5_7

the concept of a Cartan gauge: we show here how to construct a Cartan gauge starting from our definition. Sharpe's second definition is called the bundle definition of a Cartan geometry. We deal with this in the next section.

Suppose given a Cartan geometry on a generalized F-space $\pi : E \to M$. We show how the infinitesimal connection on $A\mathcal{G}$ may be used to define a Cartan gauge.

We work with local trivializations as described in Chap. 3. Let $\tau_U : U \times F \to \pi^{-1}(U)$ be a local trivialization of $\pi : E \to M$; we saw in Proposition 3.2.2 that, for each $x \in U$, τ_U induces an isomorphism $\tau_{U,x\sharp} : T_x M \oplus \mathfrak{g} \to \mathcal{A}^G_x \cong A_x\mathcal{G}$. Given $v \in T_x M$ and $a \in \mathfrak{g}$, $\tau_{U,x\sharp}(v, a)$ is the projectable vector field along E_x defined as follows: for $p \in F$, so that $\tau_U(x, p) \in E_x$,

$$\tau_{U,x\sharp}(v, a)|_{\tau_U(x,p)} = \tau_{U*}(v, a^\sharp(p)) .$$

We define a $\mathfrak{g}$-valued 1-form θ_U on U as follows: for $v \in T_x M$, $\theta_{U,x}(v)$ is the (unique) element of $\mathfrak{g}$ such that

$$\gamma_x(v) = \tau_{U,x\sharp}\big(v, \theta_{U,x}(v)\big) \in A_x\mathcal{G} .$$

Suppose given an open covering of M, and for each open set U of the covering a $\mathfrak{g}$-valued 1-form θ_U on U. Provided that the θ_U satisfy certain conditions, to be specified immediately below, this set-up defines a *Cartan structure* [36, Definition 5.1.6]; the pair (U, θ_U) is called a *Cartan gauge*. In the definition cited the group H and its Lie algebra $\mathfrak{h}$ are the subgroup G_o and its Lie algebra $\mathfrak{g}_o$ of our Cartan geometry; in this section and the next we shall use the former notation rather than the latter.

The conditions for the θ_U to define a Cartan structure are that

- for each open set U of the covering and each $x \in U$ the linear map $T_x M \to \mathfrak{g}/\mathfrak{h}$ given by

$$v \mapsto \theta_{U,x}(v) \mod \mathfrak{h}$$

  is an isomorphism;
- when U, V are open sets of the covering with $U \cap V$ nonempty, there is a function $h : U \cap V \to H$ such that on $U \cap V$

$$\theta_V = \mathrm{Ad}_{h^{-1}} \theta_U + h^* \omega_H$$

  where ω_H is the Maurer–Cartan 1-form of H.

(The Maurer–Cartan 1-form of H is an $\mathfrak{h}$-valued 1-form on the group manifold H, and $h^*\omega_H$ is its pull-back by h to $U \cap V$: thus each of the terms in the last displayed equation is a $\mathfrak{g}$-valued 1-form defined on $U \cap V$.)

In our circumstances the required functions h are provided by the fact that E is an H-bundle. That is, we can find an open covering of M such that

- over each open set U of the covering E is trivial, with trivializing map $\tau_U : U \times F \to \pi^{-1}(U)$

- for every pair U, V of open subsets of the covering with $U \cap V$ nonempty, for all $x \in U \cap V$

$$\mathrm{T}_U^{-1} \circ \mathrm{T}_V|_{\{x\}\times F} : F \to F$$

is a diffeomorphism of F which takes the form of the left action on F of an element $h(x)$ of H; there is thus a function $h : U \cap V \to H$ such that

$$\mathrm{T}_V(x, p) = \mathrm{T}_U(x, h(x)p) \text{ for all } p \in F\,.$$

From the Maurer–Cartan form, for $x \in U \cap V$, $v \in T_x M$,

$$(h^*\omega_H)_x(v) = \omega_{H,h(x)}(h_{*x}v) = l_{h(x)^{-1}*}(h_{*x}v)$$

where l denotes left multiplication in H: here $h_{*x}v \in T_{h(x)}H$ and so $l_{h(x)^{-1}*}(h_{*x}v) \in \mathfrak{h}$. Thus the compatibility formula in the second condition for a Cartan structure can be expressed as follows:

$$\theta_{V,x}(v) = \mathrm{Ad}_{h(x)^{-1}}(\theta_{U,x}(v)) + l_{h(x)^{-1}*}(h_{*x}v)\,.$$

Theorem 7.1.1 *The θ_U defined by*

$$\gamma_x(v) = \mathrm{T}_{U,x\sharp}\big(v, \theta_{U,x}(v)\big)$$

satisfy the conditions for a Cartan structure.

Proof For the first condition, we know that $A_x\mathcal{H} \cap \gamma_x(T_x M) = \{0\}$. Thus if $\theta_{U,x}(v) \in \mathfrak{h}$ then $v = 0$. So $v \mapsto \theta_{U,x}(v) \bmod \mathfrak{h}$ is injective, and by dimension it is an isomorphism.

For the second condition we need the relationship between the maps $\mathrm{T}_{U,x\sharp}$ and $\mathrm{T}_{V,x\sharp}$. We shall consider arguments $(0, a)$, $a \in \mathfrak{g}$, and $(v, 0)$, $v \in T_x M$, separately, and sum the results. It is straightforward to show that

$$\mathrm{T}_{U,x\sharp}(0, a) = \mathrm{T}_{V,x\sharp}\big(0, \mathrm{Ad}_{h(x)^{-1}}(a)\big)\,.$$

Let $t \mapsto c(t)$ be a curve in M with $c(0) = x$ and $\dot{c}(0) = v$. Then

$$\begin{aligned}
\mathrm{T}_{U,x\sharp}(v, 0)|_{\mathrm{T}_U(x,p)} &= \frac{d}{dt}\Big(\mathrm{T}_U(c(t), p)\Big)_{t=0} \\
&= \frac{d}{dt}\Big(\mathrm{T}_V(c(t), h(c(t))^{-1}p)\Big)_{t=0} \\
&= \frac{d}{dt}\Big(\mathrm{T}_V\big(c(t), \big(h(x)^{-1}h(c(t))\big)^{-1}h(x)^{-1}p\big)\Big)_{t=0}\,.
\end{aligned}$$

Now $t \mapsto h(x)^{-1}h(c(t))$ is a curve through the identity in H, and its tangent vector there (where $t = 0$) is $l_{h(x)^{-1}*}(h_{*x}v)$. Moreover,

$$\tau_V\big(c(t), \big(h(x)^{-1}h(c(t))\big)^{-1}h(x)^{-1}p\big)_{t=0} = \tau_V(x, h(x)^{-1}p) = \tau_U(x, p).$$

Thus

$$\tau_{U,x\sharp}(v, 0)\big|_{\tau_U(x,p)} = \tau_{V,x\sharp}\big(v, l_{h(x)^{-1}*}(h_{*x}v)\big)\big|_{\tau_U(x,p)}.$$

Now

$$\tau_{V,x\sharp}\big(v, \theta_{V,x}(v)\big) = \gamma_x(v) = \tau_{U,x\sharp}\big(v, \theta_{U,x}(v)\big),$$

while

$$\begin{aligned}\tau_{U,x\sharp}\big(v, \theta_{U,x}(v)\big) &= \tau_{U,x\sharp}\big(0, \theta_{U,x}(v)\big) + \tau_{U,x\sharp}(v, 0)\\ &= \tau_{V,x\sharp}\big(0, \mathrm{Ad}_{h(x)^{-1}}(\theta_{U,x}(v))\big) + \tau_{V,x\sharp}\big(v, l_{h(x)^{-1}*}(h_{*x}v)\big).\end{aligned}$$

It follows that

$$\theta_{V,x}(v) = \mathrm{Ad}_{h(x)^{-1}}(\theta_{U,x}(v)) + l_{h(x)^{-1}*}(h_{*x}v)$$

as asserted.

7.2 The Bundle Definition

In this section we discuss how our definition of a Cartan geometry is related to the principal bundle definition.

Throughout this section we assume that we are given a generalized F-space, with a group G which acts transitively on F with isotropy subgroup H, and that $\dim F = \dim M$. We let $\Pi : P \to M$ be the principal G-bundle associated with π. Further, we assume that the structure group of P is reducible from G to H: that is, that there is a sub-bundle Q of P which is invariant under the right action of $H \subset G$, and under the restricted action is a principal H-bundle.

As we mentioned in the introduction to this chapter, the key ingredient of the principal bundle definition of a Cartan geometry is a Cartan connection form. A *Cartan connection form* is a $\mathfrak{g}$-valued 1-form ϖ on Q such that

- the map $\varpi_q : T_qQ \to \mathfrak{g}$ is an isomorphism;
- $r_h^*\varpi = \mathrm{Ad}_{h^{-1}}\varpi$ for $h \in H$;
- $\varpi(a^\sharp) = a$ for $a \in \mathfrak{h}$.

(The dimension of T_qQ is $\dim M + \dim \mathfrak{h} = \dim \mathfrak{g}$.)

We shall show how to construct from a Cartan geometry on the generalized F-space $E \to M$ a Cartan connection form on the principal H-bundle $Q \to M$, and vice versa.

We first assume the existence of an infinitesimal connection γ on $A\mathcal{G}$ which satisfies the nondegeneracy condition that $A_x\mathcal{H} \cap \gamma_x(T_xM) = \{0\}$. We have seen in

Sect. 3.5 that for each $p \in P$ we can define a map $\tilde{p} : T_p P \to A_x \mathcal{G}$, where $x = \Pi(p)$, which is an isomorphism of vector spaces, and whose restriction to the vertical subspace of $T_p P$ maps it isomorphically onto $K_x \mathcal{G}$. In particular we can construct the map $\tilde{q}$ for any $q \in Q \subset P$, and the restriction of $\tilde{q}$ to $T_q Q$ is an isomorphism $T_q Q \to A_x \mathcal{H}$.

We consider the kernel projection ω of γ. For any $u \in T_q Q$, $\omega_{\Pi(q)}(\tilde{q}(u)) \in K_x \mathcal{G}$, and therefore takes the form $\tilde{q}(a_q^\sharp)$ for a unique $a \in \mathfrak{g}$. We define a map $\varpi : Q \to \mathfrak{g}$ as follows: for $u \in T_q Q$, $\varpi_q(u) = a$ where $\omega_{\Pi(q)}(\tilde{q}(u)) = \tilde{q}(a_q^\sharp)$. Evidently ϖ is a $\mathfrak{g}$-valued 1-form on Q.

Theorem 7.2.1 *The form ϖ is a Cartan connection form.*

Proof We show that ϖ has the requisite properties, in the order in which they are listed above.

Note that ϖ_q is a linear map between spaces of the same dimension. Suppose that $\varpi_q(u) = 0$. Then $\omega_x(\tilde{q}(u)) = 0$, $x = \Pi(q)$. This says that $\tilde{q}(u) \in \gamma_x(T_x M)$. But $u \in T_q Q$, and therefore $\tilde{q}(u) \in A_x \mathcal{H}$. So by the nondegeneracy assumption $\tilde{q}(u) = 0$. But $\tilde{q}$ is an isomorphism of $T_q Q$ with $A_x \mathcal{H}$, so in particular is injective: thus $u = 0$. Thus $\varpi_q : T_q Q \to \mathfrak{g}$ is an isomorphism.

Secondly, suppose that $(r_h^* \varpi)_q(u) = \varpi_{qh}(r_{h*}u) = b$ say, $b \in \mathfrak{g}$. Then

$$\omega_{\Pi(qh)}\big(\widetilde{qh}(r_{h*}u)\big) = \widetilde{qh}(b^\sharp|_{qh}) = \tilde{q}\big((\mathrm{Ad}_h\, b)^\sharp|_q\big),$$

using properties of $\tilde{q}$ from the end of Chap. 3. On the other hand, $\Pi(qh) = \Pi(q)$ and $\widetilde{qh}(r_{h*}u) = \tilde{q}(u)$, so

$$\tilde{q}\big((\mathrm{Ad}_h\, b)^\sharp|_q\big) = \tilde{q}(a_q^\sharp) \text{ where } \varpi_q(u) = a\,.$$

Thus $b = \mathrm{Ad}_{h^{-1}}\, a$, as required.

Finally, for $a \in \mathfrak{h}$, $\omega_{\Pi(q)}(\tilde{q}(a_q^\sharp)) = \tilde{q}(a_q^\sharp)$ since $\tilde{q}(a_q^\sharp) \in K_{\Pi(q)}\mathcal{G}$, so $\varpi_q(a_q^\sharp) = a$, as required. □

A similar construction, but no longer restricting things to Q, defines a $\mathfrak{g}$-valued 1-form on P. This is the connection form of an Ehresmann connection on P; moreover, the relation between the two connection forms is that described in [31] and [36, Appendix A].

We also establish a converse: we assume that Q is equipped with a Cartan connection form ϖ and construct an infinitesimal connection γ on $A\mathcal{G}$, which satisfies the nondegeneracy condition and therefore defines the structure of an infinitesimal Cartan geometry.

For any $q \in Q$,

$$\psi_q : u \mapsto \tilde{q}\big(u - \varpi_q(u)^\sharp|_q\big)$$

is a linear map $T_q Q \to A_x \mathcal{G}$, $x = \Pi(q)$. Since for $a \in \mathfrak{h}$, $\varpi(a^\sharp) = a$, ψ_q vanishes on the vertical subspace of $T_q Q$. Moreover,

$$\begin{aligned}\psi_{qh}(r_{h*}u) &= \widetilde{qh}\big(r_{h*}u - \varpi_{qh}(r_{h*}u)^\sharp|_{qh}\big) \\ &= \tilde{q}(u) - \widetilde{qh}\big((r_h^*\varpi)_q(u)^\sharp|_{qh}\big) \\ &= \tilde{q}(u) - \widetilde{qh}\big((\mathrm{Ad}_{h^{-1}}\varpi_q(u))^\sharp|_{qh}\big) \\ &= \tilde{q}(u) - \tilde{q}\big(\varpi_q(u)^\sharp|_q\big) = \psi_q(u),\end{aligned}$$

using the second property of ϖ and properties of $\tilde{q}$ from the end of Chap. 3. So if for $x \in M$, $v \in T_xM$, we set

$$\gamma_x(v) = \psi_q(u)$$

where $q \in Q$ is any point such that $\Pi(q) = x$ and $u \in T_qQ$ is any vector such that $\Pi_{*q}(u) = v$, then γ_x is well-defined and is a linear map $T_xM \to A_x\mathcal{G}$.

Theorem 7.2.2 *With this definition, γ is an infinitesimal connection on $A\mathcal{G}$, and is nondegenerate.*

Proof We need to show that γ is a section of the anchor $\rho : A\mathcal{G} \to TM$. Now $\rho \circ \tilde{q} = \Pi_{*q}$, so that

$$\rho(\gamma_x(v)) = \Pi_{*q}\big(u - \varpi_q(u)^\sharp|_q\big) = \Pi_{*q}(u) = v,$$

as required.

We show next that γ is nondegenerate, that is, that $A_x\mathcal{H} \cap \gamma_x(T_xM) = \{0\}$. Suppose that for some $v \in T_xM$, $\gamma_x(v) \in A_x\mathcal{H}$. Then for $q \in Q$ with $\Pi(q) = x$ and $u \in T_qQ$ with $\Pi_{*q}u = v$, $\psi_q(u) \in A_x\mathcal{H}$. That is, $\psi_q(u) = \tilde{q}(w)$ for some $w \in T_qQ$, because $\tilde{q} : T_qQ \to A_x\mathcal{H}$ is an isomorphism. Then

$$\tilde{q}\big(\varpi_q(u)^\sharp|_q\big) = \tilde{q}(u - w) \in A_x\mathcal{H},$$

and therefore $\varpi_q(u)^\sharp|_q \in T_qQ$. That is to say, $\varpi_q(u) \in \mathfrak{h}$: say $\varpi_q(u) = a$, $a \in \mathfrak{h}$. Now $\varpi_q(a^\sharp|_q) = a$, by the third property of Cartan connection forms. On the other hand, $\varpi_q : T_qQ \to \mathfrak{g}$ is injective, by the first property: so $u = a^\sharp|_q$. But then $v = 0$. □

We shall apply these results to the case of a model Cartan geometry. We have $F = M = G/H$, $H = G_o$; $E = M \times M$ with projection on the first factor; $P = M \times G$; $Q = G \subset P$; $A\mathcal{G} = TM \oplus_M (M \times \mathfrak{g})$; $\gamma_x(v) = (v, 0)$.

We first obtain an explicit expression for the map $\tilde{p}$, for $p \in P$, in this situation. With $p = (x, g)$, $\hat{p} : M \to \{x\} \times M$ is given by $\hat{p}(y) = (x, gy)$. The map $\tilde{p}$ is an isomorphism of vector spaces $T_pP = T_xM \oplus T_gG$ with $A_x\mathcal{G}$, which we may identify with the space of projectable vector fields along $\{x\} \times M$ of the form $(v, a^\sharp)$ with $v \in T_xM$, $a \in \mathfrak{g}$, $a^\sharp$ the fundamental vector field of the action of G on M given by $a^\sharp_y = d/dt(\exp(-ta)y)_{t=0}$, $y \in M$. Any element of T_gG takes the form a^{R}_g for some $a \in \mathfrak{g}$, where a^{R} is the right-invariant vector field on G such that $a^{\mathsf{R}}_{1_G} = a$. Thus a^{R}_g is the tangent vector at $t = 0$ to the curve $t \mapsto g(t) = \exp(ta)g$ in G. Let $t \mapsto x(t)$ be a curve in M with $x(0) = x$, $\dot{x}(0) = v$. Then $\tilde{p}(v, a^{\mathsf{R}}_g)$ is the vector field along $\{x\} \times M$ whose value at $(x, \bar{x})$ is the tangent vector at $t = 0$ to the curve

$$t \mapsto \widehat{p(t)}(\hat{p}^{-1}(\tilde{x})) = (x(t), g(t)g^{-1}\tilde{x}) = (x(t), \exp(ta)\tilde{x}),$$

which is $\left(v, -a^{\sharp}_{\tilde{x}}\right)$. So

$$\tilde{p}\left(v, a^{\mathsf{R}}_{g}\right) = (v, -a^{\sharp}).$$

The Cartan connection form ϖ is defined in this particular case as follows. It is a map $G \to \mathfrak{g}$, where we regard G as a sub-bundle of P via the injection $\iota : G \to M \times G$ with $\iota(g) = (go, g)$. For $u \in T_{\iota(g)}\iota(G)$, $\varpi_{\iota(g)}(u) = a$ where $\omega_x(\tilde{q}(u)) = \tilde{q}(0, a^{\mathsf{L}}_{g})$, $q = \iota(g) = (go, g) = (x, g)$, a^{L} the left-invariant vector field on G with $a^{\mathsf{L}}_{e} = a$ (the fundamental vector field of the right action of G on itself). Now $u = \iota_*(a^{\mathsf{R}}_{g})$ for some $a \in \mathfrak{g}$; explicitly, $u = (-a^{\sharp}_{x}, a^{\mathsf{R}}_{g})$. Then $\tilde{q}(u) = (-a^{\sharp}_{x}, -a^{\sharp})$, and

$$\omega_x(\tilde{q}(u)) = (0, -a^{\sharp}) = \tilde{q}(0, a^{\mathsf{R}}_{g}).$$

But right- and left-invariant vector fields on G are related as follows:

$$a^{\mathsf{R}}_{g} = \left(\mathrm{Ad}_{g^{-1}}(a)\right)^{\mathsf{L}}_{g}.$$

Thus

$$\varpi_{\iota(g)}\left(\iota_*(a^{\mathsf{R}}_{g})\right) = \mathrm{Ad}_{g^{-1}}(a);$$

that is to say

$$\varpi_{\iota(g)}\left(\iota_*(a^{\mathsf{L}}_{g})\right) = \varpi_{\iota(g)}\left(\iota_*((\mathrm{Ad}_{g}(a))^{\mathsf{R}}_{g})\right) = \mathrm{Ad}_{g^{-1}}(\mathrm{Ad}_{g}(a)) = a.$$

Thus $\iota^*\varpi$ is the Maurer–Cartan form of G.

In the principal bundle approach to Cartan geometries, given a homogeneous space G/H one takes for Cartan connection form the Maurer–Cartan form on G. This is the standard model geometry for the homogeneous space, the Klein geometry. The result just obtained shows that our model Cartan geometry corresponds, via the procedures explained in Sect. 3.5 and this section, to the Klein geometry for the same homogeneous space.

7.3 The Tractor Connection

In this and the following section we shall show how to define covariant derivative operators on the sections of $\mathcal{A}_o$, for any infinitesimal Cartan geometry. In each case the construction makes use of the fact that the kernel projection ω defines an isomorphism of vector bundles $\mathcal{A}_o \to \mathcal{K}$, to carry objects from one to the other. But first let us note that $\mathcal{A}_o$, as well as being a Lie algebroid, can be endowed with the structure of a Lie algebra bundle, by using ω to carry the Lie algebra bracket on $\mathcal{K}$ over to $\mathcal{A}_o$ in just this way. Thus $\sec(\mathcal{A}_o)$ is equipped with a $C^\infty(M)$-bilinear bracket, which we denote by $\{\cdot,\cdot\}$, where for $\xi, \eta \in \sec(\mathcal{A}_o)$

$$\omega \circ \{\xi, \eta\} = [\![\omega \circ \xi, \omega \circ \eta]\!] ;$$

the Jacobi identity for $[\![\cdot,\cdot]\!]$ with arguments $\omega \circ \xi, \omega \circ \eta, \omega \circ \zeta$ gives the Jacobi identity for $\{\cdot,\cdot\}$. Since ω is the identity on $\mathcal{K}_o$, the restriction of $\{\cdot,\cdot\}$ to $\mathcal{K}_o$ coincides with the Lie algebra bracket it has already, namely the restriction of the Lie algebroid bracket.

As we pointed out in Sect. 2.2, given an infinitesimal connection γ on a Lie algebroid $\mathcal{A}$, we may define the kernel derivative ∇^γ of sections κ of $\mathcal{K} \to M$ by $\nabla^\gamma_X \kappa = [\![\gamma \circ X, \kappa]\!]$. When $\mathcal{A}$ is the Lie algebroid of an infinitesimal Cartan geometry we may use the kernel projection ω to transfer ∇^γ to $\mathcal{A}_o$. Thus for any $\xi \in \sec(\mathcal{A}_o)$ and vector field X on M there is a unique element of $\sec(\mathcal{A}_o)$, which we denote by $\nabla^{\mathrm{T}}_X \xi$, such that

$$\omega \circ \nabla^{\mathrm{T}}_X \xi = \nabla^\gamma_X (\omega \circ \xi) = [\![\gamma \circ X, \omega \circ \xi]\!] .$$

Proposition 7.3.1 *The operator ∇^{T} so defined is a covariant derivative.*

Proof The only point which is perhaps not immediately obvious is establishing the property that $\nabla^{\mathrm{T}}_X(f\xi) = f\nabla^{\mathrm{T}}_X\xi + X(f)\xi$ for a function f on M: but ∇^γ is a covariant derivative, so

$$\omega \circ \nabla^{\mathrm{T}}_X(f\xi) = f\nabla^\gamma_X(\omega \circ \xi) + X(f)\omega \circ \xi = \omega \circ \left(f\nabla^{\mathrm{T}}_X\xi + X(f)\xi\right),$$

which gives the required result since the arguments of ω on each side belong to $\sec(\mathcal{A}_o)$. □

This covariant derivative corresponds more-or-less to what is known in the literature as a *tractor connection* [10] whence the notation.

Proposition 7.3.2 *The covariant derivative $\nabla^{\mathrm{T}}$is a derivation of the Lie algebra bracket on $\mathcal{A}_o$.*

Proof

$$\begin{aligned}
\omega \circ \nabla^{\mathrm{T}}_X\{\xi, \eta\} &= [\![\gamma \circ X, [\![\omega \circ \xi, \omega \circ \eta]\!]]\!] \\
&= [\![[\![\gamma \circ X, \omega \circ \xi]\!], \omega \circ \eta]\!] + [\![\omega \circ \xi, [\![\gamma \circ X, \omega \circ \eta]\!]]\!] \\
&= [\![\omega \circ \nabla^{\mathrm{T}}_X\xi, \omega \circ \eta]\!] + [\![\omega \circ \xi, \omega \circ \nabla^{\mathrm{T}}_X\eta]\!] \\
&= \omega \circ \left(\{\nabla^{\mathrm{T}}_X\xi, \eta\} + \{\xi, \nabla^{\mathrm{T}}_X\eta\}\right).
\end{aligned}$$

□

The curvature R^{T} of ∇^{T} is given by the usual formula in terms of the covariant derivative operators. For each $X, Y \in \sec(TM)$, $R^{\mathrm{T}}(X, Y)$ is a $C^\infty(M)$-linear map $\sec(\mathcal{A}_o) \to \sec(\mathcal{A}_o)$. A straightforward calculation yields the following result.

Proposition 7.3.3

$$\omega \circ \big(R^{\mathrm{T}}(X, Y)\big)\xi = [\![\omega \circ \xi, R^{\gamma}(X, Y)]\!] . \qquad \square$$

7.4 The Blaom Connection

There is another way of constructing, for an infinitesimal Cartan geometry, a covariant derivative on the sections of the Lie subalgebroid $\mathcal{A}_o \to M$. We again use the fact that the kernel projection is a linear isomorphism of $\mathcal{A}_o$ with $\mathcal{K}$. For any section ξ of $\mathcal{A}_o \to M$ and vector field X on M, $\gamma \circ [\rho \circ \xi, X] - [\![\xi, \gamma \circ X]\!]$ is a section of $\mathcal{K} \to M$. We may therefore define a section $\nabla^{\mathrm{B}}_X \xi$ of $\mathcal{A}_o \to M$ by

$$\gamma \circ [\rho \circ \xi, X] - [\![\xi, \gamma \circ X]\!] = \omega \circ \nabla^{\mathrm{B}}_X \xi .$$

Proposition 7.4.1 *The operator* ∇^{B} *so defined is a covariant derivative.*

Proof It is evident that $\nabla^{\mathrm{B}}_X \xi$ is $C^\infty(M)$-linear in X. We have

$$\begin{aligned} \omega \circ \big(\nabla^{\mathrm{B}}_X(f\xi) - f\nabla^{\mathrm{B}}_X \xi\big) &= -X(f)(\gamma \circ \rho \circ \xi) + X(f)\xi \\ &= X(f)\big(\xi - \gamma \circ \rho \circ \xi\big) = \omega \circ (X(f)\xi) , \end{aligned}$$

so that $\nabla^{\mathrm{B}}_X(f\xi) = f\nabla^{\mathrm{B}}_X \xi + X(f)\xi$. $\square$

This covariant derivative is essentially the one introduced by Blaom, and called by him, somewhat unfortunately from our point of view, a Cartan connection. We shall call it the *Blaom connection* of the infinitesimal Cartan geometry.

The condition for a section ξ of $\mathcal{A}_o$ to be an infinitesimal symmetry of an infinitesimal Cartan geometry, namely $\mathcal{L}_\xi \gamma = 0$, can be written equivalently as $\nabla^{\mathrm{B}} \xi = 0$.

The two connections we have introduced are related as follows.

Proposition 7.4.2

$$\omega \circ \big(\nabla^{\mathrm{T}}_X \xi - \nabla^{\mathrm{B}}_X \xi\big) = R^{\gamma}(X, \rho \circ \xi) .$$

Proof

$$\begin{aligned} \omega \circ \big(\nabla^{\mathrm{T}}_X \xi - \nabla^{\mathrm{B}}_X \xi\big) &= [\![\gamma \circ X, \omega \circ \xi]\!] - \gamma \circ [\rho \circ \xi, X] + [\![\xi, \gamma \circ X]\!] \\ &= \gamma \circ [X, \rho \circ \xi] - [\![\gamma \circ X, \gamma \circ \rho \circ \xi]\!] \\ &= R^{\gamma}(X, \rho \circ \xi) . \end{aligned}$$

$\square$

Blaom in fact defines a Cartan connection as one which satisfies the following condition, arising from a requirement that the corresponding map between the Lie algebroids $\tau : \mathcal{A} \to M$ and $\tau_1 : J^1\tau \to M$ be a morphism:

$$\nabla^{\mathrm{B}}_X [\![\xi_1, \xi_2]\!] - [\![\nabla^{\mathrm{B}}_X \xi_1, \xi_2]\!] - [\![\xi_1, \nabla^{\mathrm{B}}_X \xi_2]\!] = \nabla^{\mathrm{B}}_{\overline{\nabla}^{\mathrm{B}}_{\xi_2} X} \xi_1 - \nabla^{\mathrm{B}}_{\overline{\nabla}^{\mathrm{B}}_{\xi_1} X} \xi_2 ,$$

where

$$\overline{\nabla}^{\mathrm{B}}_{\xi} X = \rho \circ \nabla^{\mathrm{B}}_X \xi + [\rho \circ \xi, X] .$$

(This formula defines an operator $\overline{\nabla}^{\mathrm{B}}$ which may be considered as dual to ∇^{B}, and in the terminology of Sect. 2.5 is an $\mathcal{A}$-derivative on TM.)

Proposition 7.4.3 *The connection on $\mathcal{A}_o$ whose covariant derivative operator is ∇^{B} is indeed a Cartan connection in the sense of Blaom.*

Proof We shall show that the condition on ∇^{B} for it to be a Cartan connection in the sense of Blaom is simply the property

$$\mathcal{L}_{\xi_1}(\mathcal{L}_{\xi_2}\gamma) - \mathcal{L}_{\xi_2}(\mathcal{L}_{\xi_1}\gamma) = \mathcal{L}_{[\![\xi_1, \xi_2]\!]}\gamma$$

expressed in terms of ∇^{B}.

We shall need the following result concerning ω. Since $\omega \circ \eta = \eta - \gamma \circ \rho \circ \eta$,

$$\begin{aligned} [\![\xi, \omega \circ \eta]\!] &= [\![\xi, \eta]\!] - [\![\xi, \gamma \circ \rho \circ \eta]\!] \\ &= [\![\xi, \eta]\!] - \mathcal{L}_\xi \gamma \circ \rho \circ \eta - \gamma \circ [\rho \circ \xi, \rho \circ \eta] \\ &= \omega \circ [\![\xi, \eta]\!] + \omega \circ \nabla^{\mathrm{B}}_{\rho \circ \eta} \xi . \end{aligned}$$

Now

$$\begin{aligned} \mathcal{L}_{\xi_1}(\mathcal{L}_{\xi_2}\gamma) \circ X &= -[\![\xi_1, \omega \circ \nabla^{\mathrm{B}}_X \xi_2]\!] + \omega \circ \nabla^{\mathrm{B}}_{[\rho \circ \xi_1, X]} \xi_2 \\ &= -\omega \circ [\![\xi_1, \nabla^{\mathrm{B}}_X \xi_2]\!] - \omega \circ \nabla^{\mathrm{B}}_{\rho \circ \nabla^{\mathrm{B}}_X \xi_2} \xi_1 + \omega \circ \nabla^{\mathrm{B}}_{[\rho \circ \xi_1, X]} \xi_2 . \end{aligned}$$

We conclude, from the fact that $\mathcal{L}_{\xi_1}(\mathcal{L}_{\xi_2}\gamma) \circ X - \mathcal{L}_{\xi_2}(\mathcal{L}_{\xi_1}\gamma) \circ X = \mathcal{L}_{[\![\xi_1, \xi_2]\!]}\gamma \circ X$, that

$$-\omega \circ [\![\xi_1, \nabla^{\mathrm{B}}_X \xi_2]\!] + \omega \circ \nabla^{\mathrm{B}}_{\overline{\nabla}^{\mathrm{B}}_{\xi_1} X} \xi_2 + \omega \circ [\![\xi_2, \nabla^{\mathrm{B}}_X \xi_1]\!] - \omega \circ \nabla^{\mathrm{B}}_{\overline{\nabla}^{\mathrm{B}}_{\xi_2} X} \xi_1 = -\omega \circ \nabla^{\mathrm{B}}_X [\![\xi_1, \xi_2]\!] .$$

The result now follows from the fact that $\omega : \mathcal{A}_o \to \mathcal{K}$ is an isomorphism. □

The curvature of ∇^{B} is denoted by R^{B}. A straightforward calculation yields the following result relating R^{B} and R^γ.

Proposition 7.4.4

$$\omega \circ \big(R^{\mathrm{B}}(X,Y)\big)\xi = \mathcal{L}_\xi\big(R^\gamma(X,Y)\big) + R^\gamma(\rho \circ \nabla^{\mathrm{B}}_X\xi, Y) + R^\gamma(X, \rho \circ \nabla^{\mathrm{B}}_Y\xi). \quad \square$$

Note that if $R^\gamma = 0$ then $R^{\mathrm{B}} = 0$; but the converse is not necessarily true.

As we have already mentioned

$$\omega \circ \nabla^{\mathrm{B}}_X\xi = -\mathcal{L}_\xi\gamma \circ X.$$

Thus ξ is an infinitesimal symmetry if and only if it is a parallel section of $\mathcal{A}_o$ with respect to the Blaom connection. This observation allows us to determine the maximal dimension of the space of infinitesimal symmetries.

Theorem 7.4.5 *The infinitesimal symmetries of an infinitesimal Cartan geometry form a finite-dimensional Lie algebra* $\mathfrak{a}$ *whose dimension is at most* $\dim \mathfrak{g}$. *When that dimension is attained,* $R^{\mathrm{B}} = 0$*; the anchor* $\rho : \mathcal{A}_o \to TM$ *defines an action of* $\mathfrak{a}$ *on* M*, and* $\mathcal{A}_o$ *may be identified with the action algebroid of this action. Without further restrictions on* M *the converse holds only in a local sense: that is, if* $R^{\mathrm{B}} = 0$ *then there is a covering of* M *by open sets* U*, such that for each* U *the algebra of infinitesimal symmetries of* $\mathcal{A}_o|_U$ *has dimension* $\dim \mathfrak{g}$, ρ *defines an action of it on* U*, and* $\mathcal{A}_o|_U$ *is its action algebroid. If however* M *is connected and simply connected, and* $R^{\mathrm{B}} = 0$*, then* $\mathcal{A}_o$ *may be identified with the action algebroid of an action of the Lie algebra* $\mathfrak{a}$ *of its infinitesimal symmetries on* M.

Note that we do not assert that when the symmetry algebra has maximal dimension $\mathfrak{a} = \mathfrak{g}$: indeed, this need not necessarily be the case. Nor do we assert that the action defined by ρ is effective, though in fact this is the case, as we know for a Cartan geometry with F connected from Theorem 6.7.4, and as we shall show for infinitesimal Cartan geometries in Chap. 11.

Proof Each infinitesimal symmetry is a (global) section of $\mathcal{A}_o \to M$ which is parallel with respect to ∇^{B}. Now the set of parallel sections of a vector bundle with connection is a linear space over $\mathbb{R}$. It follows from the Blaom connection condition that if $\nabla^{\mathrm{B}}\xi_1 = \nabla^{\mathrm{B}}\xi_2 = 0$ then $\nabla^{\mathrm{B}}[\![\xi_1, \xi_2]\!] = 0$, so in this case the space of parallel sections is closed under the algebroid bracket, and forms a Lie algebra with that bracket. (We knew these facts already, but the alternative proofs are significant in the present context.) The space of parallel sections of a vector bundle with connection is finite dimensional, and its dimension is at most that of the fibre: if that dimension is attained, so that the bundle admits a basis of parallel sections, then the curvature vanishes.

Suppose that $\dim \mathfrak{a} = \dim \mathfrak{g}$. Then there is a basis of parallel sections of $\mathcal{A}_o$ made up of elements of $\mathfrak{a}$, and so $\mathcal{A}_o$ is isomorphic as a vector bundle to $M \times \mathfrak{a}$. Let us denote by $\{\cdot, \cdot\}_\mathfrak{a}$ the Lie algebra bracket of $\mathfrak{a}$, that is, the restriction of the Lie algebroid bracket. Let $\{\xi_a\}$, $a = 1, 2, \dots, \dim \mathfrak{g}$ be a basis for $\mathfrak{a}$. Then

$$\{\xi_a, \xi_b\}_\mathfrak{a} = C^c_{ab}\xi_c$$

where at first sight the coefficients might appear to be (nonconstant) functions on M: but for any vector field X on M,

$$\nabla^{\mathrm{B}}_X(C^c_{ab}\xi_c) = 0 = X(C^c_{ab})\xi_c,$$

so the coefficients are in fact constants. Now for $\xi_1, \xi_2 \in \mathfrak{a} \subset \mathrm{sec}(\mathcal{A}_o)$,

$$\rho \circ \{\xi_1, \xi_2\}_{\mathfrak{a}} = \rho \circ [\![\xi_1, \xi_2]\!] = [\rho \circ \xi_1, \rho \circ \xi_2],$$

so ρ defines an action of $\mathfrak{a}$ on M; for $\xi \in \mathfrak{a}$ we denote the corresponding fundamental vector field $\rho \circ \xi$ by $\xi^\sharp$. Any section η of $\mathcal{A}_o$ may be written $\eta = \sum_a \eta^a \xi_a$ with respect to a basis $\{\xi_a\}$ of $\mathfrak{a}$, for some functions η^a on M; and then

$$\begin{aligned}
[\![\eta, \zeta]\!] &= \sum_{a,b} [\![\eta^a \xi_a, \zeta^b \xi_b]\!] \\
&= \sum_{a,b} \big((\eta^b \xi^\sharp_b(\zeta^a) - \zeta^b \xi^\sharp_b(\eta^a))\xi_a + \eta^a \zeta^b \{\xi_a, \xi_b\}_{\mathfrak{a}}\big) \\
&= \sum_{a,b} \big(\eta^b \xi^\sharp_b(\zeta^a) - \zeta^b \xi^\sharp_b(\eta^a) + \eta^b \zeta^c C^a_{bc}\big)\xi_a,
\end{aligned}$$

which identifies the Lie algebroid bracket of $\mathcal{A}_o$ with the action algebroid bracket of $M \times \mathfrak{a}$.

For the general converse, if $R^{\mathrm{B}} = 0$ we know only that there are local bases of parallel sections of $\mathcal{A}_o$; the rest of the argument works as before, but only in the local sense.

If, on the other hand, M is connected and simply connected, and $R^{\mathrm{B}} = 0$, then the holonomy group of ∇^{B} at any point of M consists of just the identity, and parallel transport is therefore independent of the curve. Given a point $x_0 \in M$ and a basis $\{\xi_a|_0\}$, $a = 1, 2, \ldots, \dim \mathfrak{g}$, for $\mathcal{A}_{o;x_0}$, we may define a basis of global sections $\{\xi_a\}$ of $\mathcal{A}_o$ by parallel transport of $\{\xi_a|_0\}$ along arbitrary curves. Being global sections parallel with respect to ∇^{B} the ξ_a are infinitesimal symmetries, and therefore form a basis for $\mathfrak{a}$. Thus $\mathcal{A}_o$ is isomorphic as a vector bundle to $M \times \mathfrak{a}$. The remainder of the proof that $\mathcal{A}_o$ is an action algebroid proceeds as in the second paragraph of the proof overall. □

The kernel projection ω is a vector bundle isomorphism $\mathcal{A}_o \to \mathcal{K}$, but doesn't preserve brackets: recall that in particular, when $\xi, \eta \in \mathrm{sec}(\mathcal{A}_o)$ are infinitesimal symmetries then

$$\omega \circ [\![\xi, \eta]\!] - [\![\omega \circ \xi, \omega \circ \eta]\!] = R^\gamma(\rho \circ \xi, \rho \circ \eta),$$

or in other words

$$\omega \circ \big(\{\xi, \eta\}_{\mathfrak{a}} - \{\xi, \eta\}\big) = R^\gamma(\xi^\sharp, \eta^\sharp).$$

When $R^\gamma = 0$, and so $R^{\mathrm{B}} = 0$, the Lie algebra brackets on $\mathcal{A}_o \cong M \times \mathfrak{a}$ coincide; but R^{B} can be zero without R^γ vanishing, and in that case R^γ is the obstruction to equivalence of the Lie algebra structures. Let us record formally what happens when $R^\gamma = 0$.

Corollary 7.4.6 *The infinitesimal symmetries of a flat infinitesimal Cartan geometry form a finite-dimensional Lie algebra isomorphic to* $\mathfrak{g}$, *and* $\mathcal{A}_o$ *is locally an action algebroid corresponding to an action of* $\mathfrak{g}$ *on* M. *Conversely, if the infinitesimal symmetry algebra of an infinitesimal Cartan geometry is (isomorphic to)* $\mathfrak{g}$ *then the geometry is flat.*

Proof If $R^\gamma = 0$ then for any pair of infinitesimal symmetries ξ, η, $\{\xi, \eta\}_{\mathfrak{a}} = \{\xi, \eta\}$. □

An infinitesimal Cartan geometry with local symmetry algebra of maximal dimension, that is (according to the theorem) one for which $R^{\mathrm{B}} = 0$, is said to be *locally symmetric*.

7.5 The Fundamental Self-Representation

Recall from Sect. 2.6 that any transitive Lie algebroid $\mathcal{A}$ has a canonical representation (a flat $\mathcal{A}$-derivative) on its kernel $\mathcal{K}$, given by $D_\xi \kappa = [\![\xi, \kappa]\!]$, where $\xi \in \sec(\mathcal{A})$ and $\kappa \in \sec(\mathcal{K})$. In this section we shall show that for any infinitesimal Cartan geometry the Lie algebroid $\mathcal{A}_o$ is equipped with a self-representation which extends its canonical representation on $\mathcal{K}_o$. For any $\xi, \eta \in \sec(\mathcal{A}_o)$ we define $D_\xi \eta$ by

$$\omega \circ D_\xi \eta = [\![\xi, \omega \circ \eta]\!] .$$

Proposition 7.5.1 *The operator* D *so defined is a self-representation of* $\mathcal{A}_o$ *which extends its canonical representation. It is moreover a derivation of the Lie algebra bracket* $\{\cdot, \cdot\}$. *The torsion of* D *is given by*

$$\omega \circ T(\xi, \eta) = R^\gamma(\rho \circ \xi, \rho \circ \eta) + \omega \circ \{\xi, \eta\} .$$

If $\lambda \in \sec(\mathcal{K}_o)$ *then* $D_\lambda \xi = \{\lambda, \xi\}$, *which is to say that* D_λ *coincides with inner derivation by* λ *in the Lie algebraic sense.*

Proof Since $[\![\xi, \omega \circ \eta]\!] \in \sec(\mathcal{K})$, $D_\xi \eta$ is well-defined. It follows directly from the properties of $[\![\cdot, \cdot]\!]$ that D is a $\mathcal{A}_o$-derivative on $\mathcal{A}_o$. The Jacobi identity for $[\![\cdot, \cdot]\!]$ with arguments $\xi, \eta, \omega \circ \zeta$ yields the vanishing of the curvature of D, and the same identity with arguments $\xi, \omega \circ \eta, \omega \circ \zeta$ the derivation property. If $\lambda \in \sec(\mathcal{K}_o)$ then $\omega \circ \lambda = \lambda$; so $[\![\xi, \omega \circ \lambda]\!] = [\![\xi, \lambda]\!] \in \sec(\mathcal{K}_o)$, whence $\omega \circ D_\xi \lambda = [\![\xi, \lambda]\!] = \omega \circ [\![\xi, \lambda]\!]$, and so $D_\xi \lambda = [\![\xi, \lambda]\!]$. The curvature R^γ is given in terms of ω by

$$R^\gamma(\rho \circ \xi, \rho \circ \eta) = -\omega \circ [\![\xi, \eta]\!] + [\![\xi, \omega \circ \eta]\!] + [\![\omega \circ \xi, \eta]\!] - [\![\omega \circ \xi, \omega \circ \eta]\!] ;$$

when $\xi, \eta \in \sec(\mathcal{A}_o)$ the right-hand side is $\omega \circ (T(\xi, \eta) - \{\xi, \eta\})$. If $\lambda \in \sec(\mathcal{K}_o)$ then $[\![\lambda, \omega \circ \xi]\!] = [\![\omega \circ \lambda, \omega \circ \xi]\!]$, that is, $\omega \circ D_\lambda \xi = \omega \circ \{\lambda, \xi\}$. □

We call the $\mathcal{A}_o$-derivative D defined in the proposition the *fundamental self-representation* for the infinitesimal Cartan geometry.

The torsion of D must of course be distinguished from the torsion of the infinitesimal connection γ. However, the two concepts are not entirely unrelated.

Corollary 7.5.2 *If the infinitesimal connection is torsion free then*

$$T(\xi, \eta) = R^\gamma(\rho \circ \xi, \rho \circ \eta) + \{\xi, \eta\}.$$

Proof When γ is torsion free R^γ takes its values in $\mathcal{K}_o = \mathcal{A}_o \cap \mathcal{K}$, so

$$R^\gamma(\rho \circ \xi, \rho \circ \eta) = \omega \circ R^\gamma(\rho \circ \xi, \rho \circ \eta),$$

and we may dispense with the ω s. □

We have earlier introduced two ordinary covariant derivative operators on sections of $\mathcal{A}_o$, the one associated with the tractor connection, ∇^{T}, and the other with the Blaom connection, ∇^{B}. Each can be expressed in terms of the fundamental self-representation. Recall that if $\lambda \in \sec(\mathcal{K}_o)$ then $D_\lambda \xi - \{\lambda, \xi\} = 0$ (as we showed in Proposition 7.5.1), and $D_\xi \lambda - [\![\xi, \lambda]\!] = 0 = D^*_\lambda \xi$ (since D extends the canonical representation). That is to say, for $\xi, \eta, \zeta \in \sec(\mathcal{A}_o)$ with $\rho \circ \eta = \rho \circ \zeta$,

$$\begin{aligned} D_\eta \xi - \{\eta, \xi\} &= D_\zeta \xi - \{\zeta, \xi\} \\ D^*_\eta \xi &= D^*_\zeta \xi. \end{aligned}$$

Thus if for any vector field X on M and $\xi \in \sec(\mathcal{A}_o)$ we set

$$\begin{aligned} \nabla^{\mathrm{T}}_X \xi &= D_\eta \xi - \{\eta, \xi\} \\ \nabla^{\mathrm{B}}_X \xi &= D^*_\eta \xi \end{aligned}$$

where η is any section of $\mathcal{A}_o$ such that $\rho(\eta) = X$, then ∇^{T} and ∇^{B} are well-defined, and are easily seen to be covariant derivative operators on $\mathcal{A}_o$.

Proposition 7.5.3 *The operator ∇^{T} so defined is the tractor covariant derivative, while ∇^{B} is the Blaom covariant derivative.*

Proof We have

$$\begin{aligned} \omega \circ \nabla^{\mathrm{T}}_X \xi &= \omega \circ D_\eta \xi - \omega \circ \{\eta, \xi\} \\ &= [\![\eta, \omega \circ \xi]\!] - [\![\omega \circ \eta, \omega \circ \xi]\!] \\ &= [\![\gamma \circ X, \omega \circ \xi]\!]. \end{aligned}$$

On the other hand

$$\begin{aligned}
\omega \circ \nabla^{\mathrm{B}}_X \xi &= \omega \circ D_\xi \eta - \omega \circ [\![\xi, \eta]\!] \\
&= [\![\xi, \omega \circ \eta]\!] - \omega \circ [\![\xi, \eta]\!] \\
&= [\![\xi, \eta]\!] - [\![\xi, \gamma \circ X]\!] - [\![\xi, \eta]\!] + \gamma \circ [\rho \circ \xi, X] \\
&= -\mathcal{L}_\xi \gamma \circ X\,.
\end{aligned}$$

□

We summarize this discussion by listing the key properties of $\mathcal{A}_o$ that we have established.

- $\mathcal{A}_o \to M$ is a Lie algebra bundle, with standard fibre $\mathfrak{g}$;
- it is simultaneously a transitive Lie algebroid, with kernel $\mathcal{K}_o$ whose standard fibre is the Lie subalgebra $\mathfrak{g}_o$ of $\mathfrak{g}$;
- its Lie algebra bracket and its Lie algebroid bracket coincide on $\mathcal{K}_o$;
- it is equipped with a self-representation D, which extends the canonical representation, is a derivation of the Lie algebra bracket, and coincides with an inner derivation when restricted to $\mathcal{K}_o$.

The bundle $\mathcal{A}_o$, with its dual structure of Lie algebra bundle and transitive Lie algebroid and its self-representation, seems to play a role in our theory parallel to that played by the adjoint tractor bundle in the tractor calculus. With some trepidation we call it the *adjoint tractor bundle* of the infinitesimal Cartan geometry.

7.6 Reconstructing a Cartan Geometry from an Adjoint Tractor Bundle

If we start from a finite Cartan geometry (in the sense of Sect. 6.1) we obtain the structure of an adjoint tractor bundle on $A\mathcal{G}_o$. When $\mathfrak{g}$ and $\mathfrak{g}_o$ satisfy certain conditions it is possible to reverse the process, and starting with a vector bundle with the properties of an adjoint tractor bundle reconstruct a Cartan geometry from it.

Lemma 7.6.1 *Let $\mathfrak{g}$ be a Lie algebra with trivial centre such that all derivations of $\mathfrak{g}$ are inner, and let $\mathfrak{g}_o \subset \mathfrak{g}$ be a subalgebra such that the normalizer of $\mathfrak{g}_o$ in $\mathfrak{g}$ is $\mathfrak{g}_o$. Let G be the identity component of the Lie group of automorphisms of $\mathfrak{g}$ and G_o the Lie subgroup of G which preserves $\mathfrak{g}_o$. Then the Lie algebra of G is $\mathfrak{g}$, and the Lie algebra of G_o is $\mathfrak{g}_o$.* □

Theorem 7.6.2 *Assume that $\mathfrak{g}$ is a Lie algebra and $\mathfrak{g}_o$ a subalgebra of $\mathfrak{g}$ which satisfy the conditions of the lemma. Let $\mathcal{A}_o \to M$ be a vector bundle*

- *which is a Lie algebra bundle, with standard fibre $\mathfrak{g}$;*
- *which is simultaneously a transitive Lie algebroid, with kernel $\mathcal{K}_o$ whose standard fibre is the Lie subalgebra $\mathfrak{g}_o$ of $\mathfrak{g}$;*
- *whose Lie algebra bracket and Lie algebroid bracket coincide on $\mathcal{K}_o$;*
- *which is equipped with a self-representation D, which extends the canonical representation, is a derivation of the Lie algebra bracket, and coincides with inner derivation when restricted to $\mathcal{K}_o$.*

Then there is a fibre bundle $E \to M$ with standard fibre G/G_o, whose dimension is $\dim M$, *and a Cartan geometry over E, such that the Lie subalgebroid $A\mathcal{G}_o$ in the corresponding infinitesimal Cartan geometry can be identified with $\mathcal{A}_o$.*

Proof We denote by $\{\cdot, \cdot\}$ the LAB bracket of $\mathcal{A}_o$, by $[\![\cdot, \cdot]\!]$ its Lie algebroid bracket, and by ρ its anchor.

We first construct a principal bundle $P \to M$ with group G_o, using the Lie algebra bundle structure of $\mathcal{A}_o$. Consider the set P of pairs (x, φ_x) where $x \in M$ and φ_x is a Lie algebra isomorphism $\mathfrak{g} \to \mathcal{A}_{o,x}$ such that $\varphi_x(\mathfrak{g}_o) = \mathcal{K}_{o,x}$. It can be given a manifold structure by using local triviality of $\mathcal{A}_o$. The adjoint action of G_o on $\mathfrak{g}$ defines also an action of G_o to the right on P, and this action is free. So P a principal bundle over M with group G_o; and $\mathcal{A}_o$ is evidently associated with P by the adjoint action of G_o on $\mathfrak{g}$. (In fact P is simply a reduction of the bundle of frames of the vector bundle $\mathcal{A}_o$ to those frames that are adapted to the extra structure of Lie algebra etc.)

We may also construct a fibre bundle $E \to M$ with standard fibre $F = G/G_o$, namely the bundle associated with P by the action of G_o on F. Now $\mathcal{A}_o$, as a Lie algebroid, is transitive, its fibre dimension is $\dim \mathfrak{g}$, and the fibre dimension of its kernel is $\dim \mathfrak{g}_o$, so $\dim \mathfrak{g} - \dim \mathfrak{g}_o = \dim M$, and therefore $\dim F = \dim G - \dim \mathfrak{g}_o = \dim M$.

Denote by $\mathcal{G}$ the Lie groupoid of fibre morphisms of E corresponding to the action of G on F, and by $\mathcal{G}_o$ the Lie subgroupoid corresponding to the action of G_o on F. The Lie algebroid $A\mathcal{G}$ of $\mathcal{G}$ is of course transitive, with kernel $K\mathcal{G}$. The Lie algebroid bracket on $A\mathcal{G}$ will be denoted by $[\![\cdot, \cdot]\!]_{\mathcal{G}}$, the anchor by $\rho_{\mathcal{G}}$, the canonical representation of $A\mathcal{G}$ on $K\mathcal{G}$ by $D^{\mathcal{G}}$. Now for $x \in M$, $K_x\mathcal{G}$ is isomorphic as a Lie algebra to $\mathfrak{g}$.

We are now able to construct a LAB-isomorphism $\psi : \mathcal{A}_o \to K\mathcal{G}$, over the identity of M, in the following way. Fix a point $x \in M$, and let $(x, \varphi_x) \in P_x$. By definition φ_x is a Lie algebra isomorphism of the fibre $\mathcal{A}_{o,x}$ with $\mathfrak{g}$, whereas the choice of a point in the principal bundle $P \to M$ gives a diffeomorphism of the associated bundle fibre E_x with the standard fibre F, so that for each $a \in \mathfrak{g}$ the fundamental vector field $a^\sharp$ on F may be identified with a vector field on the fibre E_x and hence an element of the kernel $K_x\mathcal{G}$. We therefore obtain by composition a Lie algebra isomorphism between $\mathcal{A}_{o,x}$ and $K_x\mathcal{G}$. Choosing a different element of P_x gives a different isomorphism of $\mathcal{A}_{o,x}$ with $\mathfrak{g}$, but also a different identification of E_x with F, in such a way that ψ is well defined; it may be seen to be smooth by considering local trivializations.

Now ψ maps the kernel $\mathcal{K}_o$ of $\mathcal{A}_o$ onto $K\mathcal{G}_o$, and since it is a Lie algebra isomorphism, for any $\xi, \eta \in \sec(\mathcal{A}_o)$

$$\psi \circ \{\xi, \eta\} = [\![\psi \circ \xi, \psi \circ \eta]\!]_{\mathcal{G}} .$$

Just as in the previous section, the self-representation D may be used to define a covariant derivative operator ∇^{T} on $\mathcal{A}_o$, by

$$\nabla^{\mathrm{T}}_X \xi = D_\eta \xi - \{\eta, \xi\} \quad \text{for any } \eta \in \sec(\mathcal{A}_o) \text{ such that } \rho \circ \eta = X\,.$$

It is a derivation of the LAB bracket.

Consider, for any $\boldsymbol{a} \in A_x\mathcal{G}$ and local section ξ of $\mathcal{A}_o$ defined near x, the expression

$$\psi^{-1}\big(D^{\mathcal{G}}_{\boldsymbol{a}}(\psi \circ \xi)\big) - \nabla^{\mathrm{T}}_u \xi \quad \text{where } u = \rho_{\mathcal{G}}(\boldsymbol{a}) \in T_x M\,.$$

For any $f \in C^\infty(M)$,

$$\psi^{-1}\big(D^{\mathcal{G}}_{\boldsymbol{a}}(\psi \circ (f\xi))\big) - \nabla^{\mathrm{T}}_u(f\xi) = f(x)\big(\psi^{-1}\big(D^{\mathcal{G}}_{\boldsymbol{a}}(\psi \circ \xi)\big) - \nabla^{\mathrm{T}}_u \xi\big)\,;$$

so we may define a linear map $\Phi_x(\boldsymbol{a}) : \mathcal{A}_{o,x} \to \mathcal{A}_{o,x}$ by

$$\Phi_x(\boldsymbol{a})(\boldsymbol{b}) = \psi^{-1}\big(D^{\mathcal{G}}_{\boldsymbol{a}}(\psi \circ \xi)\big) - \nabla^{\mathrm{T}}_u \xi \quad \text{for any } \xi \text{ with } \xi(x) = \boldsymbol{b}\,.$$

Moreover, $\Phi_x(\boldsymbol{a})$ depends linearly on $\boldsymbol{a}$ and smoothly on x. Since $D^{\mathcal{G}}_{\boldsymbol{a}}$ is a derivation of the Lie algebra bracket $[\![\cdot,\cdot]\!]_{\mathcal{G}}$ on $K_x\mathcal{G}$, and ∇^{T}_u is a derivation of the Lie algebra bracket $\{\cdot,\cdot\}$ on $\mathcal{A}_{o,x}$, $\Phi_x(\boldsymbol{a})$ is a derivation of the Lie algebra $\mathcal{A}_{o,x}$. But $\mathcal{A}_{o,x}$ is isomorphic to $\mathfrak{g}$, all of whose derivations are inner: so there is an element $\varpi_x(\boldsymbol{a})$ of $\mathcal{A}_{o,x}$ such that $\Phi_x(\boldsymbol{a})(\boldsymbol{b}) = \{\varpi_x(\boldsymbol{a}), \boldsymbol{b}\}$. Moreover, $\varpi_x(\boldsymbol{a})$ is unique, and depends linearly on $\boldsymbol{a}$, because the centre of $\mathfrak{g}$ is $\{0\}$; and depends smoothly on x because $\Phi_x(\boldsymbol{a})$ does. That is, ϖ defines a vector bundle morphism $A\mathcal{G} \to \mathcal{A}_o$. Define a vector bundle morphism $\omega : A\mathcal{G} \to K\mathcal{G}$ by

$$\omega = \psi \circ \varpi\,.$$

We show that ω is the kernel projection of an infinitesimal connection on $A\mathcal{G}$ which satisfies the nondegeneracy condition. To show that it defines an infinitesimal connection it is enough to show that it is the identity on $K\mathcal{G}$. But (going back to the definition of $\Phi_x(\boldsymbol{a})$) if $\boldsymbol{a} \in K_x\mathcal{G}$ then $u = 0$ and $D^{\mathcal{G}}_{\boldsymbol{a}}(\psi \circ \xi) = [\![\boldsymbol{a}, \psi(\xi(x))]\!]_{\mathcal{G}} = \Phi_x(\boldsymbol{a})(\psi(\xi(x)))$, so (again by the fact that the centre of $\mathfrak{g}$ is $\{0\}$) $\psi(\varpi_x(\boldsymbol{a})) = \omega_x(\boldsymbol{a}) = \boldsymbol{a}$ as required. Now suppose that $\boldsymbol{a} \in A_x\mathcal{G}_o$ and $\omega_x(\boldsymbol{a}) = 0$. Then $\varpi_x(\boldsymbol{a}) = 0$, and so

$$\nabla^{\mathrm{T}}_u \xi = \psi^{-1}\big(D^{\mathcal{G}}_{\boldsymbol{a}}(\psi \circ \xi)\big)$$

for all local sections ξ of $\mathcal{A}_o$. Take for ξ a local section λ of $\mathcal{K}_o$. Then the right-hand side of the equation above belongs to $\mathcal{K}_{o,x}$, since $D^{\mathcal{G}}_{\boldsymbol{a}}(\psi \circ \lambda) \in K_x\mathcal{G}_o$. Now

$$\nabla^{\mathrm{T}}_u \lambda = (D_\eta \lambda)(x) - \{\eta(x), \lambda(x)\}$$

where η is any local section of $\mathcal{A}_o$ such that $\rho(\eta(x)) = u$. But $(D_\eta\lambda)(x) \in \mathcal{K}_{o,x}$ since D extends the canonical representation, so $\{\eta(x), \lambda(x)\} \in \mathcal{K}_{o,x}$ for all $\lambda(x) \in \mathcal{K}_{o,x}$. That is, $\{\eta(x), \mathcal{K}_{o,x}\} \subset \mathcal{K}_{o,x}$, or $\eta(x)$ belongs to the normalizer of $\mathcal{K}_{o,x}$. But $\mathcal{K}_{o,x}$ is isomorphic to $\mathfrak{g}_o$ and the normalizer of $\mathfrak{g}_o$ is $\mathfrak{g}_o$: so $\eta(x) \in \mathcal{K}_{o,x}$. But then $u = \rho(\eta(x)) = 0$, and so $\boldsymbol{a} \in K_x\mathcal{G}_o$. Thus $\omega_x(\boldsymbol{a}) = \boldsymbol{a}$, while by assumption $\omega_x(\boldsymbol{a}) = 0$;

and so $\boldsymbol{a} = 0$. It follows that the restriction of ω_x to $A_x\mathcal{G}_o$ is injective, and so by dimension it is an isomorphism $A_x\mathcal{G}_o \to K_x\mathcal{G}$.

Now $\omega : A\mathcal{G}_o \to K\mathcal{G}$ is an isomorphism of vector bundles, and $\psi : \mathcal{A}_o \to K\mathcal{G}$ is an isomorphism of vector bundles, so $\psi^{-1} \circ \omega = \varpi : A\mathcal{G}_o \to \mathcal{A}_o$ is an isomorphism of vector bundles. By its definition, for any $\sigma \in \sec(A\mathcal{G}_o)$, $\xi \in \sec(\mathcal{A}_o)$, and $\eta \in \sec(\mathcal{A}_o)$ such that $\rho \circ \eta = \rho_{\mathcal{G}} \circ \sigma$,

$$\{\varpi \circ \sigma, \xi\} = \psi^{-1} \circ D^{\mathcal{G}}_{\sigma}(\psi \circ \xi) - \left(D_{\eta}\xi - \{\eta, \xi\}\right),$$

so that

$$\{\varpi \circ \sigma - \eta, \xi\} = \psi^{-1} \circ D^{\mathcal{G}}_{\sigma}(\psi \circ \xi) - D_{\eta}\xi .$$

Again, choose $\xi = \lambda \in \sec(\mathcal{K}_o)$: the right-hand side is a section of $\mathcal{K}_o$, and so $\varpi \circ \sigma - \eta \in \sec(\mathcal{K}_o)$, as before. That is, $\rho \circ \varpi \circ \sigma = \rho \circ \eta = \rho_{\mathcal{G}} \circ \sigma$, and we conclude that $\rho \circ \varpi = \rho_{\mathcal{G}}$, and the anchors of $\mathcal{A}_o$ and $A\mathcal{G}_o$ correspond.

We may therefore choose $\eta = \varpi \circ \sigma$ in the last displayed equation, to obtain

$$D^{\mathcal{G}}_{\sigma}(\psi \circ \xi) = \psi \circ D_{\varpi\circ\sigma}\xi .$$

Both D and $D^{\mathcal{G}}$ are representations, from which fact one easily deduces that

$$D_{\varpi\circ[\![\sigma,\tau]\!]_{\mathcal{G}}} = D_{[\![\varpi\circ\sigma,\varpi\circ\tau]\!]} .$$

Note that

$$\rho \circ \varpi \circ [\![\sigma, \tau]\!]_{\mathcal{G}} = \rho_{\mathcal{G}} \circ [\![\sigma, \tau]\!]_{\mathcal{G}} = [\rho_{\mathcal{G}} \circ \sigma, \rho_{\mathcal{G}} \circ \tau] = \rho \circ [\![\varpi \circ \sigma, \varpi \circ \tau]\!] .$$

Let us set $\varpi \circ [\![\sigma, \tau]\!]_{\mathcal{G}} - [\![\varpi \circ \sigma, \varpi \circ \tau]\!] = \lambda$. Then $\lambda \in \sec(\mathcal{K}_o)$, and so $D_{\lambda}\xi = \{\lambda, \xi\} = 0$ for all $\xi \in \sec(\mathcal{A}_o)$. But this means that for any $x \in M$, $\lambda(x)$ is in the centre of the Lie algebra $\mathcal{A}_{o,x}$, which is isomorphic to $\mathfrak{g}$ whose center is $\{0\}$. Thus $\lambda = 0$,

$$\varpi \circ [\![\sigma, \tau]\!]_{\mathcal{G}} = [\![\varpi \circ \sigma, \varpi \circ \tau]\!] ,$$

and ϖ preserves the Lie algebroid brackets.

The kernel projection ω induces a LAB bracket on $A\mathcal{G}_o$, which we denote by $\{\cdot, \cdot\}_{\mathcal{G}}$:

$$\omega \circ \{\sigma, \tau\}_{\mathcal{G}} = [\![\omega \circ \sigma, \omega \circ \tau]\!]_{\mathcal{G}}.$$

That is to say,

$$\psi \circ \varpi \circ \{\sigma, \tau\}_{\mathcal{G}} = [\![\psi \circ \varpi \circ \sigma, \psi \circ \varpi \circ \tau]\!]_{\mathcal{G}} = \psi \circ \{\varpi \circ \sigma, \varpi \circ \tau\}.$$

But $\psi : \mathcal{A}_o \to K\mathcal{G}$ is an isomorphism, so

$$\varpi \circ \{\sigma, \tau\}_{\mathcal{G}} = \{\varpi \circ \sigma, \varpi \circ \tau\} ,$$

and ϖ preserves the LAB bracket also.

Thus $A\mathcal{G}_o$ and $\mathcal{A}_o$ are isomorphic as vector bundles, as Lie algebroids, and as Lie algebra bundles. □

The proof of this theorem is modelled on the proofs of Proposition 2.3 and Theorem 2.7 of [10]; see also [17].

It is the role of the adjoint action of G_o on $\mathfrak{g}$ in the construction of $\mathcal{A}_o$ at the beginning of the proof that explains the name 'adjoint tractor bundle'.

7.7 Locally Symmetric Geometries and Torsion

We note first that for any infinitesimal Cartan geometry the curvature R^{B} of the Blaom connection is easily expressed in terms of the fundamental self-representation D and its torsion.

Proposition 7.7.1

$$D_\xi T(\eta, \zeta) = R^{\mathrm{B}}(\rho \circ \eta, \rho \circ \zeta)\xi \,.$$

Proof Since $\nabla^{\mathrm{B}}_{\rho\circ\eta}\xi = D^*_\eta \xi$, evidently R^{B} is essentially just the curvature of D^*:

$$R^{\mathrm{B}}(\rho \circ \eta, \rho \circ \zeta)\xi = D^*_\eta(D^*_\zeta \xi) - D^*_\zeta(D^*_\eta \xi) - D^*_{[\![\eta,\zeta]\!]}\xi \,.$$

But $D^*_\zeta \xi = D_\zeta \xi + T(\xi, \zeta)$. Substituting this into the formula for R^{B}, and using the fact that D is a representation, together with the first Bianchi identity, yields the result. □

When the infinitesimal connection γ is torsion free we may apply D to R^γ: in the first place, R^γ takes its values in $\mathcal{K}_o \subset \mathcal{A}_o$; and secondly, since for $\lambda \in \sec(\mathcal{K}_o)$, $D_\xi \lambda \in \sec(\mathcal{K}_o)$ we may apply D to the arguments of R^γ: that is, we may consistently define $D_\xi R^\gamma$ by

$$D_\xi R^\gamma(\rho \circ \eta, \rho \circ \zeta) = D_\xi(R^\gamma(\rho \circ \eta, \rho \circ \zeta)) - R^\gamma(\rho \circ D_\xi \eta, \rho \circ \zeta) - R^\gamma(\rho \circ \eta, \rho \circ D_\xi \zeta) \,.$$

Proposition 7.7.2 *When γ is torsion free $DT = DR^\gamma$.*

Proof From Corollary 7.5.2

$$T(\eta, \zeta) = R^\gamma(\rho \circ \eta, \rho \circ \zeta) + \{\eta, \zeta\} \,.$$

But D_ξ is a derivation of the Lie algebra bracket:

$$D_\xi\{\eta, \zeta\} - \{D_\xi \eta, \zeta\} - \{\eta, D_\xi \zeta\} = 0 \,,$$

from which it follows that

$$\begin{aligned}
D_\xi T(\eta,\zeta) &= D_\xi(T(\eta,\zeta)) - T(D_\xi\eta,\zeta) - T(\eta, D_\xi\zeta)\\
&= D_\xi(R^\gamma(\rho\circ\eta,\rho\circ\zeta)) - R^\gamma(\rho\circ D_\xi\eta,\rho\circ\zeta) - R^\gamma(\rho\circ\eta,\rho\circ D_\xi\zeta)\\
&= D_\xi R^\gamma(\rho\circ\eta,\rho\circ\zeta)\,. \qquad \square
\end{aligned}$$

Corollary 7.7.3 *The necessary and sufficient condition for an infinitesimal Cartan geometry to be locally symmetric is that the torsion T of its fundamental self-representation satisfies $DT = 0$; and if its infinitesimal connection is torsion free, that its curvature R^γ satisfies $DR^\gamma = 0$.* □

It would be natural to say that a torsion-free infinitesimal Cartan geometry such that $DR^\gamma = 0$ has *constant* or *parallel curvature*; and almost as natural to extend this terminology to any infinitesimal Cartan geometry for which $DT = 0$. Then the corollary asserts that an infinitesimal Cartan geometry is locally symmetric if and only if it is of constant (parallel) curvature.

In fact any transitive Lie algebroid which is equipped with a self-representation which extends the canonical representation and is such that $DT = 0$ is locally symmetric, that is, locally an action algebroid.

Lemma 7.7.4 *Let $\mathcal{A}_o \to M$ be a transitive Lie algebroid with a self-representation D which extends the canonical representation and satisfies $DT = 0$. Then T defines a Lie algebra bracket on* $\mathrm{sec}(\mathcal{A}_o)$*, which gives $\mathcal{A}_o$ the properties of an adjoint tractor bundle, namely those listed immediately after the proof of Proposition 7.5.3.*

Proof Note first of all that T depends $C^\infty(M)$-linearly on its arguments and is skew. The first Bianchi identity for a self-representation D whose torsion satisfies $DT = 0$ is simply

$$T(T(\xi,\eta),\zeta) + T(T(\eta,\zeta),\xi) + T(T(\zeta,\xi),\eta) = 0\,,$$

which is the Jacobi identity. Thus the restriction of T to each fibre of $\mathcal{A}_o \to M$ defines a Lie algebra bracket on it. We need to show that the Lie algebras defined on different fibres are isomorphic.

We may define a covariant derivative ∇^{B} by $\nabla^{\mathrm{B}}_{\rho\circ\eta}\xi = D^*_\eta\xi$ as before, since by assumption D extends the canonical representation. The curvature of ∇^{B} vanishes as a consequence of $DT = 0$; the sections ξ such that $D^*\xi = 0$ are just those which are parallel with respect to ∇^{B}. It follows from the fact that $DT = 0$, together with the Jacobi identity, that $D^*T = 0$. Let $\{\xi_a\}$ be a local basis of sections of $\mathcal{A}_o \to M$ each of which is parallel with respect to ∇^{B}. We may write $T(\xi_a,\xi_b) = T^c_{ab}\xi_c$ for certain functions T^c_{ab} on M: then for any $\eta \in \mathrm{sec}(\mathcal{A}_o)$

$$D^*_\eta T(\xi_a,\xi_b) = 0 = D^*_\eta(T^c_{ab}\xi_c) - T(D^*_\eta\xi_a,\xi_b) - T(\xi_a, D^*_\eta\xi_b) = (\rho\circ\eta)(T^c_{ab})\xi_c\,.$$

Thus the coefficients T^c_{ab} are locally constant on M, and we have defined a local trivialization of $\mathcal{A}_o$ where the standard fibre is the Lie algebra whose structure constants are the T^c_{ab}.

So $\mathcal{A}_o$ is a Lie algebra bundle. We denote the Lie algebra bracket by $\{\cdot,\cdot\}'$, to distinguish it from previous Lie algebra brackets: so $\{\xi,\eta\}' = T(\xi,\eta)$.

It remains to show that D is a derivation of this Lie algebra bracket, that the Lie algebra bracket coincides with the Lie algebroid bracket on the kernel $\mathcal{K}_o$ of $\mathcal{A}_o$, and that for $\lambda \in \sec(\mathcal{K}_o)$, D_λ coincides with inner derivation with respect to the Lie algebra bracket.

The first of these properties is simply the condition $DT = 0$ rewritten in terms of $\{\cdot,\cdot\}'$. For the second, for $\lambda \in \sec(\mathcal{K}_o)$, $D_\xi\lambda = [\![\xi,\lambda]\!]$; so for $\lambda, \mu \in \sec(\mathcal{K}_o)$

$$\{\lambda,\mu\}' = D_\lambda\mu - D_\mu\lambda - [\![\lambda,\mu]\!] = [\![\lambda,\mu]\!] - [\![\mu,\lambda]\!] - [\![\lambda,\mu]\!] = [\![\lambda,\mu]\!] .$$

Finally, we have

$$D_\lambda\xi = \{\lambda,\xi\}' + D_\xi\lambda + [\![\lambda,\xi]\!] = \{\lambda,\xi\}' .$$

□

Corollary 7.7.5 *If $\mathcal{A}_o$ is the adjoint tractor bundle of an infinitesimal Cartan geometry, and its Lie algebra bracket in that guise is $\{\cdot,\cdot\}$, and $DT = 0$, then for $\lambda \in \sec(\mathcal{K}_o)$, $\{\lambda,\xi\}' = \{\lambda,\xi\}$.*

Proof We have

$$\{\lambda,\xi\}' = D_\lambda\xi = \{\lambda,\xi\} .$$

□

Proposition 7.7.6 *Suppose that the manifold M is connected and simply connected. Let $\mathcal{A}_o \to M$ be a transitive Lie algebroid with a self-representation D which extends the canonical representation and satisfies $DT = 0$; and let $\{\cdot,\cdot\}'$ be the Lie algebra bracket determined by T. Then the space of sections ξ of $\mathcal{A}_o$ such that $D^*\xi = 0$ is a Lie algebra $\mathfrak{a}$ with respect to $\{\cdot,\cdot\}'$ whose dimension is the fibre dimension of $\mathcal{A}_o$; moreover $\xi \to \rho\circ\xi = \xi^\sharp$ defines an action of $\mathfrak{a}$ on M, and $\mathcal{A}_o$ is the action algebroid of this action.*

Proof In view of the results of the lemma, the proof essentially follows that of the final assertion of Theorem 7.4.5. □

Note that in this situation the Lie algebra bracket on $\sec(\mathcal{A}_o)$ and the bracket of the symmetry Lie algebra $\mathfrak{a}$ coincide. There is no infinitesimal connection γ in view: but if there were, recalling that for infinitesimal symmetries

$$\omega \circ \big(\{\xi,\eta\}_{\mathfrak{a}} - \{\xi,\eta\}\big) = R^\gamma(\xi^\sharp,\eta^\sharp) ,$$

we would have $R^\gamma = 0$. It is in fact possible to construct an infinitesimal connection with this property: the first step is to define a Lie algebroid $\mathcal{A}$ for it to live on, and this we may do by using Corollary 2.6.2, taking advantage of the fact that $\mathcal{A}_o$ is a

Lie algebra bundle with bracket $\{\xi, \eta\}' = T(\xi, \eta)$ and that ∇^{B} is a derivation of this Lie algebra bracket, a result established in the following lemma.

Lemma 7.7.7 *When $DT = 0$,*

$$D^*_\xi(T(\eta, \zeta)) = T(D^*_\xi\eta, \zeta) + T(\eta, D^*_\xi\zeta),$$

so that

$$\nabla^{\mathrm{B}}_X\{\eta, \zeta\}' = \{\nabla^{\mathrm{B}}_X\eta, \zeta\}' + \{\eta, \nabla^{\mathrm{B}}_X\zeta\}'.$$

Proof By assumption,

$$D_\xi(T(\eta, \zeta)) = T(D_\xi\eta, \zeta) + T(\eta, D_\xi\zeta);$$

and $D_\xi\eta = D^*_\xi\eta + T(\xi, \eta)$. It follows that

$$\begin{aligned} D^*_\xi(T(\eta, \zeta)) &= T(D^*_\xi\eta, \zeta) + T(\eta, D^*_\xi\zeta) \\ &\quad - T(\xi, T(\eta, \zeta)) + T(T(\xi, \eta), \zeta) + T(\eta, T(\xi, \zeta)); \end{aligned}$$

but the last three terms collectively vanish by the Jacobi identity. □

Proposition 7.7.8 *Let $\mathcal{A}_o \to M$ be a transitive Lie algebroid satisfying the conditions of Lemma 7.7.4, so that in particular $DT = 0$. There is a transitive Lie algebroid $\mathcal{A}$ with flat infinitesimal connection whose kernel $\mathcal{K}$ is isomorphic as a Lie algebra bundle with $\mathcal{A}_o$ equipped with the Lie algebra bracket $\{\cdot, \cdot\}'$ defined by T. Moreover, as a Lie algebroid $\mathcal{A}_o$ may be realised as a Lie subalgebroid of $\mathcal{A}$, such that the kernel projection of the flat infinitesimal connection is a vector bundle isomorphism of $\mathcal{A}_o$ as the sub-bundle so defined with $\mathcal{A}_o$ as the kernel of $\mathcal{A}$.*

Proof By Corollary 2.6.2 we can take $\mathcal{A} = TM \oplus_M \mathcal{A}_o$ with Lie algebroid bracket

$$[\![(X, \xi), (Y, \eta)]\!]_{\mathcal{A}} = \left([X, Y], \nabla^{\mathrm{B}}_X\eta - \nabla^{\mathrm{B}}_Y\xi + \{\xi, \eta\}'\right),$$

and with infinitesimal connection

$$\gamma(X) = (X, 0).$$

Now consider $\mathcal{A}_o$ as a Lie algebroid: denote its anchor by ρ and its algebroid bracket by $[\![\cdot, \cdot]\!]$. We map $\mathcal{A}_o$ linearly and injectively into $\mathcal{A} = TM \oplus_M \mathcal{A}_o$ by the map (ρ, id). For any $\xi, \eta \in \mathrm{sec}(\mathcal{A}_o)$,

$$\begin{aligned} [\![(\rho\circ\xi, \xi), (\rho\circ\eta, \eta)]\!]_{\mathcal{A}} &= \left([\rho\circ\xi, \rho\circ\eta], \nabla^{\mathrm{B}}_{\rho\circ\xi}\eta - \nabla^{\mathrm{B}}_{\rho\circ\eta}\xi + \{\xi, \eta\}'\right) \\ &= \left(\rho\circ[\![\xi, \eta]\!], D^*_\xi\eta - D^*_\eta\xi + T(\xi, \eta)\right). \end{aligned}$$

But

$$\begin{aligned} D^*_\xi\eta - D^*_\eta\xi + T(\xi,\eta) &= D_\xi\eta - T(\xi,\eta) - D_\eta\xi + T(\eta,\xi) + T(\xi,\eta) \\ &= D_\xi\eta - D_\eta\xi - T(\xi,\eta) \\ &= [\![\xi,\eta]\!] \,. \end{aligned}$$

That is to say, (ρ, id) is a morphism of Lie algebroids. Now $\omega \circ (\rho, \mathrm{id}) = (0, \mathrm{id})$, so the restriction of ω to the image of $\mathcal{A}_o$ under (ρ, id) is an isomorphism. □

Corollary 7.7.9 *If M is connected and simply connected then the Lie algebroid $\mathcal{A}$ is isomorphic to the trivial Lie algebroid $TM \oplus_M (M \times \mathfrak{a})$ (where $\mathfrak{a}$ is the Lie algebra defined in Proposition 7.7.6).*

Proof As in the proof of Theorem 7.4.5 we may construct a basis of global sections of $\mathcal{A}_0 \to M$ which are parallel with respect to ∇^{B}, which provides a trivialization of the vector bundle $\mathcal{A}_o \to M$ and identifies it with $M \times \mathfrak{a}$. □

If we start with an infinitesimal Cartan geometry of constant, but nonzero, curvature over a connected and simply connected base we may construct by this process a related infinitesimal Cartan geometry of zero curvature, but at the expense of changing the underlying Lie algebra $\mathfrak{g}$—in fact to the action Lie algebra $\mathfrak{a}$. The action of $\mathfrak{a}$ on M is transitive. Moreover, since there are no nontrivial vertical symmetries of an infinitesimal Cartan geometry (a result we discussed in Sect. 6.6 and shall prove in full generality in Chap. 11) the action is effective. The related infinitesimal Cartan geometry we have in mind is the model infinitesimal Cartan geometry defined by the action (see Sect. 6.4), whose Lie algebroid structure is isomorphic to that of $\mathcal{A}$ as revealed by the preceding proposition and corollary. This change of algebras is related to the process called *mutation* by Sharpe [36]. Note that by Corollary 7.7.5 the two Lie algebras $\mathfrak{g}$ and $\mathfrak{a}$ have the same isotropy subalgebra: in fact the two infinitesimal Cartan geometries in effect share the same $\mathcal{A}_o$, which is the action algebroid of the action of $\mathfrak{a}$.

7.8 Comparisons Compared

We have by now mentioned quite a number of possible definitions or representations (in a non-technical sense) of a Cartan geometry (finite or infinitesimal), progressively differing slightly one from another. We list them in order of, roughly speaking, increasing generality.

1. The fibre-morphism groupoid of a fibre bundle with a homogeneous space $F = G/G_o$ as fibre (Sect. 6.1).
2. The conventional definition as formulated by Sharpe, the point in this context being that it dispenses with the group G, while retaining a Lie algebra $\mathfrak{g}$ (Sects. 7.1 and 7.2).

3. The Lie algebroid of a Lie groupoid of the kind described in item 1 (Sect. 6.2).
4. A Lie algebroid $\mathcal{A}$ of projectable vector fields on a fibre bundle with standard fibre F (Sect. 6.4).
5. The sub-bundle $\mathcal{A}_o$ of the algebroid in item 4 with its joint algebra and algebroid structure and self-representation — where we have lost sight of the fibre bundle with fibre F (Sect. 7.5).
6. A bundle $\mathcal{A}_o$ with the structures itemized in the statement of Theorem 7.6.2, regardless of where it might have come from: this could be interpreted both as what Blaom in [4] calls a classical Cartan algebroid, and as an adjoint tractor bundle in tractor theory [10] (Sects. 7.6 and 7.7).

Theorem 7.6.2 is then rather remarkable in that it closes the circle, for Lie algebras $\mathfrak{g}$ and $\mathfrak{g}_o$ with suitable properties.

Chapter 8
Infinitesimal Cartan Geometries on *TM*

Our basic approach to the study of specific infinitesimal Cartan geometries is to represent elements of $\mathcal{A}$ as projectable vector fields along the fibres of E, and therefore realise $\mathcal{K}$ as vector fields tangent to fibres, forming a representation of $\mathfrak{g}$ on each fibre. There is then an interesting class of examples in which $E = TM$ and $\mathcal{K}$ consists of vector fields on the fibres whose coefficients are polynomial in the canonical fibre coordinates. The standard examples (affine, projective, Riemannian and conformal geometries) all have realisations of this form, as indeed we have seen already in the affine and Riemannian cases. In the first four sections of this chapter we discuss each of these specific cases in turn, while in the final section we develop the general theory of such geometries.

In each of the first four sections we shall describe the structure of the relevant Lie algebroid $\mathcal{A}$, specify the infinitesimal connection γ, and obtain the infinitesimal symmetries. We shall always choose the infinitesimal connection in such a way that the soldering it defines is that given by the vertical lift construction. We pointed out in Sect. 4.5, in relation to affine connections, that when this choice is made there is a linear connection naturally associated with the affine one. This happens to be true for each of the four infinitesimal geometries treated here, and indeed for all members of this class of infinitesimal Cartan geometries: that is to say, there is in each case a linear connection to hand, which moreover will be assumed to be symmetric. We shall frequently work in terms of this associated linear connection, its covariant derivative, its curvature, and so on. In this way we can relate our results directly to the classical ones. Our first choice of reference for these results is Yano's 'The theory of Lie derivatives and its applications' [46]. Our aim in fact is to reproduce his results, but of course using our rather different methods.

In Theorem 7.4.5 we showed that the set of infinitesimal symmetries of an infinitesimal Cartan geometry is a finite-dimensional Lie algebra whose dimension is at most that of the underlying Lie algebra $\mathfrak{g}$. The argument is based on the construction of a connection on the vector bundle $\mathcal{A}_o$, with covariant derivative operator ∇^{B}, such that a section ξ of $\mathcal{A}_o$ is an infinitesimal symmetry if and only if $\nabla^{\mathrm{B}}\xi = 0$. Speaking locally, the maximal dimension is achieved exactly when the curvature R^{B} of the

M. Crampin and D. Saunders, *Cartan Geometries and their Symmetries*,
Atlantis Studies in Variational Geometry 4, DOI 10.2991/978-94-6239-192-5_8

connection vanishes. We shall use these results to obtain the maximal dimension of the infinitesimal symmetry algebra in each of the specific geometries under consideration. However, we shall not use the operator ∇^{B} or its curvature R^{B} explicitly. The conditions for a symmetry we shall obtain are equations involving certain tensorial quantities and their covariant derivatives with respect to the linear connection associated with the geometry; in effect they comprise a system of first-order partial differential equations, with as many equations as there are partial derivatives of the unknowns, and each equation expresses one partial derivative of one unknown as a linear combination of the unknowns themselves. We shall obtain conditions equivalent to the vanishing of R^{B} by deriving the integrability conditions of these equations, that is, in effect by differentiating again and using the symmetry of second-order partial derivatives to get algebraic conditions on the unknowns—though in fact we covariantly differentiate again and use the Ricci identities, taking advantage of the tensorial nature of the unknowns. We shall come back to the consideration of ∇^{B} and R^{B} in the final section.

8.1 Affine Geometry

The simplest case of this type of infinitesimal Cartan geometry is that in which $\mathcal{A}$ consists of vector fields along fibres of *TM* of the form

$$X^i \frac{\partial}{\partial x^i} + (Y^i + Y^i_j y^j)\frac{\partial}{\partial y^i}$$

with constant coefficients, and $\mathcal{A}_o$ with such vector fields for which $Y^i = 0$. For sections one takes the coefficients to be functions on M. As we showed in Sect. 4.2, we can identify $\mathcal{A}$ in this case with the Lie algebroid of the Lie groupoid of affine maps between fibres of *TM*.

In terms of coordinate fields the infinitesimal connection we require takes the form

$$\begin{aligned}\gamma\left(\frac{\partial}{\partial x^i}\right) &= \frac{\partial}{\partial x^i} - (\delta^j_i + \Gamma^j_{ik} y^k)\frac{\partial}{\partial y^j} \\ &= \left(\frac{\partial}{\partial x^i} - \Gamma^j_{ik} y^k \frac{\partial}{\partial y^j}\right) - \frac{\partial}{\partial y^i} \\ &= H_i - V_i\end{aligned}$$

say. That is, for any vector field X on M

$$\gamma(X) = X^{\mathsf{h}} - X^{\mathsf{v}}$$

where $X \mapsto X^{\mathsf{h}}$ is the horizontal lift operator of the associated linear connection as defined in Sect. 4.2. An infinitesimal Cartan geometry on $\mathcal{A}$ with such an infinitesimal connection is called an *infinitesimal Cartan affine geometry.*

The curvature of γ is given simply by

$$R^{\gamma}\left(\frac{\partial}{\partial x^k}, \frac{\partial}{\partial x^l}\right) = R^i_{jkl} y^j \frac{\partial}{\partial y^i}$$

where R^i_{jkl} are the components of the curvature tensor of the associated linear connection; there is no term independent of y because the linear connection is assumed symmetric.

We next discuss the infinitesimal symmetries of an infinitesimal Cartan affine geometry. That is, we consider the equation $\mathcal{L}_\xi \gamma = 0$: but in contrast to the discussion of symmetries of γ in the affine case in Sect. 5.5, and following the definition in Sect. 6.1, we now restrict ξ to be a section of $\mathcal{A}_o$. We may therefore write

$$\xi = X^i H_i + X^i_j y^j V_i$$

where the X^i are components of a vector field $X = \pi_* \xi$ on M and the X^i_j those of a type (1, 1) tensor field on M. We could obtain the infinitesimal symmetries of the geometry by specializing the results of Sect. 5.5, but we prefer to proceed by a direct route so as to establish a uniform approach for the whole chapter.

Proposition 8.1.1 *The section ξ of $\mathcal{A}_o$ is an infinitesimal symmetry of the infinitesimal Cartan affine geometry if and only if X is an infinitesimal affine transformation of the associated linear connection and $X^i_j = X^i_{|j}$. The map $\xi \mapsto \pi_*\xi$ is an isomorphism of the Lie algebra of infinitesimal symmetries of the infinitesimal Cartan affine geometry with the Lie algebra of infinitesimal affine transformations of the associated linear connection. The greatest dimension the symmetry algebra can attain is $n(n+1)$; it has this dimension if and only if the curvature vanishes.*

Proof We have

$$\begin{aligned}
(\mathcal{L}_\xi \gamma)\left(\frac{\partial}{\partial x^j}\right) &= [\xi, H_j - V_j] - \gamma\left(\left[X, \frac{\partial}{\partial x^j}\right]\right) \\
&= \left[X^i H_i + X^i_k y^k V_i, H_j - V_j\right] + \frac{\partial X^i}{\partial x^j}(H_i - V_i) \\
&= -\big((R^i_{klj} X^l + X^i_{k|j}) y^k + (X^i_j - X^i_{|j})\big) V_i.
\end{aligned}$$

From the terms independent of y^i we obtain $X^i_j = X^i_{|j}$, and from the terms linear in y we find (after rearranging the indices) that $X^i_{j|k} = R^i_{jkl} X^l$. These together are the conditions for X to be an infinitesimal affine transformation of the linear connection. The map $\xi \mapsto \pi_*\xi$ is evidently a linear isomorphism of the vector space of infinitesimal symmetries of the infinitesimal Cartan affine geometry with the vector space of infinitesimal affine transformations of the associated linear connection; and it preserves brackets.

The integrability conditions for the equations

$$X^i_{|j} = X^i_j, \quad X^i_{j|k} = R^i_{jkl}X^l$$

are obtained as follows. From the first set

$$\begin{aligned} X^i_{|jk} - X^i_{|kj} &= -R^i_{ljk}X^l \\ &= X^i_{j|k} - X^i_{k|l} = (R^i_{jkl} - R^i_{kjl})X^l, \end{aligned}$$

which is satisfied automatically by the first Bianchi identity. From the second

$$\begin{aligned} X^i_{j|kl} - X^i_{j|lk} &= -R^i_{mkl}X^m_j + R^m_{jkl}X^i_m \\ &= (R^i_{jkm|l} - R^i_{jlm|k})X^m + R^i_{jkm}X^m_{|l} - R^i_{jlm}X^m_{|k} \\ &= R^i_{jkl|m}X^m + R^i_{jkm}X^m_l - R^i_{jlm}X^m_k, \end{aligned}$$

using the second Bianchi identity. The integrability conditions therefore reduce to

$$R^i_{jkl|m}X^m - R^m_{jkl}X^i_m + R^i_{mkl}X^m_j + R^i_{jml}X^m_k + R^i_{jkm}X^m_l = 0.$$

This holds for all X^i and X^i_j if and only if the curvature is everywhere zero.

The fibre dimension is $n + n^2 = n(n+1)$. □

For any vector field X the quantity

$$R^i_{jkl|m}X^m - R^m_{jkl}X^i_{|m} + R^i_{mkl}X^m_{|j} + R^i_{jml}X^m_{|k} + R^i_{jkm}X^m_{|l}$$

is $(\mathcal{L}_X R)^i_{jkl}$, a component of the Lie derivative of the curvature tensor with respect to X. The derivation of the integrability conditions may also be seen as a direct proof of the following result.

Corollary 8.1.2 *An infinitesimal affine transformation leaves the curvature invariant, in the sense that* $\mathcal{L}_X R = 0$. □

As we shall show later, the same conclusion follows from the general result that an infinitesimal symmetry ξ of an infinitesimal Cartan geometry satisfies $\mathcal{L}_\xi R^\gamma = 0$ (Proposition 5.4.4).

As we pointed out in Sect. 5.5, when $X^i_j = X^i_{|j}$, $\xi = X^{\mathrm{C}}$. So the first part of Proposition 8.1.1 is the special case of Theorem 5.5.4 with $Z_2 = 0$ (and $Z_1 = X$). There can be no symmetry in the form of a translation of the fibres for a Cartan affine geometry because we assume, as part of the definition, that ξ is a section of $\mathcal{A}_o$, that is, that the zero section is preserved as well as the infinitesimal connection. Indeed there can be no vertical symmetries at all: this is evident from the conditions given in Proposition 8.1.1, since if $X^i = 0$ then $X^i_j = 0$.

In fact it will be obvious by inspection that there are no vertical symmetries in any of the specific cases dealt with in the following three sections, and we shall prove in Chap. 11 that an infinitesimal Cartan geometry can never have any vertical infinitesimal symmetries.

8.2 Projective Geometry

As we remarked in Sect. 4.4, the set of vector fields on fibres of TM of the form

$$X^i \frac{\partial}{\partial x^i} + (Y^i + Y^i_j y^j + (Y_j y^j) y^i) \frac{\partial}{\partial y^i}$$

(with constant coefficients) constitutes a Lie algebroid $\mathcal{A}$, which is formally similar to the Lie algebroid of the Lie groupoid of projective maps between fibres of the Cartan projective bundle. However, quadratic vertical vector fields on fibres of TM do not have vertically uniform flow boxes, so this Lie algebroid does not come from a Lie groupoid of fibre morphisms of TM. We may nevertheless study this collection of vector fields as the arena of an infinitesimal Cartan geometry.

For a connection we take

$$\gamma\left(\frac{\partial}{\partial x^i}\right) = \frac{\partial}{\partial x^i} - (\delta^j_i + \Gamma^j_{ik} y^k + (\Lambda_{ik} y^k) y^j) \frac{\partial}{\partial y^j};$$

here Λ_{ij} is a type (0, 2) tensor field on M, with no assumed symmetry properties. An infinitesimal Cartan geometry of this type is called an *infinitesimal Cartan projective geometry*. We regard the associated linear connection as given, or in other words the connection coefficients Γ^i_{jk} as fixed; we next show that there is a canonical choice for the tensor Λ_{ij}.

Since we now have a quadratic term in γ the curvature R^γ contains a term of homogeneity degree 2 in y as well as one of degree 1. We consider first the term in R^γ of homogeneity degree 1. It is going to be most convenient to work with components, and we denote the components of this term by $\bar{R}^i_{jkl}$, with $\bar{R}^i_{jlk} = -\bar{R}^i_{jkl}$. We find that

$$\bar{R}^i_{jkl} = R^i_{jkl} + (\Lambda_{kl} - \Lambda_{lk})\delta^i_j + \Lambda_{kj}\delta^i_l - \Lambda_{lj}\delta^i_k.$$

One may choose Λ_{ij} to make $\bar{R}^i_{jkl}$ trace-free. By taking the trace on i and l one finds that this requires that

$$\Lambda_{jk} = \frac{1}{n^2 - 1}(R_{jk} + n R_{kj}),$$

where $R_{jk} = R^i_{jik}$ is the Ricci tensor. Then $\bar{R}^i_{jkl} = P^i_{jkl}$ is the projective curvature tensor of the associated linear connection Γ^i_{jk}; all of its traces vanish, that on i and j by virtue of the first Bianchi identity. We have (as components of a $\mathcal{K}$-valued 2-form)

$$R^{\gamma}_{kl} = \left(P^i_{jkl}y^j + (\Lambda_{lj|k} - \Lambda_{kj|l})y^j y^i\right)\frac{\partial}{\partial y^i}.$$

With this choice of Λ_{ij} the connection γ is called the normal projective algebroid connection, and the geometry is called the *normal* infinitesimal Cartan projective geometry, for the given linear connection.

To obtain the conditions for a symmetry it is best to work as before in terms of the horizontal vector fields for the associated linear connection. We set

$$\xi = X^i H_i + (X^i_j y^j + (X_k y^k)y^i)V_i$$

where now the X_k are the components of a 1-form on M. In the first instance we obtain the conditions for a symmetry of an arbitrary, that is, not necessarily normal, infinitesimal Cartan projective geometry.

Proposition 8.2.1 *The conditions for ξ to be an infinitesimal symmetry of the infinitesimal Cartan projective geometry are*

$$\begin{aligned} X^i_{|j} &= X^i_j \\ X^i_{j|k} &= R^i_{jkl}X^l + \delta^i_k X_j + \delta^i_j X_k \\ X_{i|j} &= -(\Lambda_{ji|k}X^k + \Lambda_{kj}X^k_i + \Lambda_{ik}X^k_j). \end{aligned}$$

Proof These conditions are obtained by carrying out a calculation similar to, but of course more complicated than, that in the proof of Proposition 8.1.1, and setting to zero separately the terms of different homogeneity degrees. □

Corollary 8.2.2 *If ξ is a symmetry of an arbitrary infinitesimal Cartan projective geometry then its projection $\pi_*\xi$ to M is an infinitesimal projective transformation of the associated linear connection, and ξ differs by a vertical field from a symmetry of the normal infinitesimal Cartan projective geometry with the same associated linear connection. In particular, $\xi \mapsto \pi_*\xi$ is an isomorphism of the Lie algebra of infinitesimal symmetries of the normal infinitesimal Cartan projective geometry with the Lie algebra of infinitesimal projective transformations of the associated linear connection. The greatest dimension that the symmetry algebra of the normal geometry can attain is $n(n+2)$; it has this dimension if and only if the space is projectively flat.*

Proof A vector field X on M is an infinitesimal projective transformation of a linear connection if and only if there is a 1-form α such that

$$X^i_{|jk} = R^i_{jkl}X^l + \delta^i_k\alpha_j + \delta^i_j\alpha_k.$$

It then follows (by taking a further covariant derivative and a trace, and using the Bianchi identities) that α satisfies

$$\alpha_{i|j} = \Lambda_{ji|k}X^k + \Lambda_{kj}X^k_{|i} + \Lambda_{ik}X^k_{|j}$$

where

$$\Lambda_{ij} = \frac{1}{n^2 - 1}(R_{ij} + nR_{ji}),$$

the value for the normal geometry. The first assertions follow.

Again, the integrability conditions for the first set of equations in the proposition are automatically satisfied, but now because the terms $\delta^i_k X_j + \delta^i_j X_k$ are symmetric in j and k and so play no role. The further integrability conditions for the normal geometry turn out to be

$$P^i_{jkl|m} X^m - P^m_{jkl} X^i_m + P^i_{mkl} X^m_j + P^i_{jml} X^m_k + P^i_{jkm} X^m_l = 0$$

for the second set of equations, and

$$P^m_{jkl} X_m - P_{jkl|m} X^m - P_{mkl} X^m_j - P_{jml} X^m_k - P_{jkm} X^m_l = 0,$$

where $P_{jkl} = \Lambda_{jk|l} - \Lambda_{jl|k}$, for the third. The dimension is maximal if and only if $P^i_{jkl} = 0$ and $P_{jkl} = 0$. When $n \geq 3$ the former is the condition for the space to be projectively flat, and it implies the latter. When $n = 2$ the former holds automatically, and the latter is the condition for the space to be projectively flat.

The fibre dimension is $n + n^2 + n = n(n + 2)$. □

Corollary 8.2.3 *An infinitesimal projective transformation leaves the projective curvature invariant, in the sense that* $\mathcal{L}_X P = 0$.

Proof

$$(\mathcal{L}_X P)^i_{jkl} = P^i_{jkl|m} X^m - P^m_{jkl} X^i_{|m} + P^i_{mkl} X^m_{|j} + P^i_{jml} X^m_{|k} + P^i_{jkm} X^m_{|l} = 0. \quad \square$$

8.3 Riemannian Geometry

Suppose now that M is equipped with a Riemannian metric g. (In fact most of what is said below is easily extended to the case of a pseudo-Riemannian metric.) We consider those projectable vector fields ξ along fibres of TM for which

$$\xi = X^i \frac{\partial}{\partial x^i} + (Y^i + Y^i_j y^j) \frac{\partial}{\partial y^i}$$

where

$$X^k \frac{\partial g_{ij}}{\partial x^k} + g_{kj} Y^k_i + g_{ik} Y^k_j = 0.$$

The collection of such vector fields forms a Lie algebroid $\mathcal{A}$, which is evidently the Lie algebroid of the Lie groupoid of Euclidean maps between fibres of TM defined by g discussed in Sect. 4.5.

Let γ be an infinitesimal connection on $\mathcal{A}$,

$$\gamma\left(\frac{\partial}{\partial x^i}\right) = \frac{\partial}{\partial x^i} - (\delta_i^j + \Gamma_{ik}^j y^k)\frac{\partial}{\partial y^j}.$$

In order that γ takes its values in $\mathcal{A}$ we must have

$$\frac{\partial g_{ij}}{\partial x^k} + g_{lj}\Gamma_{ki}^l + g_{il}\Gamma_{kj}^l = 0.$$

Assuming that the Γ_{ij}^k are symmetric in their lower indices, this forces the connection coefficients to be those of the Levi-Civita connection, as we showed in Proposition 4.4.5. We call the corresponding infinitesimal Cartan geometry the *infinitesimal Cartan Riemannian geometry* of g.

To find the infinitesimal symmetries of an infinitesimal Cartan Riemannian geometry we take ξ to be a section of $\mathcal{A}_o$, and as before express it as

$$\xi = X^i H_i + X_j^i y^j V_i$$

for some vector field and type (1, 1) tensor field on M, where the horizontal field H_i is that of the Levi-Civita connection: but now X_j^i must be skew-symmetric with respect to g,

$$g_{kj}X_i^k + g_{ik}X_j^k = 0,$$

or (with $X_{ij} = g_{ik}X_j^k$)

$$X_{ij} + X_{ji} = 0.$$

Proposition 8.3.1 *The section ξ of $\mathcal{A}_o$ is an infinitesimal symmetry of the infinitesimal Cartan Riemannian geometry if and only if the vector field X is an infinitesimal isometry (a Killing field) of g and $X_j^i = X^i_{|j}$. The map $\xi \mapsto \pi_*\xi$ is an isomorphism of the Lie algebra of infinitesimal symmetries of the infinitesimal Cartan Riemannian geometry with the Lie algebra of infinitesimal isometries of g. The greatest dimension the symmetry algebra can attain is $\frac{1}{2}n(n+1)$; it has this dimension if and only if the space is of constant curvature.*

Proof An infinitesimal symmetry of an infinitesimal Cartan Riemannian geometry is (by definition) an infinitesimal symmetry of γ, and so Proposition 8.1.1 applies. Thus $X_j^i = X^i_{|j}$. But X_j^i is skew-symmetric with respect to g, so

$$g_{kj}X^k_{|i} + g_{ik}X^k_{|j} = 0;$$

this is one version of Killing's equation. The further condition that $X^i_{j|k} = R^i_{jkl}X^l$ is a consequence of Killing's equation.

As before the integrability conditions for the first set of equations $X^i_j = X^i_{|j}$ are satisfied automatically; those for the second set can be written

$$R_{ijkl|m}X^m + (\delta^q_i R^p_{jkl} - \delta^q_j R^p_{ikl} + \delta^q_k R^p_{lij} - \delta^q_l R^p_{kij})X_{pq} = 0$$

(where again indices have been lowered with g_{ij}). In the maximal case this must hold for every X^i and every skew X_{jk}. So $R_{ijkl|m} = 0$, and the coefficient of X_{pq} is symmetric in p and q, that is

$$\delta^p_i R^q_{jkl} - \delta^p_j R^q_{ikl} + \delta^p_k R^q_{lij} - \delta^p_l R^q_{kij} = \delta^q_i R^p_{jkl} - \delta^q_j R^p_{ikl} + \delta^q_k R^p_{lij} - \delta^q_l R^p_{kij}.$$

If one sums over i and p one obtains

$$(n-1)R_{ijkl} = g_{ik}R_{jl} - g_{il}R_{jk}$$

after bringing an index down, where $R_{ij} = R^k_{ikj}$ is the Ricci tensor, which in this case is symmetric, though it is not in general. It follows (by multiplying by g^{jl} and summing) that $nR_{ik} = Rg_{ik}$, where R is the scalar curvature, and so

$$R_{ijkl} = \frac{R}{n(n-1)}(g_{ik}g_{jl} - g_{il}g_{jk}).$$

The space is of constant curvature; $R_{ijkl|m} = 0$ automatically.

The fibre dimension is $n + \frac{1}{2}n(n-1) = \frac{1}{2}n(n+1)$. □

8.4 Conformal Geometry

For conformal geometry we start with a Riemannian metric g considered as a fibre metric, and find those projectable vector fields ξ on TM such that $\mathcal{L}_\xi g^\mathsf{v} = \lambda g^\mathsf{v}$ for some function λ on TM. The set of such vector fields is an $\mathbb{R}$-linear space closed under bracket. Throughout this section we assume that $n \geq 3$, for reasons that will soon become apparent.

Let

$$\xi = X^i \frac{\partial}{\partial x^i} + Y^i \frac{\partial}{\partial y^i};$$

we require that

$$X^k \frac{\partial g_{ij}}{\partial x^k} + g_{kj}\frac{\partial Y^k}{\partial y^i} + g_{ik}\frac{\partial Y^k}{\partial y^j} = \lambda g_{ij}$$

where Y^k and λ may depend on the y^i but the other terms are independent of them. By repeated differentiation with respect to the y^i and the taking of traces one finds that

$$\frac{\partial^2 \lambda}{\partial y^i \partial y^j} = 0,$$

so that $\lambda = \lambda_0 + \lambda_i y^i$. It follows that the dependence of Y^i on y is given explicitly by

$$Y^i = Y^i_0 + Y^i_j y^j + \tfrac{1}{2}(\lambda_j \delta^i_k + \lambda_k \delta^i_j - g_{jk} g^{il} \lambda_l) y^j y^k$$

where

$$X^k \frac{\partial g_{ij}}{\partial x^k} + g_{kj} Y^k_i + g_{ik} Y^k_j = \lambda_0 g_{ij}.$$

Vector fields along the fibres of *TM* of the form

$$X^i \frac{\partial}{\partial x^i} + \left(Y^i_0 + Y^i_j y^j + \tfrac{1}{2}(Y_j \delta^i_k + Y_k \delta^i_j - g_{jk} g^{il} Y_l) y^j y^k\right) \frac{\partial}{\partial y^i},$$

where the (constant) coefficients satisfy the immediately preceding equation for some λ_0, form a Lie algebroid. The kernel is isomorphic to the *Möbius algebra*.

To obtain an infinitesimal Cartan geometry with this algebroid we take for a connection

$$\gamma\left(\frac{\partial}{\partial x^i}\right) = \frac{\partial}{\partial x^i} - \left(\delta^j_i + \Gamma^j_{ik} y^k + \tfrac{1}{2}(\Lambda_{ik} \delta^j_l + \Lambda_{il} \delta^j_k - g_{kl} g^{jp} \Lambda_{ip}) y^k y^l\right) \frac{\partial}{\partial y^j},$$

with Γ^k_{ij} the Levi-Civita connection of g_{ij}. The resulting infinitesimal Cartan geometry is an *infinitesimal Cartan conformal geometry*. As in the projective case, this geometry is strictly infinitesimal: its Lie algebroid (an algebroid on *TM*, remember) does not come from a Lie groupoid of fibre morphisms.

The y-linear part of the curvature R^γ is given by

$$\bar{R}^i_{jkl} = R^i_{jkl} + (\Lambda_{kl} - \Lambda_{lk}) \delta^i_j + \Lambda_{kj} \delta^i_l - \Lambda_{lj} \delta^i_k - g_{jl} g^{ip} \Lambda_{kp} + g_{jk} g^{ip} \Lambda_{lp}.$$

For $n \geq 3$ one may choose Λ_{jk} so that $\bar{R}^l_{jkl} = 0$, as in the projective case: here this requires that

$$\Lambda_{jk} = \frac{1}{n-2}\left(R_{jk} - \frac{1}{2(n-1)} R g_{jk}\right);$$

the Ricci tensor R_{jk} is symmetric, and so is Λ_{jk}. With this choice $\bar{R}^i_{jkl} = C^i_{jkl}$, the conformal curvature tensor of g_{ij}. The connection γ is then called the *normal* conformal algebroid connection.

The quadratic part of the curvature, in the general case, is

$$\tfrac{1}{2}\big((\Lambda_{lm|k} - \Lambda_{km|l})\delta^i_j + (\Lambda_{lj|k} - \Lambda_{kj|l})\delta^i_m - g_{jm}g^{ip}(\Lambda_{lp|k} - \Lambda_{kl|l})\big)y^j y^m \frac{\partial}{\partial y^i}.$$

For the normal connection the coefficients $\Lambda_{lm|k} - \Lambda_{km|l}$ etc. are closely related to the components of the Cotton tensor, which are given by

$$R_{ij|k} - R_{ik|j} + \frac{1}{2(n-1)}(R_{|j}g_{ik} - R_{|k}g_{ij}) = (n-2)(\Lambda_{ij|k} - \Lambda_{ik|j}).$$

To obtain the conditions for a symmetry of an infinitesimal Cartan conformal geometry we work as before in terms of the horizontal vector fields H_i for the associated linear connection (the Levi-Civita connection of g). Consider the vector field

$$\xi = X^i H_i + \big(X^i_j y^j + \tfrac{1}{2}(Y_j\delta^i_k + Y_k\delta^i_j - g_{jk}g^{ip}Y_p)y^j y^k\big)\frac{\partial}{\partial y^i}.$$

Here Y_i are the components of a 1-form on M. We shall find it convenient to raise and lower indices with g_{ij}, so we must distinguish notationally between $g_{ij}X^j$ and the coefficient in the y-quadratic terms which is therefore denoted by Y_i; and of course we may set $Y^i = g^{ij}Y_j$. We must have

$$g_{kj}X^k_i + g_{ik}X^k_j = X_{ji} + X_{ij} = \mu g_{ij}$$

for some function μ on M.

Once again, we first obtain the conditions for a symmetry of an arbitrary infinitesimal Cartan conformal geometry. This is done by carrying out a similar calculation to previous ones.

Proposition 8.4.1 *The conditions for ξ to be an infinitesimal symmetry of the infinitesimal Cartan conformal geometry are*

$$\begin{aligned} X^i_{|j} &= X^i_j \quad \textit{where } X_{ij} + X_{ji} = \mu g_{ij} \\ X^i_{j|k} &= R^i_{jkl}X^l + \delta^i_k Y_j + \delta^i_j Y_k - g_{jk}Y^i \\ Y_{i|j} &= -(\Lambda_{ji|k}X^k + \Lambda_{kj}X^k_i + \Lambda_{ki}X^k_j). \end{aligned}$$

In particular,

$$\mu = \frac{2}{n}X^i_i, \quad Y_i = \tfrac{1}{2}\mu_{|i}.$$

Proof The final conditions involving μ are obtained by taking traces in the first two. □

Corollary 8.4.2 *If ξ is a symmetry of an arbitrary infinitesimal Cartan conformal geometry then its projection $\pi_*\xi$ to M is an infinitesimal conformal transformation, or conformal Killing field, of the metric g. Then ξ differs by a vertical field from a symmetry of the normal infinitesimal Cartan conformal geometry with the same metric. In particular, $\xi \mapsto \pi_*\xi$ is an isomorphism of the Lie algebra of infinitesimal symmetries of the normal infinitesimal Cartan conformal geometry with the Lie algebra of infinitesimal conformal transformations of g. The greatest dimension that the symmetry algebra of the normal geometry can attain is $\frac{1}{2}(n+1)(n+2)$; it has this dimension if and only if the space is conformally flat.*

Proof A vector field X on M is an infinitesimal conformal transformation of g if and only if there is a function μ such that

$$X_{i|j} + X_{j|i} = \mu g_{ij}.$$

It then follows (by taking a further covariant derivative and using the Ricci and first Bianchi identities) that

$$X_{i|jk} = R_{ijkl}X^l + \tfrac{1}{2}(\mu_{|j}g_{ik} + \mu_{|k}g_{ij} - \mu_{|i}g_{jk}).$$

A further covariant differentiation followed by the taking of appropriate traces yields

$$\mu_{|ij} = -2(\Lambda_{ij|k}X^k + \Lambda_{kj}X^k_{|i} + \Lambda_{ki}X^k_{|j})$$

where

$$\Lambda_{ij} = \frac{1}{n-2}\left(R_{ij} - \frac{1}{(n-1)}Rg_{ij}\right)$$

the value for the normal geometry. The first assertions follow.

The integrability conditions for the first set of equations in the proposition are automatically satisfied, as before. The further integrability conditions for the normal geometry are

$$C^i_{jkl|m}X^m - C^m_{jkl}X^i_m + C^i_{mkl}X^m_j + C^i_{jml}X^m_k + C^i_{jkm}X^m_l = 0$$

for the second set of equations, and

$$C^m_{jkl}Y_m - C_{jkl|m}X^m - C_{mkl}X^m_j - C_{jml}X^m_k - C_{jkm}X^m_l = 0,$$

where $C_{jkl} = \Lambda_{jk|l} - \Lambda_{jl|k}$, for the third. The dimension is maximal if and only if both equations hold pointwise for any choice of X^i, Y_i and X^j_k such that $X_{jk} + X_{kj} \propto g_{jk}$. In particular they must hold for $X^i = 0$, $Y_i = 0$ and $X^j_k = \delta^j_k$, so we must have $C^i_{jkl} = 0$ and $C_{jkl} = 0$. When $n \geq 4$ the former is the condition for the space to be conformally flat, and it implies the latter. When $n = 3$ the former holds automatically, and the latter is the condition for the space to be conformally flat.

The fibre dimension is $n + \frac{1}{2}n(n-1) + 1 + n = \frac{1}{2}(n+1)(n+2)$. □

Corollary 8.4.3 *An infinitesimal conformal transformation leaves the conformal curvature invariant, in the sense that* $\mathcal{L}_X C = 0$. □

8.5 The General Theory

In the general case the Lie algebra $\mathfrak{g}$ consists of vector fields on $\mathbb{R}^n$ whose components are polynomial in the standard coordinates. Such algebras of vector fields were analysed by Guillemin and Sternberg [26]. We require $\mathfrak{g}$ to be transitive and finite, in their terminology. Let $\mathfrak{g}^r$ be the space of elements of $\mathfrak{g}$ whose coefficients are homogeneous polynomials of degree r (here we depart slightly from the standard notation); then

$$[\mathfrak{g}^r, \mathfrak{g}^s] \subset \mathfrak{g}^{r+s-1};$$

this makes $\mathfrak{g}$ a graded Lie algebra (where the elements of grade r comprise $\mathfrak{g}^{r+1}$). Then $\mathfrak{g}$ is transitive if $\mathfrak{g}^0 \cong \mathbb{R}^n$, and finite if $\mathfrak{g}^r = \{0\}$ for all r sufficiently large. As a vector space $\mathfrak{g}$ is the direct sum of the $\mathfrak{g}^r$, and $\mathfrak{g}_o = \sum_{r>0} \mathfrak{g}^r$. So far as the algebra structure goes, note first that $\mathfrak{g}^0$ is an abelian subalgebra of $\mathfrak{g}$. Furthermore $\mathfrak{g}^1$ is a subalgebra, called the linear isotropy algebra of $\mathfrak{g}$; since $[\mathfrak{g}^1, \mathfrak{g}^0] \subset \mathfrak{g}^0$ we have a representation of $\mathfrak{g}^1$ on $\mathfrak{g}^0 \cong \mathbb{R}^n$ which can be used to identify $\mathfrak{g}^1$ with a subalgebra of $\mathfrak{gl}(n)$. Finally, note that $\mathfrak{g}_o$ is a subalgebra of $\mathfrak{g}$ (as it should be).

Let us correspondingly denote by $\mathcal{K}^r$ the sub-bundle of $\mathcal{K}$ consisting of elements whose coefficients are homogeneous of degree r: we say that such elements themselves are homogeneous of degree r. As a vector bundle $\mathcal{K}^r$ is isomorphic to a sub-bundle of $TM \otimes S^r T^*M$, the bundle of tensors of type $(1, r)$ totally symmetric in the covariant indices. For any tensor T_x at $x \in M$ of the given type, with components $T^i_{j_1 j_2 \dots j_r}$, we define a vertical vector field T^{V}_x on $T_x M$ by

$$T^{\mathsf{V}}_x = \frac{1}{r!} T^i_{j_1 j_2 \dots j_r} y^{j_1} y^{j_2} \dots y^{j_r} \frac{\partial}{\partial y^i}.$$

It is immediately apparent that T^{V}_x is well-defined, that is, that the definition does not depend on the choice of coordinates, and that it is homogeneous of degree r. Furthermore if T is a globally defined tensor field of the given type then this formula defines a global vertical vector field T^{V} on TM. The bracket of vertical vector fields induces an algebraic bracket of tensors, which we denote with braces, so that for S of type $(1, r)$ and T of type $(1, s)$, $\{S, T\}$ is of type $(1, r+s-1)$, is symmetric in the covariant indices, and satisfies

$$\{S, T\}^{\mathsf{V}} = [S^{\mathsf{V}}, T^{\mathsf{V}}].$$

This formula defines equally a bracket of tensors at a point of M and a bracket of tensor fields: with the latter interpretation the bracket is $C^\infty(M)$-bilinear. Let us denote by $\bar{\mathcal{K}}^r$ the module of tensor fields corresponding to $\mathcal{K}^r$, and $\bar{\mathcal{K}}$ the direct sum of the $\bar{\mathcal{K}}^r$. Then for each $x \in M$, $\bar{\mathcal{K}}_x$ is a Lie algebra under the bracket $\{\cdot, \cdot\}$ isomorphic to $\mathfrak{g}$, with $\bar{\mathcal{K}}^r_x$ isomorphic (as a vector space) to $\mathfrak{g}^r$.

With a view to incorporating covariant differentials, curvatures etc., we extend these notions to $\mathcal{K}$-valued p-forms. We denote by $\mathcal{K}^r \otimes \bigwedge^p T^*M$ those $\mathcal{K}$-valued p-forms whose values are homogeneous of degree r. The coefficients of an r-homogeneous $\mathcal{K}$-valued p-form define a tensor field of type $(1, r+p)$ which is symmetric in the first r covariant indices, skew in the last p; in other words, an element of $\mathcal{K}^r \otimes \bigwedge^p T^*M$ can be identified with a section of $TM \otimes S^r T^*M \otimes \bigwedge^p T^*M$. For any such tensor field Q on M and any vector fields $X_1, X_2, \dots, X_p$, we have a tensor field $Q(X_1, X_2, \dots, X_p)$ of type $(1, r)$, and an r-homogeneous vertical vector field $Q(X_1, X_2, \dots, X_p)^{\mathsf{v}}$ on *TM*: we may denote by Q^{v} the resulting type $(1, p)$ vertical tensor field.

We may use the vertical lift operation and the bracket of vertical vector fields to define an operator

$$\partial : \sec\big(TM \otimes S^r T^*M \otimes \textstyle\bigwedge^p T^*M\big) \to \sec\big(TM \otimes S^{r-1} T^*M \otimes \bigwedge^{p+1} T^*M\big)$$

by

$$\partial Q(X_1, X_2, \dots, X_{p+1})^{\mathsf{v}} = \sum_{u=1}^{p+1} (-1)^{u+1} \big[X_u^{\mathsf{v}}, Q(X_1, \dots \widehat{X_u} \dots, X_{p+1})^{\mathsf{v}}\big].$$

It is evident that ∂Q does belong to the specified space: taking the bracket with the vertical lift of a vector field on M reduces homogeneity degree by 1. Moreover, ∂ is $C^\infty(M)$-linear, and satisfies $\partial^2 = 0$.

Suppose we have a symmetric linear connection on M, with corresponding horizontal lift $X \mapsto X^{\mathsf{h}}$ and covariant derivative ∇. The well-known formula $[X^{\mathsf{h}}, Y^{\mathsf{v}}] = (\nabla_X Y)^{\mathsf{v}}$ extends to $\sec(TM \otimes S^r T^*M)$ as follows: for $T \in \sec(TM \otimes S^r T^*M)$

$$[X^{\mathsf{h}}, T^{\mathsf{v}}] = (\nabla_X T)^{\mathsf{v}}.$$

We may further extend this operation to tensor-valued forms to define a covariant differential

$$d_\nabla : \sec\big(TM \otimes S^r T^*M \otimes \textstyle\bigwedge^p T^*M\big) \to \sec\big(TM \otimes S^r T^*M \otimes \bigwedge^{p+1} T^*M\big)$$

by

$$\begin{aligned} &d_\nabla Q(X_1, X_2, \dots, X_{p+1})^{\mathsf{v}} \\ &\quad = \sum_{r=1}^{p+1} (-1)^{r+1} \big[X_r^{\mathsf{h}}, Q(X_1, \dots \widehat{X_r} \dots, X_{p+1})^{\mathsf{v}}\big] \\ &\qquad + \sum_{1 \le r, s \le p+1} (-1)^{r+s} Q([X_r, X_s], X_1, \dots \widehat{X_r} \dots \widehat{X_s} \dots, X_{p+1})^{\mathsf{v}} \end{aligned}$$

$$= \sum_{r=1}^{p+1} (-1)^{r+1} \left(\nabla_{X_r} (Q(X_1, \dots \widehat{X_r} \dots, X_{p+1})) \right)^{\mathsf{v}}$$

$$+ \sum_{1 \le r,s \le p+1} (-1)^{r+s} Q(\nabla_{X_r} X_s - \nabla_{X_s} X_r, X_1, \dots \widehat{X_r} \dots \widehat{X_s} \dots, X_{p+1})^{\mathsf{v}}$$

$$= \sum_{r=1}^{p+1} (-1)^{r+1} \left((\nabla_{X_r} Q)(X_1, \dots \widehat{X_r} \dots, X_{p+1}) \right)^{\mathsf{v}}.$$

Finally, the bracket $\{\cdot, \cdot\}$ of tensors can be extended to a bracket-cum-product of tensor-valued forms

$$\{\cdot \wedge \cdot\} : \sec\big(TM \otimes S^r T^*M \otimes \textstyle\bigwedge^p T^*M \times TM \otimes S^s T^*M \otimes \bigwedge^q T^*M\big)$$
$$\to \sec\big(TM \otimes S^{r+s-1} T^*M \otimes \textstyle\bigwedge^{p+q} T^*M\big)$$

by

$$\{P \wedge Q\}(X_1, \dots, X_r, X_{r+1}, \dots, X_{r+s})$$
$$= \frac{1}{r!s!} \sum_{\sigma} (-1)^{|\sigma|} \{P(X_{\sigma(1)}, \dots, X_{\sigma(r)}), Q(X_{\sigma(r+1)}, \dots, X_{\sigma(r+s)})\}$$

where $P \in TM \otimes S^r T^*M \otimes \bigwedge^p T^*M$, $Q \in TM \otimes S^s T^*M \otimes \bigwedge^q T^*M$, the sum is over all permutations σ of $1, 2, \dots, r+s$, and $|\sigma|$ is the sign of σ. The formula is based on one for the ordinary exterior product of an r- and an s-form, which may explain the notation.

Suppose now we have an infinitesimal Cartan geometry on TM of this type. We assume that the soldering is given by the vertical lift, as we did in distinguishing affine from generalized affine connections in Sect. 4.2, and as we do throughout this chapter. Then the infinitesimal connection γ takes the form

$$\gamma(X) = X^{\mathsf{h}} - X^{\mathsf{v}} - \Gamma(X)$$

where $X \mapsto X^{\mathsf{h}}$ is the horizontal lift operator for a linear connection, and $\Gamma(X) \in \sum_{r>1} \mathcal{K}^r$. Clearly $X^{\mathsf{h}} \in \sec(\mathcal{A})$. We assume that the linear connection is symmetric, as we have done throughout this chapter. We may write

$$\Gamma(X) = \sum_{r>1} \bar{\Gamma}_{(r)}(X)^{\mathsf{v}}, \quad \bar{\Gamma}_{(r)} \in \bar{\mathcal{K}}^r \otimes T^*M.$$

Proposition 8.5.1 *The curvature R^γ of this infinitesimal connection, considered as a $\mathcal{K}$-valued 2-form, is given by*

$$R^\gamma = \Big((R + \partial\bar{\Gamma}_{(2)} + \sum_{r>1} (d_\nabla \bar{\Gamma}_{(r)} + \partial\bar{\Gamma}_{(r+1)}) + \sum_{s,t>1} \{\bar{\Gamma}_{(s)} \wedge \bar{\Gamma}_{(t)}\} \Big)^{\mathsf{v}}$$

where R is the curvature tensor of the linear connection, considered as a tensor-valued 2-form, and ∇ its covariant derivative. In particular the term of R^γ of homogeneity degree 1 is determined by the tensor-valued 2-form

$$R + \partial\bar{\Gamma}_{(2)},$$

while that of homogeneity degree r when $r > 1$ is determined by

$$d_\nabla\bar{\Gamma}_{(r)} + \partial\bar{\Gamma}_{(r+1)} + \sum_{s+t=r+1} \{\bar{\Gamma}_{(s)} \wedge \bar{\Gamma}_{(t)}\}, \quad s, t > 1.$$

Proof We have

$$\begin{aligned} R^\gamma(X, Y) &= [X, Y]^{\mathsf{h}} - [X, Y]^{\mathsf{v}} - \Gamma([X, Y]) \\ &\quad - [X^{\mathsf{h}} - X^{\mathsf{v}} - \Gamma(X), Y^{\mathsf{h}} - Y^{\mathsf{v}} - \Gamma(Y)] \\ &= R(X, Y)^{\mathsf{v}} - [X, Y]^{\mathsf{v}} + (\nabla_X Y - \nabla_Y X)^{\mathsf{v}} \\ &\quad + [X^{\mathsf{h}}, \Gamma(Y)] - [Y^{\mathsf{h}}, \Gamma(X)] - \Gamma([X, Y]) \\ &\quad - [X^{\mathsf{v}}, \Gamma(Y)] + [Y^{\mathsf{v}}, \Gamma(X)] - [\Gamma(X), \Gamma(Y)]. \end{aligned}$$

Now $-[X, Y]^{\mathsf{v}} + (\nabla_X Y - \nabla_Y X)^{\mathsf{v}} = 0$ since the linear connection is symmetric;

$$\begin{aligned} &[X^{\mathsf{h}}, \bar{\Gamma}_{(r)}(Y)^{\mathsf{v}}] - [Y^{\mathsf{h}}, \bar{\Gamma}_{(r)}(X)^{\mathsf{v}}] - \bar{\Gamma}_{(r)}([X, Y])^{\mathsf{v}} = \left(d_\nabla\bar{\Gamma}_{(r)}(X, Y)\right)^{\mathsf{v}}; \\ &[X^{\mathsf{v}}, \bar{\Gamma}_{(r)}(Y)^{\mathsf{v}}] - [Y^{\mathsf{v}}, \bar{\Gamma}_{(r)}(X)^{\mathsf{v}}] = \left(\partial\bar{\Gamma}_{(r)}(X, Y)\right)^{\mathsf{v}}; \end{aligned}$$

and

$$[\Gamma(X), \Gamma(Y)] = \sum_{s,t>1} \{\bar{\Gamma}_{(s)}(X), \bar{\Gamma}_{(t)}(Y)\}^{\mathsf{v}}.$$

□

We next consider the infinitesimal symmetries of such an infinitesimal Cartan geometry.

Proposition 8.5.2 *In order for*

$$\xi = X^{\mathsf{h}} + \sum_{r\geq 1} X^{\mathsf{v}}_{(r)}, \quad X_{(r)} \in \bar{\mathcal{K}}^r,$$

to be an infinitesimal symmetry of the infinitesimal Cartan geometry defined above the following conditions (involving sections of $\bar{\mathcal{K}}^r \otimes T^*M$) must be satisfied:

$$\nabla X = \partial X_{(1)}, \ r = 0$$
$$\nabla X_{(1)} = R(\cdot, X) + \partial X_{(2)}, \ r = 1$$
$$\nabla X_{(r)} = -\nabla_X \bar{\Gamma}_{(r)} - \bar{\Gamma}_{(r)} \circ \nabla X + \sum_{s+t=r+1} \{\bar{\Gamma}_{(s)}, X_{(t)}\} + \partial X_{(r+1)}, \ r \geq 2.$$

Here $\partial : \sec(TM \otimes S^{r+1}T^*M) \to \sec(TM \otimes S^r T^*M \otimes T^*M)$ is given by $\partial Q(Y)^{\mathsf{V}} = [Y^{\mathsf{V}}, Q^{\mathsf{V}}]$. Notice that since $X_{(r+1)} \in \bar{\mathcal{K}}^{r+1}$, for every Y $\partial X_{(r)}(Y) \in \bar{\mathcal{K}}^r$.

Proof Compute $\mathcal{L}_\xi \gamma$ with ξ as given, and separately set to zero the terms of different homogeneity degrees. □

Now $\partial : \sec(TM \otimes S^1 T^*M) \to \sec(TM \otimes T^*M)$ is the identity. So we may write the condition for $r = 0$ as $\nabla X = X_{(1)}$.

Corollary 8.5.3 *A necessary condition for ξ to be an infinitesimal symmetry is that $\nabla X \in \bar{\mathcal{K}}^1$.* □

In fact $\partial : \sec(TM \otimes S^{r+1}T^*M) \to \sec(TM \otimes S^r T^*M \otimes T^*M)$ is scarcely distinguishable from the identity even for $r > 0$. Let us denote the components of a section S of $TM \otimes S^r T^*M \otimes T^*M$ by $S^i_{j_1 j_2 \dots j_r; k}$, symmetric in the indices $j_1, j_2, \dots, j_r$. Then for $Q \in \sec(TM \otimes S^{r+1}T^*M)$,

$$(\partial Q)^i_{j_1 j_2 \dots j_r; k} = Q^i_{j_1 j_2 \dots j_r k}.$$

Thus if (say) $\partial X_{(r+1)} = 0$ then $X_{(r+1)} = 0$.

It follows that there are no vertical infinitesimal symmetries, or in other words, the map $\xi \mapsto \pi_* \xi = X$ is injective on the Lie algebra of infinitesimal symmetries.

Let ξ be any section of $\mathcal{A}_o$, given explicitly as in Proposition 8.5.2. It is not difficult to show that

$$\begin{aligned} \nabla^{\mathrm{B}}_Y \xi = &\left(\nabla_Y X - \partial X_{(1)}(Y)\right)^{\mathsf{h}} \\ &+ \left(\nabla_Y X_{(1)} - R(Y, X) - \partial X_{(2)}(Y)\right)^{\mathsf{V}} \\ &+ \left(\sum_{r>1} (\nabla_Y X_{(r)} + \nabla_X \bar{\Gamma}_{(r)}(Y) + \bar{\Gamma}_{(r)}(X_{(1)}(Y)) \right. \\ &\left. - \sum_{s+t=r+1} \{\bar{\Gamma}_{(s)}(Y), X_{(t)}\} - \partial X_{(r+1)}(Y) \right)^{\mathsf{V}}. \end{aligned}$$

The calculation of R^{B} from this expression is straightforward in principle though evidently complicated in practice. We content ourselves with computing the horizontal component. We shall use without comment the pretty evident fact that the operators ∂ and ∇_Y commute; recall also that $\partial^2 = 0$.

Proposition 8.5.4 *For any vector fields Y, Z on M, $R^{\mathrm{B}}(Y,Z)$ is vertical (and therefore defines a section of $\mathcal{K}_o$, the kernel of $\mathcal{A}_o$).*

Proof The horizontal component of $R^{\mathrm{B}}(Y,Z)$ is the horizontal lift of

$$\begin{aligned}&\nabla_Y\big(\nabla_Z X-\partial X_{(1)}(Z)\big)-\nabla_Z\big(\nabla_Y X-\partial X_{(1)}(Y)\big)\\&-\ \partial\big(\nabla_Z X_{(1)}-R(Z,X)-\partial X_{(2)}(Z)\big)(Y)\\&+\ \partial\big(\nabla_Y X_{(1)}-R(Y,X)-\partial X_{(2)}(Y)\big)(Z)\\&-\ \big(\nabla_{[Y,Z]}X-\partial X_{(1)}([Y,Z])\big).\end{aligned}$$

This expression simplifies to $R(Y,Z)X+R(Z,X)Y-R(Y,X)Z$ which vanishes by the first Bianchi identity. □

It should be clear from these computations that the conditions for the vanishing of R^{B} are essentially the same as the integrability conditions obtained from the equations in Proposition 8.5.2. In the specific cases treated earlier, the first integrability condition is always trivially satisfied: we now see that this is true in general.

We also saw in previous sections that the second integrability condition always amounts to the vanishing of the Lie derivative, by the projection of ξ to M, of a tensor on M of curvature type. This is again always the case. It is possible to obtain this result by calculating the component of $R^{\mathrm{B}}(Y,Z)$ in $\mathcal{K}^1$ using the method of the previous proposition, but the calculation is complicated. It is easier to deduce the result from the fact that $\mathcal{L}_\xi R^\gamma=0$ when ξ is an infinitesimal symmetry (Proposition 5.4.4).

Proposition 8.5.5 *If ξ is an infinitesimal symmetry then $\mathcal{L}_X R_{(1)}=0$ where $X=\pi_*\xi$ and $R_{(1)}\in\bar{\mathcal{K}}^1\otimes\bigwedge^2 T^*M$ is the tensor field such that $R_{(1)}(Y,Z)^{\mathsf{v}}$ is the component of $R^\gamma(Y,Z)$ in $\mathcal{K}^1$.*

Proof We have

$$\mathcal{L}_\xi R^\gamma(Y,Z)=[\xi,R^\gamma(Y,Z)]-R^\gamma([X,Y],Z)-R^\gamma(Y,[X,Z]).$$

Set $R^\gamma=\sum_{r\geq1}R^{\mathsf{v}}_{(r)}$, and $\xi=X^{\mathsf{h}}+\sum_{r\geq1}X^{\mathsf{v}}_{(r)}$ as before. The $\mathcal{K}^1$ component of $\mathcal{L}_\xi R^\gamma(Y,Z)$ is

$$[X^{\mathsf{h}}+X^{\mathsf{v}}_{(1)},R_{(1)}(Y,Z)^{\mathsf{v}}]-R_{(1)}([X,Y],Z)^{\mathsf{v}}-R_{(1)}(Y,[X,Z])^{\mathsf{v}},$$

which is the vertical lift of the type $(1,1)$ tensor

$$\begin{aligned}&\nabla_X\big(R_{(1)}(Y,Z)\big)-X_{(1)}\circ R_{(1)}(Y,Z)+R_{(1)}(Y,Z)\circ X_{(1)}\\&-R_{(1)}([X,Y],Z)-R_{(1)}(Y,[X,Z])\end{aligned}$$

($X_{(1)}$ and $R_{(1)}(Y,Z)$ are type $(1,1)$ tensors, so $[X^{\mathsf{v}}_{(1)},R_{(1)}(Y,Z)^{\mathsf{v}}]$ is the vertical lift of minus their commutator). But for any vector field W on M,

$$
\begin{aligned}
&\mathcal{L}_X R_{(1)}(Y, Z)W + R_{(1)}([X, Y], Z)W + R_{(1)}(Y, [X, Z])W \\
&\quad = [X, R_{(1)}(Y, Z)W] - R_{(1)}(Y, Z)[X, W] \\
&\quad = \nabla_X\big(R_{(1)}(Y, Z)\big)W - \nabla_{R_{(1)}(Y,Z)W}X + R_{(1)}(Y, Z)\nabla_W X \\
&\quad = \nabla_X\big(R_{(1)}(Y, Z)\big)W - X_{(1)}R_{(1)}(Y, Z)W + R_{(1)}(Y, Z)X_{(1)}W
\end{aligned}
$$

where we have written $X_{(1)}$ for ∇X in the last line. But when ξ is an infinitesimal symmetry $\nabla X = X_{(1)}$ and $\mathcal{L}_\xi R^\gamma = 0$, so $\mathcal{L}_X R_{(1)}(Y, Z)W = 0$ for all vector fields Y, Z and W. □

Chapter 9
Projective Geometry: The Full Version

We discussed a version of Cartan projective geometry in Sect. 8.2, in which we took for the underlying Lie algebroid a vector bundle whose elements are certain projectable vector fields along fibres of $TM \to M$, where the kernel consists of vector fields tangent to fibres which are quadratic in the canonical fibre coordinates and which form on each fibre a copy of the Lie algebra $\mathfrak{sl}(n+1)$. We called an infinitesimal Cartan geometry defined on such a Lie algebroid an infinitesimal Cartan projective geometry. We have been at pains to point out that an infinitesimal Cartan projective geometry, so defined, is not derivable from a Cartan geometry on TM in the full sense defined in Sect. 6.1, because the Lie algebroid does not come from a Lie groupoid of fibre morphisms of TM in the way we have specified. On the other hand, all of the classical features of the projective differential geometry of a fixed linear connection may be derived from considerations of the (normal) infinitesimal Cartan projective geometry corresponding to that linear connection—though we never mentioned the idea of projective equivalence of linear connections, despite the fact (so it could be argued) that it is fundamental to the whole concept of projective differential geometry.

This situation has an air of cognitive dissonance which is crying out for resolution. We shall in the following sections show how to resolve it by describing a more sophisticated theory, consistent with the definition of a Cartan geometry in Sect. 6.1, which is based on the construction, described in Chap. 4, of a bundle over M whose standard fibre is projective space of dimension $\dim M$, together with the groupoid of projective maps between its fibres.

9.1 Cartan Projective Geometry

The bundle in question is the Cartan projective bundle $\mathrm{P}\mathcal{W}M \to M$, which was introduced in Sect. 4.3. We remind the reader that $\mathrm{P}\mathcal{W}M$ is the projectivization of a vector bundle $\mathcal{W}M \to M$ whose fibres are of dimension $n+1$, so that the fibres of $\mathrm{P}\mathcal{W}M$ are projective spaces of the same dimension as M. Moreover, $\mathcal{W}M$ admits a

M. Crampin and D. Saunders, *Cartan Geometries and their Symmetries*,
Atlantis Studies in Variational Geometry 4, DOI 10.2991/978-94-6239-192-5_9

global section e over M, depending on a choice of weight p, which in turn determines a global section $[e]$ of P$\mathcal{W}M$ (where for $x \in M$, $[e(x)]$ is the ray in $\mathcal{W}M_x$ containing $e(x)$). Thus P$\mathcal{W}M \to M$ is a generalized F-space with $F = \mathrm{P}^n$, real projective space of dimension n, and with groups $G = \mathsf{PGL}(n+1)$ and G_o the isotropy subgroup of $[(1, 0, \ldots, 0)] \in \mathrm{P}^n$. We showed further that P$\mathcal{W}M$ is not just attached, but even soldered, to M along the image of the attachment section $[e]$. Most of these constructions work for an arbitrary choice of p, but for the Cartan projective bundle we fix the value of p to be $1/(n+1)$. We shall see the justification for this choice later in this section.

We may then consider the Lie groupoid $\mathcal{G}$ of projective maps between fibres of P$\mathcal{W}M$, and its subgroupoid $\mathcal{G}_o$, the corresponding Lie algebroids $A\mathcal{G}$ and $A\mathcal{G}_o$, and their representations as Lie algebroids $\mathcal{A}^G$ and $\mathcal{A}^G_o$ of vector fields on P$\mathcal{W}M$. These structures, together with a connection (path or infinitesimal) which satisfies the transversality condition, form a Cartan geometry. As we pointed out above, P$\mathcal{W}M$ is soldered to M; we shall therefore insist, as before, that the soldering defined by the infinitesimal connection coincides with this given one.

We recall that fibre coordinates on $\mathcal{W}M$ may be taken as (w^0, w^i) and that, with respect to these coordinates, the image of the section e is just $w^0 = 1$, $w^i = 0$. We shall use the corresponding fibre coordinates (y^i) on open subsets of P$\mathcal{W}M$ around the image of the attachment section $[e]$ obtained by taking $y^i = y^i_0 = w^i/w^0$.

The Lie algebroid $\mathcal{A}^G$ consists of vector fields along the fibres of P$\mathcal{W}M \to M$ which in terms of the coordinates y^i take the form

$$X^i \frac{\partial}{\partial x^i} + \left(Y^i + Y^i_j y^j + \left(Y_j y^j\right) y^i\right) \frac{\partial}{\partial y^i};$$

$\mathcal{A}^G_o$ consists of such vector fields for which $Y^i = 0$. This is formally the same as the set-up in Sect. 8.2. However, the vector field

$$\left(Y^i + Y^i_j y^j + \left(Y_j y^j\right) y^i\right) \frac{\partial}{\partial y^i}$$

on a fibre of P$\mathcal{W}M$ is now just the coordinate representation of a vector field defined on the whole fibre, which is necessarily complete as the fibre, a projective space, is compact.

An infinitesimal connection γ on $\mathcal{A}^G$ will take the form

$$\gamma\left(\frac{\partial}{\partial x^i}\right) = \frac{\partial}{\partial x^i} - \left(A^j_i + \Pi^j_{ik} y^k + (\Lambda_{ik} y^k) y^j\right) \frac{\partial}{\partial y^j}.$$

The coefficients Π^j_{ik} no longer define a linear connection on M. To explain why, and to further illustrate the difference between the theories on P$\mathcal{W}M$ and on TM, we consider the effects on P$\mathcal{W}M$ of a coordinate transformation on M, from (x^i) to $(\hat{x}^i)$. We use the following notation, which in part was introduced in Sect. 4.3: $J^i_j = \partial \hat{x}^i / \partial x^j$ are the elements of the Jacobian matrix of the coordinate transformation, $\bar{J}^i_j$ those of its inverse, $J^k_{ij} = \partial J^k_i / \partial x^j = \partial J^k_j / \partial x^i$, and J is the Jacobian determinant. As we saw in Sect. 4.3.

$$\hat{w}^0 = w^0 - \frac{1}{n+1} \frac{\partial \log |J|}{\partial x^i} w^i, \quad \hat{w}^i = J^i_j w^j.$$

Of course the equations for the image of the section e, namely $w^0 = 1$, $w^i = 0$, are globally well-defined.

From their construction we find that

$$\hat{y}^i = \left(1 - \frac{1}{n+1} \frac{\partial \log |J|}{\partial x^k} y^k\right)^{-1} J^i_j y^j.$$

It follows that

$$\frac{\partial}{\partial x^i} = J^j_i \frac{\partial}{\partial \hat{x}^j} + \left(J^j_{il} \bar{J}^l_k \hat{y}^k + \frac{1}{n+1} \frac{\partial^2 \log |J|}{\partial x^i \partial x^k} \bar{J}^k_l \hat{y}^l \hat{y}^j\right) \frac{\partial}{\partial \hat{y}^j}$$
$$\frac{\partial}{\partial y^i} = \left(1 + \frac{1}{n+1} \frac{\partial \log |J|}{\partial x^k} \bar{J}^k_l \hat{y}^l\right) \left(J^j_i + \frac{1}{n+1} \frac{\partial \log |J|}{\partial x^i} \hat{y}^j\right) \frac{\partial}{\partial \hat{y}^j}.$$

Notice that $\hat{y}^i = 0$ if and only if $y^i = 0$ (for all i): this confirms that the origin of fibre coordinates defines a global section of $\mathcal{PWM} \to M$, which is of course the image of $[e]$. Moreover, on the image of $[e]$ we have

$$\frac{\partial}{\partial x^i} = J^j_i \frac{\partial}{\partial \hat{x}^j}, \quad \frac{\partial}{\partial y^i} = J^j_i \frac{\partial}{\partial \hat{y}^j},$$

which are the same as the transformation laws for the corresponding vector fields on the zero section of TM. This is the coordinate representation of the fact that $\mathcal{PWM}$ is naturally soldered to M along the image of $[e]$: the soldering map is just $\partial/\partial x^i \mapsto \partial/\partial y^i$. We must ensure that the soldering defined by γ agrees with the natural one, which requires that $A^j_i = \delta^j_i$ as before.

The main point of this sequence of calculations is to find the transformation rule for the coefficients Π^j_{ik}: they transform as

$$\hat{\Pi}^i_{jk} = \bar{J}^p_j \bar{J}^q_k \left(J^i_l \Pi^l_{pq} - J^i_{pq}\right) + \frac{1}{n+1} \frac{\partial \log |J|}{\partial x^l} \left(\bar{J}^l_j \delta^i_k + \bar{J}^l_k \delta^i_j\right).$$

This formula differs from that for the connection coefficients of a linear connection by the terms involving $|J|$. Notice that if Π^i_{jk} is symmetric in its lower indices so is $\hat{\Pi}^i_{jk}$. Notice further that the trace of Π^i_{jk}, that is, Π^i_{ji}, transforms as a 1-form, by Jacobi's formula for the derivative of a determinant; and in particular if the trace of Π^i_{jk} vanishes so does that of $\hat{\Pi}^i_{jk}$. This property holds only because the weight p is taken to be $1/(n+1)$, which leads to that as the value of the numerical coefficient of the terms involving $|J|$. Indeed, the choice of that value for p may be ultimately traced to this feature.

We shall assume that the infinitesimal connection γ is such that Π^i_{jk} is symmetric and trace free.

We call a Cartan geometry with all the features mentioned above a *full Cartan projective geometry* on $P\mathcal{W}M$.

The calculation of the curvature R^γ of γ, with

$$\gamma\left(\frac{\partial}{\partial x^i}\right) = \frac{\partial}{\partial x^i} - \left(\delta^j_i + \Pi^j_{ik} y^k + (\Lambda_{ik} y^k) y^j\right) \frac{\partial}{\partial y^j},$$

proceeds formally as in the *TM* version. It will be convenient to denote by $\mathfrak{R}^i_{jkl}$ the curvature 'tensor' derived from the Π^i_{jk} (we use gothic type to indicate that it is not in fact a tensor):

$$\mathfrak{R}^i_{jkl} = \frac{\partial \Pi^i_{jl}}{\partial x^k} - \frac{\partial \Pi^i_{jk}}{\partial x^l} + \Pi^i_{km}\Pi^m_{jl} - \Pi^i_{lm}\Pi^m_{jk};$$

its trace, the corresponding Ricci 'tensor' $\mathfrak{R}_{jk} = \mathfrak{R}^l_{jlk}$, is given by

$$\mathfrak{R}_{jk} = \frac{\partial \Pi^l_{jk}}{\partial x^l} - \Pi^l_{km}\Pi^m_{jl}$$

since Π^i_{jk} is trace free; note that $\mathfrak{R}_{jk}$ is symmetric. Then the coefficients of the component of R^γ which is linear in the y^i are

$$\mathfrak{R}^i_{jkl} + (\Lambda_{kl} - \Lambda_{lk})\delta^i_j + \Lambda_{kj}\delta^i_l - \Lambda_{lj}\delta^i_k.$$

If one chooses

$$\Lambda_{ij} = \frac{1}{n-1}\mathfrak{R}_{ij}$$

then the component of R^γ linear in the y^i becomes trace free. With this choice the infinitesimal Cartan projective geometry on $P\mathcal{W}M$ is said to be normal, as before.

In the normal case, for the curvature as a whole we have

$$R^\gamma_{kl} = \left(P^i_{jkl} y^j + \frac{1}{n-1}(\mathfrak{R}_{jk|l} - \mathfrak{R}_{jl|k}) y^j y^i\right) \frac{\partial}{\partial y^i},$$

where

$$P^i_{jkl} = \mathfrak{R}^i_{jkl} + \frac{1}{n-1}\left(\mathfrak{R}_{jk}\delta^i_l - \mathfrak{R}_{jl}\delta^i_k\right),$$

and where the rule indicates the formal covariant derivative with respect to the Π^i_{jk}:

$$\mathfrak{R}_{jk|l} = \frac{\partial \mathfrak{R}_{jk}}{\partial x^l} - \mathfrak{R}_{mk}\Pi^m_{jl} - \mathfrak{R}_{jm}\Pi^m_{kl}.$$

9.2 The Two Versions Compared: Projective Equivalence

We turn now to the relation between the full Cartan projective geometry and the infinitesimal version discussed in the previous chapter.

In Sect. 4.4 we discussed the injection of TM into $P\mathcal{W}M$ by making use of an Ehresmann connection on the principal $\mathbb{R}_+$ bundle $\mathcal{V}M \to M$, and we showed that the injections of TM into $P\mathcal{W}M$ are in 1–1 correspondence with such Ehresmann connections. Let ϑ be the connection 1-form of an Ehresmann connection on $\mathcal{V}M$, where the map $TM \to P\mathcal{W}M$ generated by that connection was denoted $\bar{\vartheta}$. Then as we showed in Sect. 4.4, $\bar{\vartheta}(TM)$ is an open dense subset of $P\mathcal{W}M$, and the vector fields of $\mathcal{A}^G$ when restricted to it form a Lie algebroid $\mathcal{A}^\vartheta$ of vector fields along fibres of TM which is just the Lie algebroid underlying the TM version of infinitesimal Cartan geometry. Indeed the Lie algebroids $\mathcal{A}^G$ and $\mathcal{A}^\vartheta$ are isomorphic because, as mentioned in Sect. 4.4, any quadratic vector field of the correct form on a fibre of TM may be extended from $\bar{\vartheta}(TM)$ to $P\mathcal{W}M$.

We can now see why the existence of an infinitesimal Cartan projective geometry on TM does not conflict with Theorem 7.6.2. The conditions for that result were, first, that the Lie algebra $\mathfrak{g}$ should be isomorphic with its derivation algebra $\mathsf{Der}(\mathfrak{g})$ (the Lie algebra of its automorphism group), and that the subalgebra $\mathfrak{h}$ should equal the normalizer of $\mathfrak{h}$ in $\mathfrak{g}$. For projective geometry the first of these conditions holds because $\mathfrak{sl}(n+1)$ is semisimple, and the second holds for $\mathfrak{h}$ by direct computation. It is also straightforward to check that the other conditions in the theorem hold, so we should indeed expect to be able to construct a finite Cartan geometry starting with $\mathcal{A}^\vartheta_o$—but this will be a Cartan geometry *not* based on fibre morphisms of TM.

We may see the detailed relationship between the infinitesimal geometries on $P\mathcal{W}M$ and on TM by studying the Ehresmann connection. Suppose that

$$\vartheta = \frac{1}{x^0}dx^0 + \vartheta_i dx^i,$$

where the ϑ_i must be functions on M, by invariance. Then the horizontal lift of $\partial/\partial x^i \in T_xM$ is

$$\frac{\partial}{\partial x^i} - \vartheta_i \Upsilon.$$

So for each $x \in M$ the connection induces a linear map $T_xM \to \mathcal{W}_xM$ by

$$(v^i) \mapsto (-\vartheta_j v^j, v^i)$$

where v^i are canonical fibre coordinates on TM, (w, u^i) fibre coordinates on $\mathcal{W}M$. This injects T_xM as the subspace $w = -\vartheta_j u^j$ of $\mathcal{W}_xM$, which is to be regarded as the hyperplane at infinity when we projectivize. The map

$$(v^i) \mapsto (1 - \vartheta_j v^j, v^i)$$

(i.e. the linear map followed by translation through $(1, 0, \ldots, 0)$) maps $T_x M$ to the parallel hyperplane through $(1, 0, \ldots, 0)$, and maps the origin of $T_x M$ to that point, so that the points of attachment of the fibres coincide. Thus

$$\bar{\vartheta}(x^i, y^i) = \left(x^i, \frac{v^i}{1 - \vartheta_j v^j}\right).$$

Rather than dealing explicitly with $\bar{\vartheta}$ it will be convenient to regard the v^i as new fibre coordinates on P$\mathcal{W}M$, defined in $\bar{\vartheta}(TM)$. One finds that

$$\frac{\partial}{\partial y^i} = \left(1 - \vartheta_k v^k\right)\left(\delta_i^j - \vartheta_i v^j\right)\frac{\partial}{\partial v^j}.$$

Note that when $v^i = 0$ we have $\partial/\partial y^i = \partial/\partial v^i$, so that the solderings of the two structures also coincide.

It follows that the vector field

$$\left(Y^i + Y_j^i y^j + \left(Y_j y^j\right) y^i\right)\frac{\partial}{\partial y^i},$$

belonging to $\mathcal{K}^G$, when expressed in terms of the v^i becomes

$$\left(Y^i + \left(Y_j^i - \vartheta_j Y^i - \left(\vartheta_k Y^k\right)\delta_j^i\right) v^j + \left(\left(Y_j + \vartheta_k\left(Y^k \vartheta_j - Y_j^k\right)\right) v^j\right) v^i\right)\frac{\partial}{\partial v^i}.$$

Notice that when $Y^i = 0$, that is, when the vector field belongs to $\mathcal{K}_o^G$, this simplifies to

$$\left(\left(Y_j^i v^j + \left(Y_j - \vartheta_k Y_j^k\right) v^j\right) v^i\right)\frac{\partial}{\partial v^i}.$$

Let us consider what happens to the connection γ on changing from y^i to v^i. To capture all the details we shall consider a coordinate transformation on P$\mathcal{W}M$ of the form

$$\hat{x}^i = x^i, \quad \hat{y}^i = v^i = \frac{y^i}{1 + \vartheta_j y^j}.$$

We find that

$$\frac{\partial}{\partial x^i} = \frac{\partial}{\partial \hat{x}^i} - \left(\frac{\partial \vartheta_k}{\partial x^i} v^k\right) v^j \frac{\partial}{\partial v^j},$$

while the formula for $\partial/\partial y^i$ is as before.

The vector field

$$\gamma\left(\frac{\partial}{\partial x^i}\right) = \frac{\partial}{\partial x^i} - \left(\delta_i^j + \Pi^j_{ik} y^k + (\Lambda_{ik} y^k) y^j\right) \frac{\partial}{\partial y^j},$$

when expressed in terms of $\hat{x}^i$ and v^i, takes the form

$$\frac{\partial}{\partial \hat{x}^i} - \left(\delta_i^j + \Gamma^j_{ik} v^k + (\Phi_{ik} v^k) v^j\right) \frac{\partial}{\partial v^j},$$

where

$$\Gamma_i^{jk} = \Pi^i_{jk} - \vartheta_k \delta^i_j - \vartheta_j \delta^i_k$$
$$\Phi_{ij} = \Lambda_{ij} + \frac{\partial \vartheta_j}{\partial x^i} - \Pi^k_{ij} \vartheta_k + \vartheta_i \vartheta_j.$$

We consider first the relationship between Γ^i_{jk} and Π^i_{jk}. We know that the Γ^i_{jk} are the connection coefficients of a linear connection on M; and this can easily be confirmed by considering the effects of a coordinate transformation on M on the terms on the right-hand side. The linear connection is moreover symmetric. Now the difference of two Ehresmann connection forms on $\mathcal{V}M$ is the pullback to $\mathcal{V}M$ of a 1-form on M, so the linear connections determined from the same infinitesimal Cartan projective geometry on P$\mathcal{W}M$ by two different Ehresmann connections are related as follows:

$$\hat{\Gamma}^k_{ij} = \Gamma^k_{ij} + \left(\alpha_i \delta^k_j + \alpha_j \delta^k_i\right)$$

where the α_i are the components of a 1-form. The linear connection with connection coefficients $\hat{\Pi}^i_{jk}$ is said to be obtained from the connection with coefficients Γ^i_{jk} by a projective change of connection; two symmetric linear connections related in this way are said to be *projectively equivalent*, and the set of linear connections projectively equivalent to any one linear connection is a projective equivalence class of linear connections. We have shown that from an infinitesimal Cartan projective geometry on P$\mathcal{W}M$ we can recover an infinitesimal Cartan projective geometry on *TM* by injecting *TM* into P$\mathcal{W}M$ using an Ehresmann connection on $\mathcal{V}M \to M$, and that the linear connections for the infinitesimal geometries obtained in this way from a fixed infinitesimal Cartan projective geometry on P$\mathcal{W}M$ are projectively equivalent and form a projective equivalence class as the Ehresmann connection varies.

Suppose, conversely, that we are given a projective equivalence class of linear connections on M. By taking a trace in the equation for projective equivalence and writing $\Gamma_i = \Gamma^j_{ij}$, $\hat{\Gamma}_i = \hat{\Pi}^j_{ij}$, we obtain $\hat{\Gamma}_i = \Gamma_i + (n+1)\alpha_i$, so that the quantities

$$\Pi^k_{ij} = \Gamma^k_{ij} - \frac{1}{n+1}\left(\Gamma_i \delta^k_j + \Gamma_j \delta^k_i\right)$$

are projectively invariant, that is, unchanged under a projective change of connection, and therefore associated with the whole equivalence class of projectively related linear connections rather than with any individual one. These quantities were introduced by T.Y. Thomas [41, 42], who called them collectively the projective connection; to use that terminology in the present context would be to invite confusion, so we have adopted another. Douglas [21] used a generalized form of the same quantities, and called them collectively the *fundamental descriptive invariant* of a projective equivalence class of connections; this is the term we shall use. Notice that the fundamental descriptive invariant is symmetric and trace free. Then the normal infinitesimal Cartan projective geometry on $P\mathcal{W}M$ with infinitesimal connection γ given by

$$\gamma\left(\frac{\partial}{\partial x^i}\right) = \frac{\partial}{\partial x^i} - \left(\delta_i^j + \Pi^j_{ik} y^k + (\Lambda_{ik} y^k) y^j\right)\frac{\partial}{\partial y^j}$$

combines the whole projective equivalence class of linear connections into a single geometric object. The individual linear connections in the class may be recovered by means of the Ehresmann connection construction: in order to obtain the linear connection with coefficients Γ^i_{jk} we must use the Ehresmann connection with

$$\vartheta_i = -\frac{1}{n+1}\Gamma_i.$$

We can further develop the theory of projective equivalence of linear connections from the standpoint of the Cartan theory of projective connections by considering the relationship between Φ_{ij} and Λ_{ij}. We restrict our attention to the normal infinitesimal Cartan projective geometry on $P\mathcal{W}M$. It will be convenient to express things where possible in terms of Γ^i_{jk} and $\vartheta_i = -\Gamma_i/(n+1)$. We find that

$$\begin{aligned}\mathfrak{R}_{ij} &= \frac{\partial \Gamma^k_{ij}}{\partial x^k} + \frac{\partial \vartheta_i}{\partial x^j} + \frac{\partial \vartheta_j}{\partial x^i} - \Gamma^k_{il}\Gamma^l_{jk} - 2\vartheta_k \Gamma^k_{ij} kij + (n-1)\vartheta_i\vartheta_j \\ &= (n-1)\Lambda_{ij}\end{aligned}$$

and

$$\Phi_{ij} = \Lambda_{ij} + \frac{\partial \vartheta_j}{\partial x^i} - \Gamma^k_{ij}\vartheta_k - \vartheta_i\vartheta_j.$$

It follows that

$$(n-1)\Phi_{ij} = \frac{\partial \Gamma^k_{ij}}{\partial x^k} + \frac{\partial \vartheta_i}{\partial x^j} + n\frac{\partial \vartheta_j}{\partial x^i} - \Gamma^k_{il}\Gamma^l_{jk} - (n-1)\vartheta_k\Gamma^k_{ij}.$$

Now the Ricci tensor of the linear connection with coefficients Γ^i_{jk} is given by

$$R_{ij} = \frac{\partial \Gamma^k_{ij}}{\partial x^k} + (n+1)\frac{\partial \vartheta_i}{\partial x^j} - \Gamma^k_{il}\Gamma^l_{jk} - (n+1)\vartheta_k\Gamma^k_{ij}.$$

Notice that

$$R_{ij} - R_{ji} = (n+1)\left(\frac{\partial \vartheta_i}{\partial x^j} - \frac{\partial \vartheta_j}{\partial x^i}\right).$$

It follows that

$$\begin{aligned}(n-1)\Phi_{ij} &= R_{ij} - n\left(\frac{\partial \vartheta_i}{\partial x^j} - \frac{\partial \vartheta_j}{\partial x^i}\right)\\ &= \frac{1}{n+1}(R_{ij} + nR_{ji}),\end{aligned}$$

and therefore

$$\Phi_{ij} = \frac{1}{n^2-1}(R_{ij} + nR_{ji}).$$

That is to say, the infinitesimal projective connections on *TM* derived from a normal infinitesimal projective connection on $\mathrm{P}\mathcal{W}M$ are themselves normal.

Notice that the condition for the Ricci tensor R_{ij} to be symmetric is that the Ehresmann connection 1-form is closed. We know that Ehresmann connections on $\mathcal{V}M$ with closed, even exact, connection forms exist (they may be obtained from global sections of $\mathcal{V}M \to M$, as explained in Sect. 4.4). So in each projective class of linear connections we can find particular connections for which the Ricci tensor is symmetric.

Recall that the curvature of the normal geometry is given by

$$\left(P^i_{jkl}y^j + \frac{1}{n-1}(\mathfrak{R}_{jk|l} - \mathfrak{R}_{jl|k})y^j y^i\right)\frac{\partial}{\partial y^i}.$$

This belongs to $\mathcal{K}^G_o$, so in terms of the v^i it becomes

$$\left(P^i_{jkl}v^j + \left(\frac{1}{n-1}(\mathfrak{R}_{jk|l} - \mathfrak{R}_{jl|k}) - P^m_{jkl}\vartheta_m\right)v^j v^i\right)\frac{\partial}{\partial v^i}.$$

It follows from consideration of the linear parts that P^i_{jkl} is a tensor on M, and that it is projectively invariant. It is of course the *projective curvature tensor* common to each of the linear connections of the projective equivalence class determined by the normal infinitesimal Cartan projective geometry on $\mathrm{P}\mathcal{W}M$.

We note finally that an infinitesimal symmetry of a normal infinitesimal Cartan projective geometry on $\mathrm{P}\mathcal{W}M$ restricts to an infinitesimal symmetry of any of the corresponding normal infinitesimal projective geometries on *TM*; and conversely an infinitesimal symmetry of a normal infinitesimal projective geometry on *TM* extends to an infinitesimal symmetry of the normal infinitesimal projective geometry on $\mathrm{P}\mathcal{W}M$ which contains it. Thus all of the normal infinitesimal projective geometries on *TM* whose linear connections belong to the same projective class have the same infinitesimal symmetry algebra, which is the infinitesimal symmetry algebra of the normal infinitesimal Cartan projective connection on $\mathrm{P}\mathcal{W}M$ which subsumes them.

9.3 Linear Connections and Radius Vector Fields

In order to continue our study of projective geometry, we need the concept of a radius vector field, and so we make a small diversion to consider some properties of classical linear connections (covariant derivatives) on TM.

In the previous chapter we discussed infinitesimal Cartan geometries on $TM \to M$ where TM was considered as attached to M along the image of the zero section and soldered to it by means of the vertical lift construction. Ordinary linear connections found no place in that story: they are not Cartan connections, basically because $\mathsf{GL}(n)$ does not act transitively on $\mathbb{R}^n$—the origin is unmoved. But the group does act transitively on $\mathbb{R}^n - \{0\}$ (which we shall, from now on, denote by $\mathbb{R}^n_\circ$), and this observation leads to the consideration of a class of Cartan geometries over the slit tangent bundle $T^\circ M$ which are linear, in that they are based on the Lie groupoid $\mathcal{G}$ of linear fibre morphisms.

We require an additional ingredient, a section ς of $T^\circ M \to M$, that is to say a nowhere vanishing vector field Z on M, to take as attachment section. Then $T^\circ M$ is soldered to M along the image of ς, again via the vertical lift construction: or to put it another way, via the fact that the tangent space to a vector space at any point of it, not just the origin, is canonically identified with the vector space itself. Let $G_p \subset G = \mathsf{GL}(n)$ be the isotropy subgroup of any nonzero element $p \in \mathbb{R}^n$, say $p = (1, 0, \ldots, 0)$: then $G/G_p = \mathbb{R}^n_\circ$. Let $\mathcal{G}_p$ be the Lie subgroupoid of $\mathcal{G}$ corresponding to G_p and determined by ς. Let $A\mathcal{G}$ and $A\mathcal{G}_p$ be the Lie algebroids of $\mathcal{G}$ and $\mathcal{G}_p$: we can regard any element of $A\mathcal{G}$ as a projectable fibre-linear vector field on a fibre of $T^\circ M \to M$. Let $\mathfrak{s}$ be the soldering map. We suppose that M is equipped with a symmetric linear connection, with connection coefficients Γ^i_{jk} and covariant derivative operator ∇. Then we can define an infinitesimal connection γ on $A\mathcal{G}$ by

$$\gamma\left(\frac{\partial}{\partial x^i}\right) = \frac{\partial}{\partial x^i} - \Gamma^j_{ik} y^k \frac{\partial}{\partial y^j}.$$

Proposition 9.3.1 *The infinitesimal connection γ satisfies the nondegeneracy condition for an infinitesimal Cartan geometry if and only if the type* $(1, 1)$ *tensor field ∇Z is everywhere nonsingular when regarded as a linear map of the tangent space; and γ respects the soldering $\mathfrak{s}$ if and only if* $\nabla Z = \mathrm{id}$.

We shall call a vector field Z such that $\nabla Z = \mathrm{id}$ a *radius vector field*; this reflects the fact that the vector field $x^i \partial/\partial x^i$ on $\mathbb{R}^n_\circ$ has this property where ∇ is the standard covariant derivative operator. The terminology comes from Whitehead [45], although in his paper a radius vector field was also required to be an infinitesimal affine transformation of ∇, a condition satisfied by the standard example on $\mathbb{R}^n_\circ$.

Proof It will be convenient to express any section ξ of $A\mathcal{G}$ in the form

$$\xi = X^i H_i + Y^i_j y^j \frac{\partial}{\partial y^i}$$

where

$$H_i = \gamma\left(\frac{\partial}{\partial x^i}\right) = \frac{\partial}{\partial x^i} - \Gamma^j_{ik} y^k \frac{\partial}{\partial y^j},$$

the X^i are the components of a vector field X on M and the Y^i_j those of a type $(1, 1)$ tensor field Y. Then the kernel projection of γ is given by

$$\omega(\xi) = Y^i_j y^j \frac{\partial}{\partial y^i}.$$

The image of the attachment section is $y^i = Z^i(x)$, and ξ is tangent to it if and only if

$$Y^i_j Z^j = X^j\left(\frac{\partial Z^i}{\partial x^j} + \Gamma^i_{jk} Z^k\right),$$

that is, if and only if $Y(Z) = \nabla_X Z$. But then on the attachment section, for any ξ tangent to it,

$$\omega(\xi) = Y^i_j Z^j \frac{\partial}{\partial y^i} = (\nabla_X Z)^i \frac{\partial}{\partial y^i} = (\nabla Z)^i_j X^j \frac{\partial}{\partial y^i}.$$

The soldering map defined by γ is thus

$$X^i \frac{\partial}{\partial x^i} \mapsto (\nabla Z)^i_j X^j \frac{\partial}{\partial y^i};$$

this is an isomorphism if and only if ∇Z is nonsingular, and it agrees with the vertical lift if and only if $\nabla Z = \text{id}$. □

Corollary 9.3.2 *The infinitesimal connection γ defines an infinitesimal Cartan geometry on $A\mathcal{G}$ which respects the soldering $\mathfrak{s}$ if and only if Z is a radius vector field with respect to ∇.* □

We assume henceforth that Z is a radius vector field.

Evidently $K\mathcal{G}_p$ consists of those vector fields $Y^i_j y^j \partial/\partial y^i$ for which $Y^i_j Z^j = 0$. It is a straightforward consequence of the definition and the assumed symmetry of the linear connection that $R^i_{jkl} Z^j = 0$ (where of course R is the curvature). The following result is immediate.

Proposition 9.3.3 *The infinitesimal connection γ is torsion free, in the sense that R^γ takes its values in $K\mathcal{G}_p$.* □

We next consider the infinitesimal symmetries of this Cartan geometry. It will continue to be be convenient to express sections ξ of $A\mathcal{G}$ in the form given in the proof of the proposition. The condition for ξ to be a section of $A\mathcal{G}_p$ is $\nabla_X Z = Y(Z)$, so in view of the fact that Z is a radius vector field, the condition is simply $X = Y(Z)$.

Proposition 9.3.4 *A section ξ of $A\mathcal{G}_p$ is an infinitesimal symmetry of the infinitesimal Cartan geometry if and only if X is an infinitesimal affine transformation of ∇ and $\mathcal{L}_X Z = 0$; and $Y = \nabla X$. The greatest dimension the symmetry algebra can have is n^2; it has this dimension if and only if the curvature of the linear connection vanishes.*

Proof The condition for a symmetry, $\mathcal{L}_\xi \gamma = 0$, just gives

$$Y^i_{j|k} = R^i_{jkl} X^l.$$

Now $R^i_{jkl} Z^j = 0$, so $Y^i_{j|k} Z^j = 0$. On the other hand,

$$Y^i_{j|k} Z^j = \left(Y^i_j Z^j\right)_{|k} - Y^i_k = X^i_{|k} - Y^i_k.$$

Thus $Y = \nabla X$ and

$$X^i_{|jk} = R^i_{jkl} X^l,$$

which is the condition for X to be an infinitesimal affine transformation of ∇. Furthermore,

$$\mathcal{L}_X Z = [X, Z] = \nabla_X Z - \nabla_Z X = X - Y(Z) = 0,$$

so if ξ is a symmetry then X is an infinitesimal affine transformation of ∇ and $\mathcal{L}_X Z = 0$. As a vector field, ξ is X^{C}, the complete lift of X to $T^\circ M$.

Conversely, if these conditions are satisfied and we set $Y = \nabla X$ then $Y(Z) = \nabla_Z X = \nabla_X Z = X$, so ξ is a section of $A\mathcal{G}_p$; and $Y^i_{j|k} = X^i_{|jk} = R^i_{jkl} X^l$, so ξ is a symmetry.

To obtain the integrability conditions it will be convenient to write the condition $\mathcal{L}_\xi \gamma = 0$ in the form

$$Y^i_{j|k} = R^i_{jkl} Y^l_m Z^m.$$

By repeated covariant differentiation, the use of the Ricci and second Bianchi identities, and the fact that $Y^i_{j|k} Z^j = 0$, we obtain

$$Y^p_q \left(R^i_{jkl|p} Z^q - R^q_{jkl} \delta^i_p + R^i_{pkl} \delta^q_j + R^i_{jpl} \delta^q_k + R^i_{jkp} \delta^q_l\right) = 0.$$

The fibre dimension of $K\mathcal{G}$ is n^2, and this is the greatest dimension the Lie algebra of symmetries can have. When this dimension is attained it must be the case that

$$R^i_{jkl|p} Z^q - R^q_{jkl} \delta^i_p + R^i_{pkl} \delta^q_j + R^i_{jpl} \delta^q_k + R^i_{jkp} \delta^q_l = 0.$$

Take the trace on j and q. Since $R^i_{jkl}Z^j = 0$ and Z is a radius vector field, $R^i_{jkl|p}Z^j = -R^i_{pkl}$. It follows that

$$nR^i_{pkl} = (R_{kl} - R_{lk})\delta^i_p.$$

A further trace on i and k yields $R_{pl} = 0$, so $R^i_{pkl} = 0$.

In the flat case, with $M = \mathbb{R}^n_\circ$, we must have $Z = x^i\partial/\partial x^i$ since it must be nonvanishing, and the infinitesimal symmetries are just the linear vector fields. □

9.4 Thomas-Whitehead Connections

We now return to projective geometry.

The infinitesimal Cartan projective geometry discussed in the first two sections is defined over the Cartan projective bundle $\pi : \mathrm{P}\mathcal{W}M \to M$. This bundle is obtained from $T^\circ\mathcal{V}M$, the slit tangent bundle of the volume bundle $\nu : \mathcal{V}M \to M$, by a two-step process: first we construct from $T\mathcal{V}M$ a vector bundle $\mathcal{W}M \to M$ of fibre dimension $n+1$, and then we projectivize the fibres of $\mathcal{W}M$ (having deleted the zero section). We have the following commutative diagram.

$$\begin{array}{ccc} T^\circ\mathcal{V}M & \xrightarrow{\bar{\chi}} & \mathrm{P}\mathcal{W}M \\ \downarrow{\scriptstyle \tau^\circ_{\mathcal{V}M}} & & \downarrow{\scriptstyle \pi} \\ \mathcal{V}M & \xrightarrow[\nu]{} & M \end{array}$$

Here $\bar{\chi} : T^\circ\mathcal{V}M \to \mathrm{P}\mathcal{W}M$ is the composition of the restriction to $T^\circ\mathcal{V}M$ of the fibre-linear map $\chi : T\mathcal{V}M \to \mathcal{W}M$ with projectivization $\mathrm{p} : \mathcal{W}M \to \mathrm{P}\mathcal{W}M$. The pair $(\bar{\chi}, \nu)$ is a compatible fibre bundle map from $T^\circ\mathcal{V}M \to \mathcal{V}M$ to $\mathrm{P}\mathcal{W}M \to M$, according to the definition in Sect. 3.3.

There are two canonical vector fields on $T^\circ\mathcal{V}M$: Υ^c, the complete lift to $T^\circ\mathcal{V}M$ of the fundamental vector field Υ on $\mathcal{V}M$; and $\tilde{\Delta}$, the Liouville field. The distribution $\langle\Upsilon^\mathrm{c}, \tilde{\Delta}\rangle$ is integrable, and its leaves are just the fibres of $\bar{\chi} : T^\circ\mathcal{V}M \to \mathrm{P}\mathcal{W}M$; the integral curves of Υ^c are the fibres of $\chi : T\mathcal{V}M \to \mathcal{W}M$.

The vector field Υ defines a global section of $\tau^\circ_{\mathcal{V}M}$. Moreover, $T^\circ\mathcal{V}M$ is soldered to $\mathcal{V}M$ along the image of Υ, with soldering map $\tilde{\mathfrak{s}}$, given by

$$\tilde{\mathfrak{s}}(v) = v^\mathrm{V}|_{\Upsilon(z)} \quad \text{where } v \in T_z\mathcal{V}M, z \in \mathcal{V}M$$

as described in Sect. 9.3.

The vector field Υ^c on $T^\circ\mathcal{V}M$ is tangent to the image of the section Υ (this is an obvious property of any complete lift in relation to the section defined by its projection). It follows that for any $z_1, z_2 \in \mathcal{V}M$ with $\nu(z_1) = \nu(z_2)$, $(\bar{\chi}\circ\Upsilon)(z_1) = (\bar{\chi}\circ\Upsilon)(z_2)$, and therefore

$$x \mapsto (\bar{\chi} \circ \Upsilon)(z) \quad \text{for any } z \text{ wit } \nu(z) = x \in \mathrm{M}$$

is well-defined, and a section of π. It is indeed the canonical attachment section $[e]$ of $\mathrm{P}\mathcal{W}M \to M$.

We next consider the soldering. We note first that

$$\Upsilon(z)^{\mathsf{V}}|_{\Upsilon(z)} = \tilde{\Delta}|_{\Upsilon(z)},$$

and that indeed for any $v \in T_z\mathcal{V}M$

$$v^{\mathsf{V}}|_{\Upsilon(z)} \in \langle \tilde{\Delta}|_{\Upsilon(z)} \rangle \quad \text{if and only if} \quad v \in \langle \Upsilon(z) \rangle$$

(this is an obvious property of the vertical lift operator). Since $\pi_* \circ \bar{\chi}_* = \nu_* \circ \tau_{\mathcal{V}M*}$,

$$\bar{\chi}_* : V_{\Upsilon(z)} T^\circ \mathcal{V}M \to V_{[e(x)]} \mathrm{P}\mathcal{W}M, \quad x = \nu(z),$$

and the kernel of the restriction of $\bar{\chi}_*$ to $V_{\Upsilon(z)} T^\circ \mathcal{V}M$ is the 1-dimensional subspace spanned by $\tilde{\Delta}|_{\Upsilon(z)}$. Now

$$\bar{\chi}_* \circ \tilde{\mathfrak{s}} : T_z\mathcal{V}M \to V_{[e(x)]} \mathrm{P}\mathcal{W}M$$

is a linear map between vector spaces of dimension $n + 1$ and n. But

$$\begin{aligned} (\bar{\chi}_* \circ \tilde{\mathfrak{s}})(v) = 0 \quad & \text{if and only if} \quad \tilde{\mathfrak{s}}(v) \in \langle \tilde{\Delta}|_{\Upsilon(z)} \rangle \\ & \text{if and only if} \quad v^{\mathsf{V}}|_{\Upsilon(z)} \in \langle \tilde{\Delta}|_{\Upsilon(z)} \rangle \\ & \text{if and only if} \quad v \in \langle \Upsilon(z) \rangle. \end{aligned}$$

That is, the kernel of $\bar{\chi}_* \circ \tilde{\mathfrak{s}} : T_z\mathcal{V}M \to V_{[e(x)]} \mathrm{P}\mathcal{W}M$ is the 1-dimensional subspace spanned by $\Upsilon(z)$, and so $\bar{\chi}_* \circ \tilde{\mathfrak{s}}$ induces an isomorphism of $T_{\nu(z)}M$ with $V_{[e(x)]}\mathrm{P}\mathcal{W}M$, and therefore a soldering of $\mathrm{P}\mathcal{W}M$ to M along the image of $[e]$: this is just the soldering $\mathfrak{s}$ defined in Sect. 6.4.

As we pointed out in Sect. 9.3, $G = \mathsf{GL}(n + 1)$ acts transitively on $\mathbb{R}^{n+1}_{\mathrm{o}}$, and the latter space may be identified with G/G_p where G_p is the isotropy group of, say, $p = (1, 0, \ldots, 0)$.[1] Then $\tau^\circ_{\mathcal{V}M} : T^\circ\mathcal{V}M \to \mathcal{V}M$ is a generalized F-space with $F = \mathbb{R}^{n+1}_{\mathrm{o}}$ and with attachment section Υ. Let $\mathcal{G}^{\mathcal{V}}$ be the Lie groupoid of fibre morphisms of this bundle and $\mathcal{G}^{\mathcal{V}}_p$ the Lie subgroupoid defined by Υ. Let $A\mathcal{G}^{\mathcal{V}}$ and $A\mathcal{G}^{\mathcal{V}}_p$ be the corresponding Lie algebroids. Let $\mathcal{G}$ and $\mathcal{G}_p$ be the groupoids of projective maps between the fibres of the Cartan projective bundle $\mathrm{P}\mathcal{W}M \to M$ and $A\mathcal{G}$ and $A\mathcal{G}_p$ the corresponding Lie algebroids. We may regard elements of $A\mathcal{G}^{\mathcal{V}}$ as fibre-linear projectable vector fields along fibres of $T^\circ\mathcal{V}M \to \mathcal{V}M$, and elements of $A\mathcal{G}^{\mathcal{V}}_p$ as elements of $A\mathcal{G}^{\mathcal{V}}$ which are tangent to the image of Υ. Let $\boldsymbol{a}$ be an element of $A_z\mathcal{G}^{\mathcal{V}}$

[1] In this section we call the fixed point of $\mathbb{R}^{n+1}_{\mathrm{o}}$, the standard fibre, p rather than o to avoid confusion with $0 \in \mathbb{R}^{n+1}$.

with $\tau^{\circ}_{\mathcal{V}M*}\boldsymbol{a} = \bar{\boldsymbol{a}}$. Then (considered as a vector field along the fibre) $\boldsymbol{a}$ is invariant under scaling, and so $\bar{\chi}_*\boldsymbol{a}$ is well-defined as a vector field along $\mathrm{P}_x\mathcal{W}M$, the fibre of $\mathrm{P}\mathcal{W}M$ over $x = \nu(z) \in M$, and $\pi_*(\bar{\chi}_*\boldsymbol{a}) = \nu_*\bar{\boldsymbol{a}}$. Moreover, if $s \mapsto \mu_{s*}$ is the 1-parameter group on $T^{\circ}\mathcal{V}M$ whose infinitesimal generator is Υ^{c} then $\bar{\chi}_* \circ \mu_{s*} = \bar{\chi}_*$. It follows that $\bar{\chi}_*$ induces a vector bundle morphism $A\mathcal{G}^{\mathcal{V}} \to A\mathcal{G}$ over $\nu : \mathcal{V}M \to M$. Sections of these Lie algbroids may be considered as vector fields on the relevant bundles. By Theorem 3.3.5 a section of $A\mathcal{G}^{\mathcal{V}}$ is projectable under $\bar{\chi}$ if and only if as a vector field it is $\bar{\chi}$-related to a vector field on $\mathrm{P}\mathcal{W}M$, and since the bracket of $\bar{\chi}$-related vector fields is $\bar{\chi}$-related to their bracket we see that $(\bar{\chi}_*, \nu)$ is a Lie algebroid morphism. Moreover, the attachment sections correspond, which means that $(\bar{\chi}_*, \nu)$ induces a Lie algebroid morphism $A\mathcal{G}^{\mathcal{V}}_p \to A\mathcal{G}_p$.

Given the close relationship between $T^{\circ}\mathcal{V}M$ and $\mathrm{P}\mathcal{W}M$ revealed by the analysis above, it is natural to ask whether the infinitesimal connection γ of an infinitesimal Cartan projective geometry lifts to a structure on $T^{\circ}\mathcal{V}M$. The question may be answered as follows.

Theorem 9.4.1 *Given an infinitesimal Cartan projective geometry on $A\mathcal{G}$ whose infinitesimal connection γ respects the canonical soldering $\mathfrak{s}$, there is a unique infinitesimal connection $\tilde{\gamma}$ on $A\mathcal{G}^{\mathcal{V}}$ which projects onto γ and respects the canonical soldering $\tilde{\mathfrak{s}}$.*

Proof We show first that if such an infinitesimal connection $\tilde{\gamma}$ exists, it is unique. Suppose that $\tilde{\gamma}_1$ and $\tilde{\gamma}_2$ are two such infinitesimal connections. Then for any $v \in T^{\circ}\mathcal{V}M$, it must be the case that

$$\tau^{\circ}_{\mathcal{V}M*}\tilde{\gamma}_1(v) = \tau^{\circ}_{\mathcal{V}M*}\tilde{\gamma}_2(v) \quad \text{and} \quad \bar{\chi}_*\tilde{\gamma}_1(v) = \bar{\chi}_*\tilde{\gamma}_2(v).$$

It follows that

$$\tilde{\gamma}_2 = \tilde{\gamma}_1 + \theta \otimes \tilde{\Delta}$$

for some 1-form θ on $\mathcal{V}M$. The corresponding kernel projections $\tilde{\omega}_1$ and $\tilde{\omega}_2$ are related as follows:

$$\tilde{\omega}_2 = \tilde{\omega}_1 - \tau^{\circ *}_{\mathcal{V}M}\theta \otimes \tilde{\Delta}.$$

But $\tilde{\Delta}$ does not vanish on the image of the attachment section Υ (it coincides with Υ^{v} there), so if both connections respect the soldering we must have $\theta = 0$ and the two connections in fact coincide.

We next show that there is such a connection on any coordinate neighbourhood in $\mathcal{V}M$ adapted to the bundle structure $\mathcal{V}M \to M$. We denote the coordinates on $T^{\circ}\mathcal{V}M$ by (x^a, u^a) with $a = 0, 1, 2, \ldots, n$. For $u^0 \neq 0$, $\bar{\chi}$ is given by

$$\bar{\chi}(x^0, x^i, u^0, u^i) = (x^i, y^i) \quad \text{where } y^i = \frac{x^0 u^i}{u^0}.$$

Suppose that

$$\gamma\left(\frac{\partial}{\partial x^i}\right) = \frac{\partial}{\partial x^i} - \left(\delta_i^j + \Pi_{ik}^j y^k + (\Lambda_{ik} y^k) y^j\right)\frac{\partial}{\partial y^j};$$

the coefficient δ_i^j occurs because γ is assumed to respect the soldering $\mathfrak{s}$. Then we define $\tilde{\gamma}$ on the coordinate vector fields as follows:

$$\tilde{\gamma}\left(\frac{\partial}{\partial x^0}\right) = \frac{\partial}{\partial x^0} - \frac{u^i}{x^0}\frac{\partial}{\partial u^i}$$

$$\tilde{\gamma}\left(\frac{\partial}{\partial x^i}\right) = \frac{\partial}{\partial x^i} + x^0 \Lambda_{ij} u^j \frac{\partial}{\partial u^0} - \left(\frac{u^0}{x^0}\delta_i^j + \Pi_{ik}^j u^k\right)\frac{\partial}{\partial u^j}.$$

(The coefficients Λ_{ij} and Π^i_{jk} are to be regarded as functions of the x^i.) Now

$$\bar{\chi}_*\left(\left.\frac{\partial}{\partial x^0}\right|_{(x^a,u^a)}\right) = \left(\frac{u^i}{u^0}\right)\left.\frac{\partial}{\partial y^i}\right|_{(x^i,y^i)}$$

$$\bar{\chi}_*\left(\left.\frac{\partial}{\partial x^i}\right|_{(x^a,u^a)}\right) = \left.\frac{\partial}{\partial x^i}\right|_{(x^i,y^i)}$$

$$\bar{\chi}_*\left(\left.\frac{\partial}{\partial u^0}\right|_{(x^a,u^a)}\right) = -\left(\frac{x^0 u^i}{(u^0)^2}\right)\left.\frac{\partial}{\partial y^i}\right|_{(x^i,y^i)}$$

$$\bar{\chi}_*\left(\left.\frac{\partial}{\partial u^i}\right|_{(x^a,u^a)}\right) = \left(\frac{x^0}{u^0}\right)\left.\frac{\partial}{\partial y^i}\right|_{(x^i,y^i)}.$$

We note that

$$\bar{\chi}_*(\Upsilon^{\mathrm{c}}) = \bar{\chi}_*\left(x^0\frac{\partial}{\partial x^0} + u^0\frac{\partial}{\partial u^0}\right) = 0$$

$$\bar{\chi}_*(\tilde{\Delta}) = \bar{\chi}_*\left(u^0\frac{\partial}{\partial u^0} + u^i\frac{\partial}{\partial u^i}\right) = 0,$$

as expected. Now $\{\Upsilon, \partial/\partial x^i\}$ is a local basis of vector fields on $\mathcal{V}M$ projectable to M. By straightforward calculations we find that

$$\bar{\chi}_*\tilde{\gamma}(\Upsilon) = \bar{\chi}_*\tilde{\gamma}\left(x^0\frac{\partial}{\partial x^0}\right) = 0, \quad \bar{\chi}_*\tilde{\gamma}\left(\frac{\partial}{\partial x^i}\right) = \gamma\left(\frac{\partial}{\partial x^i}\right).$$

So this connection projects onto γ, at least over the coordinate neighbourhood. To establish the soldering condition, we observe that the image of the attachment section Υ is given by $u^0 = x^0$, $u^i = 0$. The vector fields Υ^{c} and $\partial/\partial x^i$ are tangent to it. Now

$$\tilde{\omega}(\Upsilon^{\mathrm{c}}) = u^0 \frac{\partial}{\partial u^0} + u^i \frac{\partial}{\partial u^i} = \tilde{\Delta}$$

$$\tilde{\omega}\left(\frac{\partial}{\partial x^i}\right) = -x^0 \Lambda_{ij} u^j \frac{\partial}{\partial u^0} + \left(\frac{u^0}{x^0}\delta_i^j + \Pi_{ik}^j u^k\right)\frac{\partial}{\partial u^j},$$

and on the attachment section

$$\tilde{\omega}(\Upsilon^{\mathrm{c}}) = x^0 \frac{\partial}{\partial u^0} = \Upsilon^{\mathrm{v}}$$

$$\tilde{\omega}\left(\frac{\partial}{\partial x^i}\right) = \frac{\partial}{\partial u^i} = \left(\frac{\partial}{\partial x^i}\right)^{\mathrm{v}}.$$

Thus $\tilde{\gamma}$ respects the soldering $\tilde{\mathfrak{s}}$. But this means that there is a globally defined connection with the required properties, because the connections on coordinate neighbourhoods defined above must agree on overlaps by uniqueness. □

Corollary 9.4.2

$$\tilde{\gamma}(\Upsilon) = \Upsilon^{\mathrm{c}} - \tilde{\Delta}.$$ □

We have defined $\tilde{\gamma}$ as an infinitesimal connection on a Lie algebroid: but it may evidently also be thought of as the horizontal lift operator of an ordinary linear connection on $\mathcal{V}M$. We now restrict our attention to the case in which the infinitesimal Cartan projective connection is normal. Since Λ_{ij} is then symmetric, we see that the linear connection on $\mathcal{V}M$ defined by $\tilde{\gamma}$ is symmetric. This symmetric linear connection is known as the *Thomas-Whitehead connection*, or *TW-connection*, corresponding to the Cartan projective geometry, because it is the central feature of the theory of projective differential geometry developed by T.Y. Thomas and J.H.C. Whitehead (see [41, 42, 45], and for more recent accounts, [19, 27, 35]). It may be seen from the formula for $\tilde{\gamma}$ in the proof of Theorem 9.4.1 that the connection coefficients $\tilde{\Gamma}_{ab}^c$ of the TW-connection are

$$\tilde{\Gamma}_{i0}^k = \tilde{\Gamma}_{0i}^k = (x^0)^{-1}\delta_i^k$$

$$\tilde{\Gamma}_{ij}^0 = -x^0 \Lambda_{ij}$$

$$\tilde{\Gamma}_{ij}^k = \Pi_{ij}^k$$

with the remaining coefficients $\tilde{\Gamma}_{0k}^0$, $\tilde{\Gamma}_{j0}^0$, $\tilde{\Gamma}_{00}^k$ and $\tilde{\Gamma}_{00}^0$ equal to zero.

Corollary 9.4.3 *The fundamental vector field Υ on $T^\circ\mathcal{V}M$ is a radius vector field with respect to the TW-connection, and $A\mathcal{G}^{\mathcal{V}}$ equipped with the infinitesimal connection $\tilde{\gamma}$ is the infinitesimal Cartan geometry corresponding to the TW-connection and the radius vector field Υ according to Corollary 9.3.2.*

Proof Let us denote the covariant derivative operator of the TW-connection by $\tilde{\nabla}$. Then

$$\tilde{\nabla}_{\partial/\partial x^0}\Upsilon = \tilde{\nabla}_{\partial/\partial x^0}\left(x^0\frac{\partial}{\partial x^0}\right) = \frac{\partial}{\partial x^0} + x^0\tilde{\Gamma}^a_{00}\frac{\partial}{\partial x^a} = \frac{\partial}{\partial x^0},$$
$$\tilde{\nabla}_{\partial/\partial x^i}\Upsilon = \tilde{\nabla}_{\partial/\partial x^i}\left(x^0\frac{\partial}{\partial x^0}\right) = x^0\tilde{\Gamma}^a_{i0}\frac{\partial}{\partial x^a} = \frac{\partial}{\partial x^i}.$$

Thus Υ is indeed a radius vector field. Since $\tilde{\gamma}$ respects the soldering $\tilde{\mathfrak{s}}$ along the attachment section determined by Υ, Corollary 9.3.2 applies. □

The case where $M = \mathrm{P}^n$ illuminates this result, as we show next.

We denote by S^n the n-sphere, the quotient of $\mathbb{R}^{n+1}_{\mathrm{o}}$ by the action generated by the Liouville field $\Delta = x^a\partial/\partial x^a$ on $\mathbb{R}^{n+1}_{\mathrm{o}}$; of course P^n is obtained from S^n by identifying diametrically opposite points. The vector field Δ on $\mathbb{R}^{n+1}_{\mathrm{o}}$ is a radius vector field (and as we pointed out earlier is indeed the prime example of a radius vector field). We shall show that in this case $\mathcal{V}M$ can be identified with $\mathbb{R}^{n+1}_{\mathrm{o}}$ with diametrically opposite points identified, which we denote by $\bar{\mathbb{R}}^{n+1}_{\mathrm{o}}$; and that Υ is just the image of Δ under this identification.

Let p be the projection $\mathbb{R}^{n+1}_{\mathrm{o}} \to S^n$. Let Ω be the standard volume form on $\mathbb{R}^{n+1}$. Then for any point $v \in \mathbb{R}^{n+1}$, $v \neq 0$, we can define an n-covector θ at $p(v) \in S^n$ as follows: let w_i, $i = 1, 2, \ldots, n$ be any n elements of $T_{p(v)}S^n$, and let u_i be any n elements of $T_v\mathbb{R}^{n+1}$ such that $p_{v*}u_i = w_i$; set

$$\theta(w_1, w_2, \ldots, w_n) = \Omega_v(\Delta_v, u_1, u_2, \ldots, u_n).$$

Adding a multiple of Δ_v to any u_i doesn't change the value of the right-hand side, so θ is well-defined. We denote by $\varphi : \mathbb{R}^{n+1}_{\mathrm{o}} \to \bigwedge^n T^*S^n$ the map so defined. If we take $s \in \mathbb{R}$, $s > 0$, and carry out the same construction but starting at sp then the right-hand side gets multiplied by s^{n+1}: that is to say, $\varphi(sv) = s^{n+1}\varphi(v)$. If we now identify diametrically opposite points throughout we obtain a diffeomorphism of $\bar{\mathbb{R}}^{n+1}_{\mathrm{o}}$ with $\mathcal{V}\mathrm{P}^n$; and under the succession of maps $\mathbb{R}^{n+1}_{\mathrm{o}} \to \bar{\mathbb{R}}^{n+1}_{\mathrm{o}} \to \mathcal{V}\mathrm{P}^n$, Δ gets mapped to Υ, which after all is the infinitesimal generator of the action on $\mathcal{V}\mathrm{P}^n$.

Finally, we consider curvatures. The curvature tensor $\tilde{R}$ of the TW-connection is easily calculated, and one finds that

$$\tilde{R}^i_{jkl} = P^i_{jkl}, \quad \tilde{R}^0_{jkl} = -\frac{x^0}{n-1}(\mathfrak{R}_{jk|l} - \mathfrak{R}_{jl|k}),$$

all other components being zero. When this is expressed as the curvature of the Cartan connection $\tilde{\gamma}$ one obtains

$$R^{\tilde{\gamma}}_{kl} = P^i_{jkl}u^j\frac{\partial}{\partial u^i} - \frac{x^0}{n-1}(\mathfrak{R}_{jk|l} - \mathfrak{R}_{jl|k})u^j\frac{\partial}{\partial u^0};$$

this projects to $\mathrm{P}\mathcal{W}M$ to give

$$\left(P^i_{jkl}y^j + \frac{1}{n-1}(\mathfrak{R}_{jk|l} - \mathfrak{R}_{jl|k})y^j y^i\right)\frac{\partial}{\partial y^i},$$

the curvature R^γ_{kl} of the infinitesimal Cartan projective geometry.

9.5 Infinitesimal Symmetries of TW-Connections and Infinitesimal Cartan Projective Geometries

By Proposition 9.3.4, as a projectable vector field on $T^\circ\mathcal{V}M$ an infinitesimal symmetry of the infinitesimal Cartan geometry corresponding to a TW-connection is the complete lift of a vector field X on $\mathcal{V}M$ which satisfies $[\Upsilon, X] = 0$, and which is an infinitesimal affine transformation of the connection. Now $[\Upsilon^c, X^c] = [\Upsilon, X]^c = 0$, and evidently $[\tilde{\Delta}, X^c] = 0$, so if X^c is an infinitesimal symmetry of the TW-connection then it descends to a projectable vector field on $\mathrm{P}\mathcal{W}M$; and since by definition X^c is a section of $A\mathcal{G}^{\mathcal{V}}_p$ its image is a section of $A\mathcal{G}_p$. Thus by the general theory of Lie algebra morphisms and projectability, in view of the relation between the TW-connection and the infinitesimal Cartan projective connection the projectable vector field that X^c defines on $\mathrm{P}\mathcal{W}M$ is an infinitesimal symmetry of the Cartan projective geometry. This is an example of the projectability of Cartan geometries and their symmetries described in Sect. 6.8.

By referring to Proposition 8.2.1 one sees that there are no vertical infinitesimal projective symmetries, an instance of the general result to be proved in Chap. 11. We may realise the Lie algebra of infinitesimal projective symmetries as a (finite dimensional) Lie algebra of vector fields on M: call it $\mathfrak{p}$. We may equally realise the Lie algebra of infinitesimal symmetries of the Cartan geometry of the TW-connection as a Lie algebra of vector fields on $\mathcal{V}M$: call it $\mathfrak{a}$. We have a Lie algebra homomorphism $\mathfrak{a} \to \mathfrak{p}$ defined as follows: for any $X \in \mathfrak{a}$, take the section of $A\mathcal{G}^{\mathcal{V}}_p$ defined by X^c, project it first to $A\mathcal{G}_p$, and then to M: call it φ.

Theorem 9.5.1 *The fundamental vector field Υ on $\mathcal{V}M$ belongs to $\mathfrak{a}$, and is a flat infinitesimal symmetry of the TW-connection. When $H^1(M) = 0$ the Lie algebra homomorphism $\varphi : \mathfrak{a} \to \mathfrak{p}$ is surjective, and its kernel is the 1-dimensional subspace spanned by Υ.*

Proof As we remarked in the proof of Theorem 9.4.1, Υ^c is tangent to the section of $T^\circ\mathcal{V}M$ that Υ defines, so Υ^c is a section of $A\mathcal{G}^{\mathcal{V}}_p$. A straightforward calculation using the defining formulae from the theorem shows that $\mathcal{L}_{\Upsilon^c}\tilde{\gamma} = 0$; the fact that Υ is flat may be seen easily by inspection of the curvature tensor $\tilde{R}$ of a TW-connection as the components $\tilde{R}^a_{bcd}$ vanish when any of the indices b, c, d is zero. Evidently $\varphi(\Upsilon) = 0$.

Suppose that ξ, a section of $A\mathcal{G}_p \to M$, is an infinitesimal symmetry of the projective geometry. We may consider ξ as defining a vector field $\hat{X}$ on $\mathrm{P}\mathcal{W}M$ projectable to M, and over any coordinate neighbourhood U of M we may express this vector field in the form

$$\hat{X} = X^i H_i + \left(X^i_j y^j + (X_k y^k)\, y^i\right) \frac{\partial}{\partial y^i}$$

where

$$H_i = \frac{\partial}{\partial x^i} - \Pi^k_{ij} y^j \frac{\partial}{\partial y^j}.$$

Then $X^i \partial/\partial x^i$ is the coordinate representation of a (globally defined) vector field X on M which is an element of $\mathfrak{p}$, and every element of $\mathfrak{p}$ defines such a section of $A\mathcal{G}_p \to M$. The coefficients X^i_j and X_k are functions on $U \subset M$; we make no claims about their tensorial properties. The necessary and sufficient conditions for ξ to be an infinitesimal symmetry of the projective geometry are just those given in Proposition 8.2.1, mutatis mutandis, and are indeed derived by almost identical calculations. They are

$$\begin{aligned}
X^i_{|j} &= X^i_j \\
X^i_{j|k} &= \mathfrak{R}^i_{jkl} X^l + \delta^i_k X_j + \delta^i_j X_k \\
X_{i|j} &= -(\Lambda_{ji|k} X^k + \Lambda_{kj} X^k_i + \Lambda_{ik} X^k_j);
\end{aligned}$$

here again the rule represents formal covariant differentiation with respect to the Π^i_{jk}, so that for example

$$X_{i|j} = \frac{\partial X_i}{\partial x^j} - \Pi^k_{ij} X_k.$$

We are assuming that the projective geometry is normal, so that in particular both Π^k_{ij} and Λ_{ij} are symmetric in i and j. The important point is that in consequence

$$\frac{\partial X_j}{\partial x^i} = \frac{\partial X_i}{\partial x^j}.$$

Let us assume that U is contractible: then there is a function f on U such that

$$X_i = -\frac{\partial f}{\partial x^i}$$

(sign chosen for convenience). Consider now the vector field

$$\bar{X} = f \Upsilon^{\mathrm{c}} + X^i \frac{\partial}{\partial x^i} - x^0 X_i u^i \frac{\partial}{\partial u^0} + \left(X^i_j - \Pi^i_{kj} X^k\right) u^j \frac{\partial}{\partial u^i}$$

defined on $T^\circ \mathcal{V}M|_U$. Then $\bar{X}$ is projectable to $\mathrm{P}\mathcal{W}M|_U$, and projects onto $\hat{X}$. It is also projectable to $\mathcal{V}M|_U$, and projects onto

$$\tilde{X} = f \Upsilon + X^i \frac{\partial}{\partial x^i}.$$

Evidently $\bar{X}$ is fibre linear, so defines a local section $\tilde{\xi}$ of $A\mathcal{G}^{\mathcal{V}}$; and it is tangent to the submanifold $u^0 = x^0$, $u^i = 0$, the image of the attachment section, so $\tilde{\xi}$ is actually a local section of $A\mathcal{G}_p^{\mathcal{V}}$. We show that $\tilde{\xi}$ is a local infinitesimal symmetry of the TW-connection $\tilde{\gamma}$. Firstly, by Corollary 9.4.2

$$\mathcal{L}_{\tilde{\xi}}\tilde{\gamma}(\Upsilon) = [\bar{X}, \Upsilon^{\mathrm{c}} - \tilde{\Delta}] - \tilde{\gamma}([\tilde{X}, \Upsilon]) = 0,$$

since $\bar{X}$ is projectable to $\mathrm{P}\mathcal{W}M$. Consider

$$\mathcal{L}_{\tilde{\xi}}\tilde{\gamma}\left(\frac{\partial}{\partial x^i}\right) = \psi_i \text{ say.}$$

Since $\tilde{\xi}$ projects to an infinitesimal symmetry of the projective geometry, whose infinitesimal connection is the projection to $\mathrm{P}\mathcal{W}M$ of $\tilde{\gamma}$, the projection of ψ_i to $\mathrm{P}\mathcal{W}M$ must be zero. But $\mathcal{L}_{\tilde{\xi}}\tilde{\gamma}$ takes its values in $K\mathcal{G}^{\mathcal{V}}$ by Lemma 2.2.3, so we must have

$$\psi_i = \Lambda_i\tilde{\Delta}$$

for some function Λ_i on U. To find Λ_i one may simply operate with these vertical vector fields on u^0 and evaluate the results on the attachment section. Clearly $\Lambda_i\tilde{\Delta}(u^0) \doteq x^0\Lambda_i$ (with $\doteq$ indicating equality on the attachment section). To find $\psi_i(u^0)$ the following results are useful:

$$\tilde{\gamma}\left(\frac{\partial}{\partial x^i}\right)(u^0) = x^0\Lambda_{ij}u^j, \quad \tilde{\gamma}(\Upsilon)(u^0) = 0, \quad \bar{X}(u^0) = fu^0 - x^0X_iu^i;$$

and on the attachment section, with $u^0 = x^0$, $u^i = 0$,

$$\tilde{\gamma}\left(\frac{\partial}{\partial x^i}\right) \doteq \frac{\partial}{\partial x^i} - \frac{\partial}{\partial u^i}, \quad \bar{X} \doteq fx^0\left(\frac{\partial}{\partial x^0} + \frac{\partial}{\partial u^0}\right) + X^i\frac{\partial}{\partial x^i}.$$

It follows that

$$\begin{aligned}
\psi_i(u^0) &= \left[\bar{X}, \tilde{\gamma}\left(\frac{\partial}{\partial x^i}\right)\right](u^0) + \frac{\partial f}{\partial x^i}\tilde{\gamma}(\Upsilon)(u^0) + \frac{\partial X^j}{\partial x^i}\tilde{\gamma}\left(\frac{\partial}{\partial x^j}\right)(u^0) \\
&\doteq \left(fx^0\left(\frac{\partial}{\partial x^0} + \frac{\partial}{\partial u^0}\right) + X^i\frac{\partial}{\partial x^i}\right)(x^0\Lambda_{ij}u^j) \\
&\quad - \left(\frac{\partial}{\partial x^i} - \frac{\partial}{\partial u^i}\right)(fu^0 - x^0X_iu^i) \\
&\doteq -x^0\left(\frac{\partial f}{\partial x^i} + X_i\right) = 0.
\end{aligned}$$

Thus $\Lambda_i = 0$ and $\tilde{X}$ is a local symmetry of the TW-connection which projects onto X. Any other local symmetry of the TW-connection projecting to the same X differs from this by a constant multiple of Υ.

We now consider the global situation. This involves a simple application of Čech cohomology. Let $\mathfrak{U} = \{U_\lambda : \lambda \in \Lambda\}$ be a covering of M by contractible coordinate neighbourhoods U_λ, indexed by Λ, such that every nonempty finite intersection of the sets of the covering is also contractible: in other words, let $\mathfrak{U}$ be a good covering of M by coordinate neighbourhoods. Now given $X \in \mathfrak{p}$, for each $\lambda \in \Lambda$ we can find a local symmetry of the TW-connection, $\tilde{X}_\lambda$, which projects onto the restriction of X to U_λ. For any $\lambda, \kappa \in \Lambda$ with $U_\lambda \cap U_\kappa \neq \emptyset$ there is a constant $k_{\lambda\kappa}$ such that $\tilde{X}_\lambda - \tilde{X}_\kappa = k_{\lambda\kappa}\Upsilon$. Clearly for any $\lambda, \kappa, \mu \in \Lambda$ with $U_\lambda \cap U_\kappa \cap U_\mu \neq \emptyset$,

$$k_{\lambda\kappa} - k_{\lambda\mu} + k_{\kappa\mu} = 0.$$

So k is a 1-cocycle in the Čech cochain complex for the covering $\mathfrak{U}$ with values in $\mathbb{R}$. But since $H^1(M) = 0$ by assumption, by the equivalence of Čech and de Rham cohomology k must be a coboundary. That is to say, for each $\lambda \in \Lambda$ there is a constant k_λ such that $k_{\lambda\kappa} = k_\lambda - k_\kappa$ for any $\lambda, \kappa \in \Lambda$ with $U_\lambda \cap U_\kappa \neq \emptyset$. But then $\tilde{X}_\lambda - k_\lambda\Upsilon = \tilde{X}_\kappa - k_\kappa\Upsilon$. So for each coordinate neighbourhood of the covering we can find a symmetry of the TW-connection which projects to X, such that these local symmetries agree on overlaps. They therefore define a global vector field $\tilde{X}$ on $\mathcal{V}M$ which belongs to $\mathfrak{a}$ and satisfies $\varphi(\tilde{X}) = X$. □

When the symmetry algebras have maximal dimension, $\dim \mathfrak{a} = (n+1)^2$ (Proposition 9.3.4), while $\dim \mathfrak{p} = n(n+2)$ (as in Corollary 8.2.2); and of course $\dim \mathfrak{p} = \dim \mathfrak{a} - 1$.

9.6 Infinitesimal Affine Transformations of a TW-Connection

We showed in the previous section that an infinitesimal symmetry of the infinitesimal Cartan geometry correspondng to a TW-connection is the complete lift of a vector field X on $\mathcal{V}M$ which is an infinitesimal affine transformation of the connection and which also satisfies $[\Upsilon, X] = 0$. It so happens that the Lie algebra of infinitesimal affine transformations of a TW-connection (without the additional condition) has an interesting structure, in which the infinitesimal symmetries of the Cartan geometry have a natural place. We examine the infinitesimal affine transformations of a TW-connection in this section.

The only nonzero connection coefficients $\tilde{\Gamma}^a_{bc}$, $a, b, c = 0, 1, 2, \dots, n$, of a TW-connection are these:

$$\tilde{\Gamma}^i_{0i} = \tilde{\Gamma}^i_{i0} = \frac{1}{x^0}, \quad \tilde{\Gamma}^0_{ij} = -x^0\Lambda_{ij}, \quad \tilde{\Gamma}^i_{jk} = \Pi^i_{jk}$$

(no sum over i). The conditions on the coefficients X^a of a vector field X on $\mathcal{V}M$ for it to be an infinitesimal affine transformation of the TW-connection are

$$\frac{\partial^2 X^a}{\partial x^b \partial x^c} - \frac{\partial X^a}{\partial x^d}\tilde{\Gamma}^d_{bc} + \frac{\partial X^d}{\partial x^b}\tilde{\Gamma}^a_{dc} + \frac{\partial X^d}{\partial x^c}\tilde{\Gamma}^a_{db} + X^d \frac{\partial \tilde{\Gamma}^a_{bc}}{\partial x^d} = 0$$

(see Sect. 5.5). We consider the equations obtained from this by taking certain particular values for a, b and c.

First of all, $(a, b, c) = (0, 0, 0)$ gives

$$\frac{\partial^2 X^0}{\partial (x^0)^2} = 0.$$

Secondly, with $(a, b, c) = (i, 0, 0)$ we obtain

$$\frac{\partial^2 X^i}{\partial (x^0)^2} + \left(\frac{2}{x^0}\right)\frac{\partial X^i}{\partial x^0} = 0.$$

It follows that

$$X^0 = \alpha^0 x^0 + \beta^0, \quad X^i = \alpha^i + \frac{1}{x^0}\beta^i$$

for certain functions α^a, β^a on M. Thus

$$X = \left(\alpha^0 x^0 \frac{\partial}{\partial x^0} + \alpha^i \frac{\partial}{\partial x^i}\right) + \left(\beta^0 \frac{\partial}{\partial x^0} + \frac{1}{x^0}\beta^i \frac{\partial}{\partial x^i}\right) = A + B,$$

say. Now $[\Upsilon, X] = -B$, so B is a well-defined vector field on $\mathcal{V}M$, and so is $A = X - B$. Moreover, $[\Upsilon, A] = 0$. Note that Υ, which is of course an infinitesimal affine transformation, is given by $\alpha^0 = 1$, $\alpha^i = \beta^a = 0$.

Next, $(a, b, c) = (0, i, 0)$ gives

$$\frac{\partial \beta^0}{\partial x^i} - \Lambda_{ij}\beta^j = 0,$$

while $(a, b, c) = (i, j, 0)$ gives

$$\frac{\partial \beta^i}{\partial x^j} + \Pi^i_{jk}\beta^k + \beta^0 \delta^i_j = 0.$$

Bearing in mind the explicit form of the vector field B,

$$B = B^a \frac{\partial}{\partial x^a} = \beta^0 \frac{\partial}{\partial x^0} + \frac{1}{x^0}\beta^i \frac{\partial}{\partial x^i},$$

we see that

$$\frac{\partial B^0}{\partial x^0} + \tilde{\Gamma}^0_{0a} B^a = \frac{\partial \beta^0}{\partial x^0} = 0$$
$$\frac{\partial B^0}{\partial x^i} + \tilde{\Gamma}^0_{ia} B^a = \frac{\partial \beta^0}{\partial x^i} - \Lambda_{ij}\beta^j = 0$$
$$\frac{\partial B^i}{\partial x^0} + \tilde{\Gamma}^i_{0a} B^a = -\frac{1}{(x^0)^2}\beta^i + \frac{1}{(x^0)^2}\beta^i = 0$$
$$\frac{\partial B^i}{\partial x^j} + \tilde{\Gamma}^i_{ja} B^a = \frac{1}{x^0}\left(\frac{\partial \beta^i}{\partial x^j} + \Pi^i_{jk}\beta^k + \beta^0\delta^i_j\right) = 0.$$

That is to say, B is parallel with respect to the TW-connection.

We denote the Lie algebra of affine transformations of the TW-connection, considered as an algebra of vector fields on $\mathcal{V}M$, by $\mathfrak{A}$. The elements of $\mathfrak{A}$ of type A satisfy $[\Upsilon, A] = 0$, while those of type B satisfy $[\Upsilon, B] = -B$. Denote by $\mathfrak{a}$ the subspace of $\mathfrak{A}$ consisting elements of type A and by $\mathfrak{b}$ the subspace of elements of type B; then $\mathfrak{A} = \mathfrak{a} \oplus \mathfrak{b}$ as vector spaces. Now for $A_1, A_2 \in \mathfrak{a}$, $[\Upsilon, [A_1, A_2]] = 0$ by the Jacobi identity, so $\mathfrak{a}$ is a subalgebra of $\mathfrak{A}$. For $B_1, B_2 \in \mathfrak{b}$, $[\Upsilon, [B_1, B_2]] = -2[B_1, B_2]$ by Jacobi, so we must have $[B_1, B_2] = 0$, that is, $\mathfrak{b}$ is Abelian. Finally, for $A \in \mathfrak{a}$, $B \in \mathfrak{b}$, $[\Upsilon, [A, B]] = -[A, B]$, so that $[\mathfrak{a}, \mathfrak{b}] \in \mathfrak{b}$. Thus $\mathfrak{b}$ is an Abelian ideal of $\mathfrak{A}$, and $\mathfrak{A}/\mathfrak{b} = \mathfrak{a}$. Moreover, $\mathfrak{a}$ has the same meaning as it had in the previous section.

We note that the maximum dimension of $\mathfrak{A}$ is $(n+1)(n+2)$, and that of $\mathfrak{a}$ is $n(n+2)+1$. If these dimensions are attained, the dimension of $\mathfrak{b}$ is $(n+1)(n+2) - n(n+2) - 1 = n+1 = \dim(\mathcal{V}M)$.

The structure of $\mathfrak{A}$, together with the fact that if $B \in \mathfrak{b}$ then B is parallel, suggests that $\mathfrak{b}$ plays the role of translations. To throw more light on this we look at the case of projective space. The TW-connection should be the standard flat connection on $\mathbb{R}^{n+1}_0$; however, in the coordinates x^a not all of the connection coefficients vanish (we have $\tilde{\Gamma}^i_{0j} = \tilde{\Gamma}^i_{j0} = (x^0)^{-1}\delta^i_j$), which indicates that these are not Cartesian coordinates. If we take the x^i to be affine coordinates on P^n, related to Cartesian coordinates y^a by $x^i = y^i/y^0$, then the construction of the volume bundle gives $x^0 = y^0$, so Cartesians are given in terms of the x^a by

$$y^0 = x^0, \quad y^i = x^0 x^i.$$

It is easy to see that the connection coefficients do indeed vanish in these coordinates.

In terms of the x^a the equations for the affine transformations of the TW-connection are

$$\frac{\partial^2 \alpha^i}{\partial x^j \partial x^k} + \frac{\partial \alpha^0}{\partial x^j}\delta^i_k + \frac{\partial \alpha^0}{\partial x^k}\delta^i_j = 0, \quad \frac{\partial^2 \alpha^0}{\partial x^j \partial x^k} = 0$$

for elements of $\mathfrak{a}$, and

$$\frac{\partial \beta^i}{\partial x^j} + \beta^0\delta^i_j = 0, \quad \frac{\partial \beta^0}{\partial x^i} = 0$$

for elements of $\mathfrak{b}$. Thus

$$\alpha^0 = K^0_i x^i + K^0_0, \quad \alpha^i = -K^0_j x^j x^i + K^i_j x^j + K^j_0,$$

where all the coefficients are constants, and

$$x^0\alpha^0 \frac{\partial}{\partial x^0} + \alpha^i \frac{\partial}{\partial x^i} = \left(K^0_j y^j + K^0_0 y^0\right) \frac{\partial}{\partial y^0} + \left(K^i_j y^j + K^i_0 y^0\right) \frac{\partial}{\partial y^i}.$$

For the β^a, we see that $\beta^0 = C^0$ is constant, and $\beta^i = -C^0 x^i + C^i$ where C^i is constant; moreover

$$\beta^0 \frac{\partial}{\partial x^0} + \frac{1}{x^0}\beta^i \frac{\partial}{\partial x^i} = C^0 \frac{\partial}{\partial y^0} + C^i \frac{\partial}{\partial y^i}.$$

Thus $\mathfrak{a}$ consists of the linear transformations and $\mathfrak{b}$ of the translations.

Note finally that if $\mathfrak{b}$ has its maximum dimension, namely $n + 1$, then there is a basis of vector fields on $\mathcal{V}M$ whose elements are parallel and commute pairwise. Thus the TW-connection is flat, as is the corresponding infinitesimal Cartan projective geometry.

9.7 The Projective Tractor Connections

We now show how the TW-connection, a linear connection on $T\mathcal{V}M$, induces linear connections which we call *projective tractor connections* on both $\widetilde{\mathcal{W}}M$ and $\mathcal{W}M$, which play the role of projective tractor bundles.

For any vector field X on $\mathcal{V}M$ we denote by X^{h} its horizontal lift to $T\mathcal{V}M$ relative to the TW-connection. Now Υ is an infinitesimal symmetry of the TW-connection, which means that $[\Upsilon^{\mathsf{c}}, X^{\mathsf{h}}] = [\Upsilon, X]^{\mathsf{h}}$. Moreover, $[\tilde{\Delta}, X^{\mathsf{h}}] = 0$ because the connection is linear. So $[\tilde{\Upsilon}, X^{\mathsf{h}}] = [\Upsilon, X]^{\mathsf{h}}$. Finally, by Corollary 9.4.2, the horizontal lift of Υ to $T\mathcal{V}M$ with respect to the TW-connection is given by $\Upsilon^{\mathsf{h}} = \tilde{\gamma}(\Upsilon) = \Upsilon^{\mathsf{c}} - \tilde{\Delta} = \tilde{\Upsilon}$. This means that the horizontal distribution descends to $\widetilde{\mathcal{W}}M$. For if X is any vector field on $\mathcal{V}M$ projectable to M, so that $[\Upsilon, X] = f\Upsilon$ for some function f on $\mathcal{V}M$, then $[\tilde{\Upsilon}, X^{\mathsf{h}}] = f\tilde{\Upsilon}$, and so X^{h} projects to $\widetilde{\mathcal{W}}M$. Moreover, changing X by the addition of a scalar multiple of Υ changes X^{h} by the addition of the same scalar multiple of $\Upsilon^{\mathsf{h}} = \tilde{\Upsilon}$, which doesn't affect the outcome.

So $\widetilde{\mathcal{W}}M$ inherits a horizontal distribution, linear in the fibre coordinates, and therefore a linear connection, which we may regard as a covariant derivative operator on local sections of $\widetilde{\mathcal{W}}M \to M$. We denote this covariant derivative by ∇: it is the tractor connection of Bailey et al. [2].

The induced horizontal lift of the ith coordinate vector field on M to $\widetilde{\mathcal{W}}M$, in the case of the normal geometry, is easily seen to be

$$\frac{\partial}{\partial x^i} + \mathfrak{P}_{ij}\tilde{w}^j \frac{\partial}{\partial \tilde{w}^0} - \left(\delta_i^j \tilde{w}^0 + \Pi_{ik}^j \tilde{w}^k\right) \frac{\partial}{\partial \tilde{w}^j},$$

so the covariant derivative of any local section $(\varsigma^0, \varsigma^i)$ of $\widetilde{\mathcal{W}}M \to M$ is given by

$$\nabla_{\partial/\partial x^i} \begin{pmatrix} \varsigma^0 \\ \varsigma^j \end{pmatrix} = \begin{pmatrix} \dfrac{\partial \varsigma^0}{\partial x^i} - \mathfrak{P}_{ij}\varsigma^j \\ \dfrac{\partial \varsigma^j}{\partial x^i} + \Pi_{ik}^j \varsigma^k + \varsigma^0 \delta_i^j \end{pmatrix}.$$

The bundle dual to $\widetilde{\mathcal{W}}M$ is $J^1\nu$, where $\nu : \mathcal{D}M \to M$ is the bundle of scalar densities of weight $-1/(n+1)$ (see Lemma 4.3.2). The covariant derivative ∇ determines a covariant derivative on sections of $J^1\nu \to M$, where

$$\nabla_i \begin{pmatrix} \varsigma_0 \\ \varsigma_j \end{pmatrix} = \begin{pmatrix} \dfrac{\partial \varsigma_0}{\partial x^i} - \varsigma_i \\ \dfrac{\partial \varsigma_j}{\partial x^i} - \Pi_{ij}^k \varsigma_k + \mathfrak{P}_{ij}\varsigma_0 \end{pmatrix}.$$

In terms of jets, the connection defines a splitting of

$$0 \to J^1\nu \otimes_M T^*M \to J^1\nu_1 \to J^1\nu \to 0$$

where $J^1\nu_1$ is the first jet bundle of $\nu_1 : J^1\nu \to M$. In terms of fibre coordinates (x_0, x_i) on $J^1\nu$, and $(x_0, x_i, x_{0i}, x_{ij})$ on $J^1\nu_1$, the splitting map is given by

$$x_{0i} = x_i, \quad x_{ij} = \Pi_{ij}^k x_k - \mathfrak{P}_{ij} x_0,$$

so that it is a second-order connection taking its values in $J^2\nu \subset J^1\nu_1$ and therefore defines a splitting of

$$0 \to \mathcal{D}M \otimes_M S^2T^*M \to J^2\nu \to J^1\nu \to 0.$$

We go back to the TW-connection. Denote its covariant derivative operator by $\tilde{\nabla}$. We show that with the aid of an Ehresmann connection on $\mathcal{V}M$ we may construct from $\tilde{\nabla}$ a covariant derivative operator on sections of $\mathcal{W}M \to M$, which is a variant of the tractor connection, and also is closely related to the ordinary covariant derivative of that symmetric linear connection in the projective class which is determined by ϑ.

Lemma 9.7.1 *The vector field Υ is an infinitesimal affine transformation of $\tilde{\nabla}$: that is to say, for every vector field Y on $\mathcal{V}M$,*

$$\mathcal{L}_\Upsilon \circ \tilde{\nabla}_Y - \tilde{\nabla}_Y \circ \mathcal{L}_\Upsilon = \tilde{\nabla}_{[\Upsilon, Y]}.$$

□

Lemma 9.7.2 *Let Y be any vector field on $\mathcal{V}M$ which is invariant by Υ, that is, for which $[\Upsilon, Y] = 0$. Let $\hat{Y}$ be the section of $T\mathcal{V}M \to M$ corresponding to Y. Then $\chi \circ \hat{Y}$ is well-defined, and is a section of $\mathcal{W}M \to M$; and every section of $\mathcal{W}M \to M$ is obtained from a Υ-invariant vector field on $\mathcal{V}M$ in this way.* □

Lemma 9.7.3 *Let ϑ be the connection form of an Ehresmann connection on $\mathcal{V}M$, and for any vector field X on M let X^ϑ be the horizontal lift of X to $\mathcal{V}M$ relative to ϑ. Then X^ϑ is Υ-invariant.* □

Lemma 9.7.4 *For any vector field X on M and any Υ-invariant vector field Y on $\mathcal{V}M$, the vector field*

$$\tilde{\nabla}_{X^\vartheta} Y$$

is Υ-invariant.

Proof From Lemma 9.7.1,

$$[\Upsilon, \tilde{\nabla}_{X^\vartheta} Y] = \tilde{\nabla}_{X^\vartheta}[\Upsilon, Y] + \tilde{\nabla}_{[\Upsilon, X^\vartheta]} Y = 0,$$

since both Y and X^ϑ are Υ-invariant. □

For any vector field X on M define an operator ∇^ϑ_X on sections of $\mathcal{W}M \to M$ as follows: for a section σ let Y_σ be the Υ-invariant vector field on $\mathcal{V}M$ such that $\chi \circ \hat{Y}_\sigma = \sigma$, and let $\nabla^\vartheta_X \sigma$ be the section of $\mathcal{W}M \to M$ corresponding to the Υ-invariant vector field $\tilde{\nabla}_{X^\vartheta} Y_\sigma$.

Proposition 9.7.5 ∇^ϑ *is a covariant derivative of sections of $\mathcal{W}M \to M$.*

Proof For any function f on M,

$$\tilde{\nabla}_{(fX)^\vartheta} Y_\sigma = \tilde{\nabla}_{fX^\vartheta} Y_\sigma = f\tilde{\nabla}_{X^\vartheta} Y_\sigma,$$

while

$$\tilde{\nabla}_{X^\vartheta} Y_{f\sigma} = \tilde{\nabla}_{X^\vartheta}(fY_\sigma) = f\tilde{\nabla}_{X^\vartheta} Y_\sigma + (X^\vartheta f)Y_\sigma = f\tilde{\nabla}_{X^\vartheta} Y_\sigma + (Xf)Y_\sigma.$$

When expressed in terms of ∇^ϑ these are the required properties. □

We can easily obtain an explicit expression for ∇^ϑ from an expression for $\tilde{\nabla}$ in terms of the local basis of vector fields $\{\Upsilon, H_i\}$ where

$$H_i = \left(\frac{\partial}{\partial x^i}\right)^\vartheta = \frac{\partial}{\partial x^i} - \vartheta_i \Upsilon$$

is the horizontal lift of the ith coordinate field on M relative to ϑ. Now Υ is a radius vector field, and $\tilde{\nabla}$ is symmetric, whence

$$\tilde{\nabla}_{H_i}\Upsilon = \tilde{\nabla}_\Upsilon H_i = H_i.$$

We know that

$$\tilde{\nabla}_{\partial/\partial x^i}\left(\frac{\partial}{\partial x^j}\right) = -\mathfrak{P}_{ij}\Upsilon + \Pi^k_{ij}\frac{\partial}{\partial x^k},$$

from which we obtain

$$\begin{aligned}
\tilde{\nabla}_{H_i}H_j &= \tilde{\nabla}_{\partial/\partial x^i}H_j - \vartheta_i\tilde{\nabla}_\Upsilon H_j \\
&= \tilde{\nabla}_{\partial/\partial x^i}\left(\frac{\partial}{\partial x^j}\right) - \tilde{\nabla}_{\partial/\partial x^i}(\vartheta_j\Upsilon) - \vartheta_i H_j \\
&= -\mathfrak{P}_{ij}\Upsilon + \Pi^k_{ij}\frac{\partial}{\partial x^k} - \frac{\partial\vartheta_j}{\partial x^i}\Upsilon - \vartheta_j\frac{\partial}{\partial x^i} - \vartheta_i H_j \\
&= -\left(\mathfrak{P}_{ij} + \frac{\partial\vartheta_j}{\partial x^i} - \Pi^k_{ij}\vartheta_k + \vartheta_i\vartheta_j\right)\Upsilon + \left(\Pi^k_{ij} - \vartheta_i\delta^k_j - \vartheta_j\delta^k_i\right)H_k.
\end{aligned}$$

But

$$\Pi^k_{ij} - \vartheta_i\delta^k_j - \vartheta_j\delta^k_i = \Gamma^k_{ij}$$

are the components of a symmetric linear connection on M, representative of the projective class determined by Π^k_{ij}; and if we set

$$\mathfrak{P}_{ij} + \frac{\partial\vartheta_j}{\partial x^i} - \Pi^k_{ij}\vartheta_k + \vartheta_i\vartheta_j = P_{ij}$$

then P_{ij} is the projective Schouten tensor of Γ^k_{ij}. So if e is the canonical global section of $\mathcal{W}M \to M$, which is determined by Υ, and e_i is the local section determined by H_i, then (writing ∇^ϑ_i for $\nabla^\vartheta_{\partial/\partial x^i}$)

$$\nabla^\vartheta_i e = e_i, \quad \nabla^\vartheta_i e_j = -P_{ij}e + \Gamma^k_{ij}e_k.$$

So if σ is any local section of $\mathcal{W}M \to M$, and we set $\sigma = \sigma^0 e + \sigma^j e_j$, then

$$\begin{aligned}
\nabla^\vartheta_i\sigma &= \left(\frac{\partial\sigma^0}{\partial x^i}\right) + \sigma_0 e_i + \left(\frac{\partial\sigma^j}{\partial x^i}\right)e_j - P_{ij}\sigma^j e + \Gamma^k_{ij}\sigma^j e_k \\
&= \left(\sigma^0_{|i} - P_{ij}\sigma^j\right)e + \left(\sigma^j_{|i} + \delta^j_i\sigma_0\right)e_j,
\end{aligned}$$

where the bar denotes covariant derivative with respect to Γ^k_{ij}, and σ^0 is considered as a scalar (so that $\sigma^0_{|i}$ is just the partial derivative) and σ^j are the components of a contravariant vector. In other terms,

$$\nabla^\vartheta_i\begin{pmatrix}\sigma^0\\ \sigma^j\end{pmatrix} = \begin{pmatrix}\sigma^0_{|i} - P_{ij}\sigma^j\\ \sigma^j_{|i} + \delta^j_i\sigma^0\end{pmatrix}.$$

This is all expressed in terms of the basis $\{e, e_i\}$ of local sections of $\mathcal{W}M \to M$, coming from the basis $\{\Upsilon, H_i\}$ of Υ-invariant local vector fields on $\mathcal{V}M$. But we could work with coordinate vector fields $\partial/\partial x^i$ on $\mathcal{V}M$, which are also Υ-invariant. Let us denote the corresponding local sections of $\mathcal{W}M \to M$ by ∂_i. Then

$$\tilde{\nabla}_{H_i}\Upsilon = \frac{\partial}{\partial x^i} - \vartheta_i\Upsilon, \quad \tilde{\nabla}_{H_i}\left(\frac{\partial}{\partial x^j}\right) = \left(\Pi^k_{ij} - \vartheta_i\delta^k_j\right)\frac{\partial}{\partial x^k} - \mathfrak{P}_{ij}\Upsilon,$$

from which it follows that

$$\nabla^{\vartheta}_i e = -\vartheta_i e + \partial_i, \quad \nabla^{\vartheta}_i \partial_j = -\mathfrak{P}_{ij}e + \left(\Pi^k_{ij} - \vartheta_i\delta^k_j\right)\partial_k.$$

This doesn't look very promising. We can, however, induce from this a covariant derivative on sections of $\widetilde{\mathcal{W}}M = \mathcal{W}M \otimes \mathcal{D}M \to M$, as follows.

First, we make a general remark about linear connections on vector bundles and tensor products. Let $E \to M$, $E' \to M$ be vector bundles over M, each equipped with a linear connection, whose covariant derivative operators are ∇ and ∇' respectively. Then there is a unique linear connection on $E \otimes E' \to M$ which satisfies Leibnitz' rule with respect to the tensor product, that is, whose covariant derivative $\nabla^{\otimes}$ satisfies

$$\nabla^{\otimes}(\sigma \otimes \sigma') = (\nabla\sigma) \otimes \sigma' + \sigma \otimes (\nabla'\sigma')$$

for all local sections σ of $E \to M$ and σ' of $E' \to M$. (One may construct $\nabla^{\otimes}$ by taking local bases of sections $\{\sigma_a\}$ and $\{\sigma'_\alpha\}$, using the formula above to define $\nabla^{\otimes}(\sigma_a \otimes \sigma'_\alpha)$, extending by the usual rules, and showing that the result is unchanged if one changes the original local sections.)

Now the Ehresmann connection induces a linear connection on the bundle of scalar densities of weight $1/(n+1)$, $\mathcal{D}M \to M$, which is associated with the principal bundle $\mathcal{V}M \to M$, with covariant derivative which we shall denote by $\nabla^{\vartheta'}$. We now use the local section x of $\mathcal{D}M \to M$ defined by $x^0 = 1$: then

$$\nabla^{\vartheta'}_i \mathsf{x} = \vartheta_i \mathsf{x}.$$

Let us denote by $\nabla^{\otimes}$ the covariant derivative on $\mathcal{W}M \otimes \mathcal{D}M \to M$ constructed from ∇^{ϑ} and $\nabla^{\vartheta'}$ as described above. Then we obtain

$$\begin{aligned}
\nabla^{\otimes}_i(e \otimes \mathsf{x}) &= (-\vartheta_i e + \partial_i) \otimes \mathsf{x} + e \otimes (\vartheta_i \mathsf{x}) = \partial_i \otimes \mathsf{x} \\
\nabla^{\otimes}_i(\partial_j \otimes \mathsf{x}) &= \mathfrak{P}_{ij} e \otimes \mathsf{x} + \left(\Pi^k_{ij} - \vartheta_i\delta^k_j\right)\partial_k \otimes \mathsf{x} + \partial_j \otimes (\vartheta_i \mathsf{x}) \\
&= -\mathfrak{P}_{ij} e \otimes \mathsf{x} + \Pi^k_{ij}\partial_k \otimes \mathsf{x},
\end{aligned}$$

which seems much more satisfactory. In particular, $\nabla^{\otimes}$ does not depend on ϑ: in fact it coincides with ∇ introduced at the beginning of this section, as one sees by expressing it in terms of components with respect to the given basis:

$$\nabla_i^{\otimes}\begin{pmatrix}\varsigma^0\\ \varsigma^j\end{pmatrix} = \begin{pmatrix}\dfrac{\partial\varsigma^0}{\partial x^i} - \mathfrak{P}_{ij}\varsigma^j\\ \dfrac{\partial\varsigma^j}{\partial x^i} + \Pi^j_{ik}\varsigma^k + \varsigma^0\delta^j_i\end{pmatrix}.$$

Chapter 10
Conformal Geometry: The Full Version

In the previous chapter we have described how the full version of projective geometry can be related to a finite Cartan geometry based on the Lie groupoid of fibre morphisms of a bundle of projective spaces, the bundle $P\mathcal{W}M \to M$, and how this is related to the infinitesimal geometry on TM studied in Sect. 8.2. There is a similar relationship between infinitesimal conformal geometry on TM (Sect. 8.4) and a finite geometry, but now the fibres of the corresponding bundle will be spheres rather than projective spaces. In this chapter we shall explain how the action of a suitable group gives rise to such a sphere as a homogeneous space, and how the corresponding Cartan geometry can be constructed.

10.1 The Möbius Group

It follows from the results of Sect. 8.4 that vector fields on $\mathbb{R}^n$ of the form

$$\left(Y^i + Y^i_j y^j + \tfrac{1}{2}(Z_j\delta^i_k + Z_k\delta^i_j - \delta_{jk}Z^i)y^j y^k\right)\frac{\partial}{\partial y^i}$$

where the δ_{jk} are components of the Euclidean scalar product ($\delta_{ii} = 1$, $\delta_{ij} = 0$ for $i \neq j$), the coefficients are constants, $\delta_{ik}Y^k_j + \delta_{kj}Y^k_i = \mu\delta_{ij}$ for some constant μ, and $Z_i = \delta_{ij}Z^j$, form a Lie algebra. (The more general case treated there, in which the basis is not necessarily orthonormal, concerns vector fields

$$\left(Y^i + Y^i_j y^j + \tfrac{1}{2}(Z_j\delta^i_k + Z_k\delta^i_j - h_{jk}Z^i)y^j y^k\right)\frac{\partial}{\partial y^i}$$

where $h_{ij} = e_i \cdot e_j$, $h_{ik}Y^k_j + h_{kj}Y^k_i = \mu h_{ij}$, and $Z_i = h_{ij}Z^j$.) Since such vector fields are not in general complete, however, this Lie algebra cannot be realised as the algebra of infinitesimal generators of a Lie group acting on $\mathbb{R}^n$. We now describe how to construct a Lie group with this Lie algebra, and a manifold F of dimension

M. Crampin and D. Saunders, *Cartan Geometries and their Symmetries*,
Atlantis Studies in Variational Geometry 4, DOI 10.2991/978-94-6239-192-5_10

n on which it acts transitively and effectively. (The method may easily be adapted to deal with metrics of other signatures.)

Consider $\mathbb{R}^{n+2}$, with standard basis which we denote by $\{e_0, e_i, e_\infty\}$ with $i = 1, 2, \ldots, n$. We equip $\mathbb{R}^{n+2}$ with a scalar product of signature $(n+1, 1)$ by setting $e_0 \cdot e_\infty = -1$, $e_i \cdot e_j = \delta_{ij}$, all other products zero. This choice is usual, and is made for calculational convenience: the standard signature $(n+1, 1)$ scalar product, corresponding to the bilinear form $\operatorname{diag}(1, 1, \ldots, 1, -1)$, is obtained by changing the basis to

$$f_0 = \frac{1}{\sqrt{2}}(e_0 - e_\infty), \quad f_i = e_i, \quad f_\infty = \frac{1}{\sqrt{2}}(e_0 + e_\infty).$$

The basis $\{f^0, f^i, f^\infty\}$ can be said to be orthonormal. We call the original basis $\{e^0, e^i, e^\infty\}$, and any other with respect to which the scalar product takes the same form, quasi-orthonormal.

The manifold F in question is the set of lines in $\mathbb{R}^{n+2}$ which are null with respect to this scalar product: thus each point of F is a line $\{tz : t \in \mathbb{R}\}$ where $z = z^0 e_0 + z^i e_i + z^\infty e_\infty \in \mathbb{R}^{n+2}$ with $\delta_{ij} z^i z^j = 2z^0 z^\infty$. In terms of orthonormal coordinates (that is, coordinates relative to an orthonormal basis $\{f_0, f_i, f_\infty\}$) the condition for a line to be null becomes

$$(w^0)^2 + (w^1)^2 + \cdots + (w^n)^2 = (w^\infty)^2,$$

where $z = w^0 f_0 + w^i f_i + w^\infty f_\infty$. The hyperplane $w^\infty = 1$ inherits a Euclidean structure, and each null line intersects it in a unique point of the sphere

$$(w^0)^2 + (w^1)^2 + \cdots + (w^n)^2 = 1\,,$$

from which we see that F can be identified with the unit sphere of dimension n.

It is well known that the group $\mathsf{O}(n+1, 1)$ acts transitively on the set of null lines of the standard scalar product of signature $(n+1, 1)$. Let us denote by $\bar{G}$ the orthogonal group of the scalar product defined in the previous paragraph: it is conjugate to $\mathsf{O}(n+1, 1)$ in $\mathsf{GL}(n+2)$, and acts transitively on F. An $(n+2)\times(n+2)$ matrix ϕ with coefficients ϕ^A_B, $A, B = 0, 1, 2, \ldots, n, \infty$, belongs to $\bar{G}$ if and only if

$$\begin{aligned}
\delta_{ij}\phi^i_0\phi^j_0 - 2\phi^0_0\phi^\infty_0 &= 0\\
\delta_{ij}\phi^i_\infty\phi^j_\infty - 2\phi^0_\infty\phi^\infty_\infty &= 0\\
\delta_{ij}\phi^i_0\phi^j_\infty - \phi^0_0\phi^\infty_\infty - \phi^\infty_0\phi^0_\infty &= -1\\
\delta_{jk}\phi^j_0\phi^k_i - \phi^\infty_0\phi^0_i - \phi^0_0\phi^\infty_i &= 0\\
\delta_{jk}\phi^j_\infty\phi^k_i - \phi^0_\infty\phi^\infty_i - \phi^\infty_\infty\phi^0_i &= 0\\
\delta_{kl}\phi^k_i\phi^l_j - 2\phi^0_i\phi^\infty_j &= \delta_{ij}\,.
\end{aligned}$$

Correspondingly, an element a of the Lie algebra $\mathfrak{g}$ of $\bar{G}$, considered as a matrix, has coefficients a^B_C where

$$a^\infty_0 = a^0_\infty = 0, \quad a^0_0 + a^\infty_\infty = 0,$$
$$a^\infty_i = \delta_{ij} a^j_0, \quad a^0_i = \delta_{ij} a^j_\infty, \quad \delta_{jk} a^k_i + \delta_{ik} a^k_j = 0.$$

We can introduce local coordinates (y^i) on F, over the open set in which $z^0 \neq 0$, by setting $y^i = z^i/z^0$. Note that $z^\infty/z^0 = \frac{1}{2}\delta_{ij} y^i y^j$. An element ϕ of $\bar{G}$ induces a transformation of F whose coordinate representation is

$$y^i \circ \phi = \frac{\phi^i_0 + \phi^i_j y^j + \frac{1}{2}\delta^i_\infty h_{jk} y^j y^k}{\phi^0_0 + \phi^0_j y^j + \frac{1}{2}\phi^0_\infty \delta_{jk} y^j y^k}.$$

Using this formula we find that the fundamental vector field $a^\sharp$ of any $a \in \mathfrak{g}$ arising from the action of $\bar{G}$ on F is given in coordinate form as follows:

$$a^\sharp = \left(Y^i + Y^i_j y^j + \tfrac{1}{2}(Z_j \delta^i_k + Z_k \delta^i_j - \delta_{jk} Z^i) y^j y^k\right) \frac{\partial}{\partial y^i}$$

with

$$Y^i = a^i_0, \quad Y^i_j = a^i_j - a^0_0 \delta^i_j, \quad Z_i = -a^0_i;$$

notice that $\delta_{ik} Y^k_j + \delta_{kj} Y^k_i = -2a^0_0 \delta_{ij}$.

The vector e_0 spans a null line; that is, it determines a point of F, which is the origin of the coordinates (y^i). We denote by $\bar{G}_o$ the subgroup of $\bar{G}$ which is the isotropy group of the line $\langle e_0 \rangle$. Then $\phi \in \bar{G}_o$ if and only if $\phi^i_0 = \phi^\infty_0 = 0$. It follows from the requirement that $\delta_{jk}\phi^j_0\phi^k_i - \phi^\infty_0\phi^0_i - \phi^0_0\phi^\infty_i = 0$ that $\phi^\infty_i = 0$ also (since $\phi^0_0 \neq 0$). Thus $\bar{G}_o$ consists of those elements of $\bar{G}$ which are, in an obvious sense, block upper triangular (we give the explicit representation shortly). To put it another way, we have a filtration of $\mathbb{R}^{n+2}$, $\langle e_0 \rangle \subset \langle e_0, e_i \rangle \subset \mathbb{R}^{n+2}$, and $\bar{G}_o$ preserves this filtration. The same results may be expressed in somewhat more geometrical terms as follows: any element of $\bar{G}$ which fixes the null line $\langle e_0 \rangle$ also fixes its orthogonal subspace $\langle e_0 \rangle^\perp$, which contains e_0 since it is null; this is the subspace $\langle e_0, e_i \rangle$, which happens also to be the tangent hyperplane to the null cone along the null line $\langle e_0 \rangle$.

The Lie algebra $\mathfrak{g}_o$ of $\bar{G}_o$ consists of those $a \in \mathfrak{g}$ with $a^i_0 = a^\infty_0 = a^\infty_i = 0$. In terms of the representation by vector fields on F this amounts precisely to $Y^i = 0$, as expected.

The action of the group $\bar{G}_o$ on $\langle e_0, e_i \rangle$ is obtained by restricting one's attention to the components ϕ^a_b with $a, b = 0, 1, 2, \ldots, n$; we have $\phi^i_0 = 0$, and so $\delta_{kl}\phi^k_i\phi^l_j = \delta_{ij}$. That is, the restriction of $\bar{G}_o$ to $\langle e_0, e_i \rangle$ consists of $(n+1) \times (n+1)$ matrices of the form

$$\begin{pmatrix} k & \theta \\ 0 & \Phi \end{pmatrix}$$

with $k \in \mathbb{R}$, $k \neq 0$; $\theta \in \mathbb{R}^{n*}$; $\Phi \in \mathsf{O}(n)$. This group of matrices is called the *Weyl group* of (h_{ij}); we denote it by $\mathsf{W}(n)$. In fact $\bar{G}_o$ is completely determined by its action on $\langle e_0, e_i\rangle$: from the conditions satisfied by an element of $\bar{G}$ we see that with $\phi_0^i = \phi_0^\infty = \phi_i^\infty = 0$

$$\begin{aligned} \delta_{ij}\phi^i_\infty\phi^j_\infty - 2\phi^0_\infty\phi^\infty_\infty &= 0 \\ \phi^0_0\phi^\infty_\infty &= 1 \\ \delta_{jk}\phi^j_\infty\phi^k_i - \phi^\infty_\infty\phi^0_i &= 0 \end{aligned}$$

from which the remaining components ϕ^∞_∞, ϕ^i_∞ and ϕ^0_∞ may be determined in turn. In a more accessible notation we may write the $(n+2)\times(n+2)$ matrix as

$$\begin{pmatrix} k & \theta & p \\ 0 & \Phi & v \\ 0 & 0 & k^{-1} \end{pmatrix}$$

where

$$v = k^{-1}\Phi\theta^T, \quad p = \tfrac{1}{2}k^{-1}|\theta|^2.$$

We thus have an injective homomorphism

$$\begin{pmatrix} k & \theta \\ 0 & \Phi \end{pmatrix} \mapsto \begin{pmatrix} k & \theta & p \\ 0 & \Phi & v \\ 0 & 0 & k^{-1} \end{pmatrix}$$

of $\mathsf{W}(n)$ into $\bar{G}$, whose image is $\bar{G}_o$. Alternatively we may regard this map as determining a representation of $\mathsf{W}(n)$ on $\mathbb{R}^{n+2}$ by matrices which leave invariant the scalar product of signature $(n+1, 1)$ and the filtration $\langle e_0\rangle \subset \langle e_0, e_i\rangle \subset \mathbb{R}^{n+2}$.

We observed earlier that the group $\bar{G}$ acts transitively on F: but it clearly cannot act effectively since $-I$ fixes each null line. We now show that this is the only element of $\bar{G}$, other than I itself, with this property. We know that if $\phi \in \bar{G}$ fixes $\langle e_0\rangle$ then it belongs to $\bar{G}_o$ and takes the block upper triangular form above. If further $\phi \in \bar{G}_o$ fixes $\langle e_\infty\rangle$ then $p = 0$ and $v = 0$, whence $\theta = 0$. Now for any nonzero $z \in \mathbb{R}^n$ we can choose z^0 and z^∞ such that (z^0, z, z^∞) is null: then

$$\phi\begin{pmatrix} z^0 \\ z \\ z^\infty \end{pmatrix} = \begin{pmatrix} k & 0 & 0 \\ 0 & \Phi & 0 \\ 0 & 0 & k^{-1} \end{pmatrix}\begin{pmatrix} z^0 \\ z \\ z^\infty \end{pmatrix} = \begin{pmatrix} kz^0 \\ \Phi z \\ k^{-1}z^\infty \end{pmatrix} = t\begin{pmatrix} z^0 \\ z \\ z^\infty \end{pmatrix}$$

for some $t \in \mathbb{R}$. Then Φ satisfies $\Phi z = tz$ for all $z \in \mathbb{R}^n$, which is possible only if $\Phi = \pm I$ (with $t = \pm 1$), since it is orthogonal. But then $\phi = \pm I$.

The group $G = \bar{G}/\{\pm I\}$ acts transitively and effectively on F. This group is known as the *Möbius group* [32, 36]. Then $F = G/G_o$ where $G_o = \bar{G}_o/\{\pm I\}$.

10.2 The Generalized F-space

To obtain a Cartan geometry from this construction we first define a generalized F-space, by the following somewhat indirect method.

Recall that in Sect. 4.4 we defined the vector bundles $\tau : \mathcal{W}M \to M$, and $\nu^p : \mathcal{D}^p M \to M$, the line bundle of scalar densities of weight p over M; we also defined $\widetilde{\mathcal{W}}^p M \cong \mathcal{W}M \otimes \mathcal{D}^p M$ (Lemma 4.3.1). We shall be concerned here mainly with the case $p = 1/n$, and to simplify notation we shall drop the superscript p except where it takes some other value.

The following observation will explain the reason behind the use of densities of weight $1/n$. Suppose given a Riemannian metric g on M; recall from Lemma 4.5.6 that there is an associated Riemannian volume form ς_g, a section of $\mathcal{D}M \to M$, given by $\varsigma_g^0 = |\det g_{ij}|^{1/2n}$. If we set[1]

$$\mathsf{g}_{ij} = \frac{g_{ij}}{|\det g_{ij}|^{1/n}}$$

then conformally equivalent Riemannian metrics g give the same g. However, g is not itself a tensor, but rather a section of $S^2T^*M \otimes \mathcal{D}^{-2/n}M$; it is, essentially, $g \otimes \varsigma_g^{-2}$, and indeed g defines a fibre metric on $TM \otimes \mathcal{D}M \to M$.

We have the exact sequence

$$0 \to \mathcal{E} \to \mathcal{W}M \to TM \to 0$$

where $\mathcal{E} = \langle e \rangle$ is the kernel of the anchor map $\rho : \mathcal{W}M \to TM$; it has a global section e corresponding to the global vector field Υ on the volume bundle $\mathcal{V}M$ (Sect. 4.3). We therefore also have an exact sequence

$$0 \to \mathcal{E} \otimes \mathcal{D}M \to \widetilde{\mathcal{W}}M \to TM \otimes \mathcal{D}M \to 0\,.$$

We suppose given a fibre metric g on $TM \otimes \mathcal{D}M$. We can choose a local basis of sections $\{\zeta_0, \zeta_i\}$ of $\widetilde{\mathcal{W}}M \to M$, or local frame, as follows. Let $\{\zeta_0\}$ be a local section of $\mathcal{E} \otimes \mathcal{D}M \to M$, and let the ζ_i project onto a local basis of sections of $TM \otimes \mathcal{D}M$ which is orthonormal with respect to g. Any other local frame $\{\hat{\zeta}_0, \hat{\zeta}_i\}$ of $\widetilde{\mathcal{W}}M \to M$ of the same form is related to $\{\zeta_0, \zeta_i\}$ by

$$\hat{\zeta}_0 = k\zeta_0\,, \quad \hat{\zeta}_i = \theta_i\zeta_0 + \Phi_i^j\zeta_j$$

with coefficients which are local functions on M, k being nonvanishing and the Φ_i^j entries in an orthogonal matrix. As $(n+1) \times (n+1)$ matrices these frame transformations take the form

$$\begin{pmatrix} k & \theta \\ 0 & \Phi \end{pmatrix},$$

[1] In this chapter we shall distinguish between the two symbols g and g.

that is, they belong to $\mathsf{W}(n)$. This shows that for a given fibre metric g on $TM \otimes \mathcal{D}M$ we can identify $\widetilde{\mathcal{W}}M$ as a vector bundle associated with a principal bundle whose group is $\mathsf{W}(n)$.

So we have a principal bundle with group $\mathsf{W}(n)$; and any (linear) representation of $\mathsf{W}(n)$ will define a vector bundle by the associated bundle construction. In particular the representation

$$\begin{pmatrix} k & \theta \\ 0 & \Phi \end{pmatrix} \mapsto \begin{pmatrix} k & \theta & p \\ 0 & \Phi & v \\ 0 & 0 & k^{-1} \end{pmatrix}$$

of $\mathsf{W}(n)$ on $\mathbb{R}^{n+2}$ determines a vector bundle of rank $(n+2)$ over M, which is equipped with a fibre metric of signature $(n+1, 1)$ and a filtration by rank $(n+1)$ and rank 1 sub-bundles. Following [10] we call this bundle the *conformal tractor bundle*, and denote it by $\mathcal{T}$ and its fibre metric by h. The rank 1 sub-bundle is denoted by $\mathcal{T}_0$. Its fibres are null lines with respect to h. The rank $(n+1)$ sub-bundle is $\mathcal{T}_0^\perp$; $\mathcal{T}_0 \subset \mathcal{T}_0^\perp \subset \mathcal{T}$. We may identify $\mathcal{T}_0^\perp$ with $\widetilde{\mathcal{W}}M$ and $\mathcal{T}_0$ with $\mathcal{E} \otimes \mathcal{D}M$.

We may introduce fibre coordinates (z^0, z^i, z^∞) on $\mathcal{T}$ such that $\mathcal{T}_0^\perp$ is given by $z^\infty = 0$. We may then identify (z^0, z^i) as fibre coordinates $(\tilde{w}^0, \tilde{w}^i)$ on $\widetilde{\mathcal{W}}M$ coming from coordinates (x^i) on M. We cannot of course assume that the coordinates are quasi-orthonormal: we do however require that the fibre metric, expressed as a quadratic form in these coordinates, is given by $\mathsf{g}_{ij} z^i z^j - 2z^0 z^\infty$. We say that such coordinates are adapted to h and call them adapted coordinates. We shall derive the coordinate transformation formulæ for adapted coordinates.

Under a change of coordinates $(x^i) \to (\hat{x}^i)$ we must have

$$\begin{aligned} \hat{z}^0 &= |J|^{-1/n}\left(z^0 - \frac{1}{n}\psi_i z^i + \varphi^0 z^\infty\right) \\ \hat{z}^i &= |J|^{-1/n} J^i_j \left(z^j + \varphi^j z^\infty\right) \\ \hat{z}^\infty &= \varphi^\infty z^\infty \end{aligned}$$

for some local functions φ^A. We require that

$$\hat{\mathsf{g}}_{ij}\hat{z}^i\hat{z}^j - 2\hat{z}^0\hat{z}^\infty = \mathsf{g}_{ij}z^i z^j - 2z^0 z^\infty .$$

We must therefore have

$$\begin{aligned} \varphi^\infty &= |J|^{1/n} \\ \varphi^j &= -\frac{1}{n}\mathsf{g}^{jk}\psi_k \\ \varphi^\infty &= \frac{1}{2n^2}(\mathsf{g}^{ij}\psi_i\psi_j) , \end{aligned}$$

and the transformation formulæ are

$$\hat{z}^0 = |J|^{-1/n}\left(z^0 - \frac{1}{n}\psi_i z^i + \frac{1}{2n^2}(\mathsf{g}^{ij}\psi_i\psi_j)z^\infty\right)$$
$$\hat{z}^i = |J|^{-1/n} J^i_j\left(z^j - \frac{1}{n}\mathsf{g}^{jk}\psi_k z^\infty\right)$$
$$\hat{z}^\infty = |J|^{1/n} z^\infty$$

where $\psi_i = \partial \log |J|/\partial x^i$. Here $\mathsf{g}_{ik}\mathsf{g}^{kj} = \delta^j_i$, and the g^{ij} are the components of a section of $S^2T^*M \otimes \mathcal{D}^{2/n}M$. These formulæ are very similar to, though not quite identical with, the formulæ proposed by Thomas in [43]. It appears that the main difference is that Thomas has in effect used a different weighting.

In terms of these coordinates, $\mathcal{T}_0$ is spanned by $(1, 0, 0)$, while as noted earlier $\mathcal{T}_0^\perp$ is $z^\infty = 0$. The fibre metric h on $\mathcal{T}$ defines a fibrewise quadratic form on $\mathcal{T}_0^\perp$ by restriction, but it is not a metric since $\mathcal{T}_0$ is null: in fact the coordinate expression of the restriction is just $\mathsf{g}_{ij}z^iz^j$. But from the identification of $\mathcal{T}_0^\perp$ with $\widetilde{\mathcal{W}}M$ we have a projection $\mathcal{T}_0^\perp \to TM \otimes \mathcal{D}M$, namely $(z^0, z^i) \mapsto (z^i)$, whose kernel is just $\mathcal{T}_0$: it follows that by restriction and projection h determines a fibre metric on $TM \otimes \mathcal{D}M$, whose coordinate expression can again be written $\mathsf{g}_{ij}z^iz^j$. Now let τ be any nonvanishing section of $\mathcal{D}M \to M$: then for any section ζ of $TM \otimes \mathcal{D}M \to M$, we can write $\zeta = X \otimes \tau$ where X is a section of $TM \to M$ (a vector field on M); and we have $\mathsf{g}_{ij}\zeta^i\zeta^j = \tau^2\mathsf{g}_{ij}X^iX^j = g_{ij}X^iX^j$ where the g_{ij} are the components of a genuine Riemannian metric on M. Moreover, the Riemannian metrics derived from different sections τ are conformally related. Conversely, given a metric g on M, if we set

$$\mathsf{g}_{ij} = \frac{g_{ij}}{|\det g_{ij}|^{1/n}}$$

then as we pointed out before, the g_{ij} are the components of a section of $S^2T^*M \otimes \mathcal{D}^{(-2/n)}M$, and $g_{ij} = \tau^2\mathsf{g}_{ij}$ with $\tau = |\det g_{ij}|^{1/2n}$; moreover, conformally equivalent Riemannian metrics g give the same g.

Given a conformal equivalence class of Riemannian metrics on M we may form, first the tractor bundle $\mathcal{T}$ with its fibre metric h, and then a fibre bundle $E \to M$ whose fibre E_x is the space of null lines in $\mathcal{T}_x$. This is the required generalized F-space. It has a global section, whose image is the bundle of null lines $\mathcal{T}_0$, given in terms of fibre coordinates by $y^i = 0$, and it is canonically soldered to M along this section.

10.3 The Cartan Connection

The tractor bundle $\mathcal{T} \to M$ is a vector bundle and is equipped with a fibre metric h. We may therefore form the Lie groupoid $\mathcal{G}^\mathcal{T}$ of (linear) fibre morphisms of $\mathcal{T}$ which are orthogonal with respect to h. The corresponding Lie algebroid $A\mathcal{G}^\mathcal{T}$ may be represented by projectable vector fields ξ on $\mathcal{T}$ such that $\mathcal{L}_\xi \hat{h} = 0$ where $\hat{h}$ is the fibre-quadratic function on $\mathcal{T}$ defined by h, as we showed in Proposition 4.5.2 and Lemma 4.5.3. We shall write $\mathcal{A}^{\mathcal{T}}$ for this Lie algebroid of vector fields.

With respect to fibre coordinates (z^A), $A, B = 0, 1, 2, \ldots, \infty$, and with

$$\xi = X^i \frac{\partial}{\partial x^i} + Y^A_B z^B \frac{\partial}{\partial z^A},$$

the condition $\mathcal{L}_\xi \hat{h} = 0$ is

$$X^i \frac{\partial h_{AB}}{\partial x^i} + h_{AC} Y^C_B + h_{CB} Y^C_A = 0 .$$

Recalling that $h_{0\infty} = h_{\infty 0} = -1$, $h_{ij} = \mathrm{g}_{ij}$, $h_{AB} = 0$ otherwise, and taking particular values for A and B in turn, we find

$$\begin{array}{ll}
(AB) = (00) & Y^\infty_0 = 0 \\
(AB) = (0i) & Y^\infty_i = \mathrm{g}_{ij} Y^J_0 \\
(AB) = (0\infty) & Y^0_0 + Y^\infty_\infty = 0 \\
(AB) = (ij) & X^k \dfrac{\partial \mathrm{g}_{ij}}{\partial x^k} + \mathrm{g}_{kj} Y^k_i + \mathrm{g}_{ik} Y^k_j = 0 \\
(AB) = (i\infty) & Y^0_i = \mathrm{g}_{ij} Y^j_\infty \\
(AB) = (\infty\infty) & Y^0_\infty = 0 .
\end{array}$$

With the exception of the case $(AB) = (ij)$ these conditions just reproduce those for the Lie algebra, while the case $(AB) = (ij)$ reproduces, mutatis mutandis, the condition for a Riemannian structure.

We are actually interested only indirectly in $\mathcal{T}$ and $\mathcal{G}^{\mathcal{T}}$. Our primary interest is the bundle $E \to M$ whose fibre E_x over $x \in M$ is the space of lines through the origin in $\mathcal{T}_x$ which are null with respect to h_x. We take for $\mathcal{G}$ the set of all Möbius transformations between fibres of E. The elements of $\mathcal{G}^{\mathcal{T}}$ map fibres of E to fibres of E, and as such induce Möbius transformations: that is, $\mathcal{G}$ is essentially the restriction of $\mathcal{G}^{\mathcal{T}}$ to E. Likewise, the elements of $\mathcal{A}^{\mathcal{T}}$ are tangent to null cones, and therefore on restriction coincide with elements of $\mathcal{A}^G \cong A\mathcal{G}$ (where G is the Möbius group).

We introduce local coordinates (y^i) on E, in a neighbourhood of the image of the attachment section, by setting $y^i = z^i/z^0$. Then $\mathcal{A}^G$ consists of vector fields along fibres of $E \to M$ which when expressed in terms of the y^i take the form

$$X^i \frac{\partial}{\partial x^i} + \left(Y^i + Y^i_j y^j + \tfrac{1}{2}(Z_j \delta^i_k + Z_k \delta^i_j - \mathrm{g}_{jk} Z^i) y^j y^k\right) \frac{\partial}{\partial y^i}$$

with $\mathrm{g}_{ik} Y^k_j + \mathrm{g}_{kj} Y^k_i = \mu \mathrm{g}_{ij}$; $\mathcal{A}^G_o$ consists of those for which $Y^i = 0$.

We can now address the question of the choices of infinitesimal connections $\tilde{\gamma}$ for $\mathcal{A}^{\mathcal{T}}$ and γ for $\mathcal{A}^{G}$. We put

$$\tilde{\gamma}\left(\frac{\partial}{\partial x^i}\right) = \frac{\partial}{\partial x^i} - \Pi^A_{iB} z^B \frac{\partial}{\partial z^A};$$

for each i, $-\Pi^A_{iB}$ must satisfy the conditions for $\tilde{\gamma}$ to take its values in $\mathcal{A}^{\mathcal{T}}$: that is,

$$\Pi^0_{i\infty} = \Pi^\infty_{i0} = 0\,, \quad \Pi^0_{ij} = \mathfrak{g}_{jk}\Pi^k_{i\infty}\,, \quad \Pi^\infty_{ij} = \mathfrak{g}_{jk}\Pi^k_{i0}\,,$$

while

$$\frac{\partial \mathfrak{g}_{jk}}{\partial x^i} = \mathfrak{g}_{lk}\Pi^l_{ij} + \mathfrak{g}_{jl}\Pi^l_{ik}\,.$$

The infinitesimal connection $\tilde{\gamma}$ on $\mathcal{A}^{\mathcal{T}}$ defines an infinitesimal connection γ on $\mathcal{A}^{G}$. In terms of the fibre coordinates y^i, using the translation rules given in the previous section we find that $\gamma(\partial/\partial x^i)$ is

$$\frac{\partial}{\partial x^i} - \left(\Pi^j_{i0} + (\Pi^j_{ik} - \Pi^0_{i0}\delta^j_k)y^k - \tfrac{1}{2}(\Pi^0_{ik}\delta^j_l + \Pi^0_{il}\delta^j_k - \mathfrak{g}_{kl}\mathfrak{g}^{jm}\Pi^0_{im})y^k y^l\right)\frac{\partial}{\partial y^j}\,.$$

In order that this infinitesimal connection respects the soldering we must take $\Pi^j_{i0} = \delta^j_i$. We assume that Π^i_{jk} is symmetric, which forces

$$\Pi^k_{ij} = \tfrac{1}{2}\mathfrak{g}^{kl}\left(\frac{\partial \mathfrak{g}_{jl}}{\partial x^i} + \frac{\partial \mathfrak{g}_{il}}{\partial x^j} - \frac{\partial \mathfrak{g}_{ij}}{\partial x^l}\right).$$

If in addition we take $\Pi^0_{i0} = 0$ we ensure that γ is torsion-free. Finally, we take $\Pi^0_{ij} = \mathfrak{P}_{ij}$, the Schouten 'tensor' of $\mathfrak{g}_{ij}$, so that γ is normal.

For the record we quote the coordinate transformation formulæ for the key coefficients:

$$\begin{aligned} J^l_i J^m_j \hat{\Pi}^k_{lm} &= J^k_l\left(\Pi^l_{ij} + \frac{1}{n}(\delta^l_i\psi_j + \delta^l_j\psi_i - \mathfrak{g}_{ij}\mathfrak{g}^{lm}\psi_m)\right) - J^k_{ij}\,; \\ J^j_i \hat{\Pi}^k_{j0} &= J^k_j \Pi^j_{i0}\,; \\ \hat{\Pi}^0_{i0} &= \bar{J}^j_i \Pi^0_{j0}\,. \end{aligned}$$

These confirm the validity of the choices we have made.

Let g_{ij} be any metric of the conformal class. By expressing $\mathfrak{g}_{ij}$ in terms of g_{ij} we find that

$$\begin{aligned} \Pi^k_{ij} &= \Gamma^k_{ij} + \delta^k_j\omega_i + \delta^k_i\omega_j - \mathfrak{g}_{ij}\mathfrak{g}^{kl}\omega_l \\ &= \Gamma^k_{ij} + \delta^k_j\omega_i + \delta^k_i\omega_j - g_{ij}g^{kl}\omega_l \end{aligned}$$

where the Γ_{ij}^k are the Levi-Civita connection coefficients of g_{ij}, and where

$$\omega_i = \frac{\partial f}{\partial x^i} = -\frac{1}{n}\Gamma_i, \quad f = -\frac{1}{2n}\log|\det g_{jk}|;$$

that is, in any coordinate neighbourhood we may formally regard $|\det g_{ij}|^{-1/n}$ as a conformal factor taking g_{ij} to $\mathfrak{g}_{ij}$, and then Π_{ij}^k is related to Γ_{ij}^k in the corresponding way. It follows that

$$\begin{aligned}\mathfrak{P}_{ij} &= P_{ij} + \nabla_i\omega_j - \omega_i\omega_j + \tfrac{1}{2}g_{ij}g^{kl}\omega_k\omega_l \\ &= P_{ij} + \frac{\partial\omega_j}{\partial x^i} - \Gamma_{ij}^k\omega_k - \omega_i\omega_j + \tfrac{1}{2}\mathfrak{g}_{ij}\mathfrak{g}^{kl}\omega_k\omega_l,\end{aligned}$$

where P_{ij} is the Schouten tensor of g_{ij}.

The expression for $\gamma(\partial/\partial x^i)$ for the normal infinitesimal Cartan conformal connection γ is

$$\frac{\partial}{\partial x^i} - \left(\delta_i^j + \Pi_{ik}^j y^k + \tfrac{1}{2}\left(\mathfrak{P}_{ik}\delta_l^j + \mathfrak{P}_{il}\delta_k^j - \mathfrak{g}_{kl}\mathfrak{g}^{jm}\mathfrak{P}_{im}\right)y^k y^l\right)\frac{\partial}{\partial y^j}.$$

We call the Cartan geometry with this infinitesimal connection the *full Cartan conformal geometry* of the conformal class of metrics.

The fact that Π_{ij}^k and $\mathfrak{P}_{ij}$ are formally related to their counterparts Γ_{ij}^k and P_{ij} for any metric in the conformal class as if by a conformal transformation means that the y-linear component of the curvature of γ is the conformal curvature tensor C_{ijk}^l of the conformal class, just as in Sect. 8.4. The quadratic part of the curvature of γ is

$$\frac{1}{2(n-2)}\left(\mathfrak{C}_{lmk}\delta_j^i + \mathfrak{C}_{ljk}\delta_m^i - \mathfrak{g}_{jm}\mathfrak{g}^{ip}\mathfrak{C}_{lpk}\right)y^j y^m\frac{\partial}{\partial y^i}$$

where $\mathfrak{C}_{ijk}$ is the Cotton 'tensor' formed from Π_{ij}^k. By a similar argument using the known conformal transformation properties of the Cotton tensor we find that

$$\mathfrak{C}_{ijk} = C_{ijk} + (n-2)\omega_l C_{ijk}^l$$

where C_{ijk} is the Cotton tensor of the metric g_{ij}.

10.4 The Conformal Tractor Connection

The original infinitesimal connection $\tilde{\gamma}$ takes its values in $\mathcal{A}^{\mathcal{T}}$. But we may think of it as defining a horizontal distribution on $T\mathcal{T}$, and therefore an ordinary linear connection on the tractor bundle $\mathcal{T}$ (which is of course a vector bundle). In this guise we call it the *conformal tractor connection*.

Let us denote the corresponding covariant derivative operator by $\tilde{\nabla}$. Then for any section ζ of $\mathcal{T} \to M$, with coordinate representation ζ^A, and with the choices above for the Π^A_{iB}, we have

$$\tilde{\nabla}_i \begin{pmatrix} \zeta^0 \\ \zeta^j \\ \zeta^\infty \end{pmatrix} = \begin{pmatrix} \partial\zeta^0/\partial x^i - \mathfrak{P}_{ij}\zeta^j \\ \partial\zeta^j/\partial x^i + \Pi^j_{ik}\zeta^k + \delta^j_i\zeta^0 - \mathsf{g}^{jk}\mathfrak{P}_{ik}\zeta^\infty \\ \partial\zeta^\infty/\partial x^i + \mathsf{g}_{ij}\zeta^j \end{pmatrix}.$$

This is essentially the conformal tractor connection of Bailey et al. [2]. It defines a covariant derivative on the dual bundle $\mathcal{T}^*$, where

$$\tilde{\nabla}_i \begin{pmatrix} \zeta_0 \\ \zeta_j \\ \zeta_\infty \end{pmatrix} = \begin{pmatrix} \partial\zeta_0/\partial x^i - \zeta_i \\ \partial\zeta_j/\partial x^i - \Pi^k_{ij}\zeta_k + \mathfrak{P}_{ij}\zeta_0 - \mathsf{g}_{ij}\zeta_\infty \\ \partial\zeta_\infty/\partial x^i + \mathsf{g}^{jk}\mathfrak{P}_{ik}\zeta_j \end{pmatrix}.$$

It is a curious feature of this connection that the expression for $\partial\zeta_\infty/\partial x^i$ is actually determined by the other two expressions, in the following sense: if a local section $\zeta = (\zeta_0, \zeta_j, \zeta_\infty)$ is such that $(\tilde{\nabla}_i\zeta)_0 = (\tilde{\nabla}_i\zeta)_j = 0$, then $(\tilde{\nabla}_i\zeta)_\infty = 0$ also, that is, ζ is parallel. This is seen as follows. We denote the formal covariant derivative with respect to $\Pi^{\,i}_{jk}$ with a bar; by construction, $\mathsf{g}_{ij|k} = 0$. We then have, by assumption,

$$\mathsf{g}_{jk}\zeta_\infty = \zeta_{k|j} + \mathfrak{P}_{jk}\zeta_0\,, \qquad \zeta_{0|i} = \zeta_i\,,$$

and $\zeta_{k|j} = \zeta_{j|k}$. By differentiating again we obtain

$$\mathsf{g}_{jk}\zeta_{\infty|i} - \mathsf{g}_{ik}\zeta_{\infty|j} = -\mathfrak{R}^l_{kij}\zeta_l + \mathfrak{P}_{jk}\zeta_i - \mathfrak{P}_{ik}\zeta_j + (\mathfrak{P}_{jk|i} - \mathfrak{P}_{ik|j})\zeta_0$$

where $\mathfrak{R}^l_{ijk}$ is the curvature 'tensor' of $\Pi^{\,i}_{jk}$. It follows that

$$(n-1)\zeta_{\infty|i} = \mathsf{g}^{jk}\big(-\mathfrak{R}^l_{kij}\zeta_l + \mathfrak{P}_{jk}\zeta_i - \mathfrak{P}_{ik}\zeta_j + (\mathfrak{P}_{jk|i} - \mathfrak{P}_{ik|j})\zeta_0\big).$$

The Schouten 'tensor' is given by

$$\mathfrak{P}_{jk} = \frac{1}{n-2}\left(\mathfrak{R}_{jk} - \frac{1}{2(n-1)}\mathfrak{R}\mathsf{g}_{jk}\right)$$

(see Sects. 4.5 and 8.4). The curvature 'tensor' satisfies the second Bianchi identity, from which one deduces that

$$\mathsf{g}^{jk}(\mathfrak{P}_{jk|i} - \mathfrak{P}_{ik|j}) = 0.$$

Moreover, $\mathsf{g}^{jk}\mathfrak{P}_{jk} = \mathfrak{R}/2(n-1)$, while $\mathsf{g}^{jk}\mathfrak{R}^l_{kij} = \mathsf{g}^{lm}\mathfrak{R}_{im}$, whence

$$\mathsf{g}^{jk}(-\mathfrak{R}^l_{kij}\zeta_l + \mathfrak{P}_{jk}\zeta_i) = \mathsf{g}^{jm}\left(-\mathfrak{R}_{im} + \frac{1}{2(n-1)}\mathfrak{R}\mathsf{g}_{im}\right)\zeta_j$$
$$= -(n-2)\mathsf{g}^{jk}\mathfrak{P}_{ik}\zeta_j\,,$$

so that

$$\frac{\partial\zeta_\infty}{\partial x^i} = -\mathsf{g}^{jk}\mathfrak{P}_{ik}\zeta_j$$

as claimed.

Let $\tau^* : \mathcal{T}^* \to M$ denote the dual of the tractor bundle. The conformal tractor connection determines a splitting of

$$0 \to \mathcal{T}^* \otimes_M T^*M \to J^1\tau^* \to \mathcal{T}^* \to 0\,,$$

which is given by

$$z_{0i} = z_i\,, \quad z_{ij} = \Pi^k_{ij}z_k + \mathfrak{P}_{ij}z_0 - \mathsf{g}_{ij}z_\infty\,, \quad z_{\infty i} = -\mathfrak{P}_{ik}\mathsf{g}^{jk}z_j\,.$$

By the previous result one can in effect ignore the last of these equations. Now $\mathcal{T}$ contains $\widetilde{\mathcal{W}}M$ as a codimension 1 sub-bundle, given in coordinates by $z^\infty = 0$. Let us denote by $\widetilde{\mathcal{W}}^\circ M \subset \mathcal{T}^*$ the fibrewise annihilator of $\widetilde{\mathcal{W}}M$: it is a sub-bundle of $\mathcal{T}^*$ of fibre dimension 1, given in coordinates by $z_0 = z_i = 0$. Then $\mathcal{T}^*/\widetilde{\mathcal{W}}^\circ M$ is isomorphic to $\widetilde{\mathcal{W}}^*M$, which is $J^1\nu$ (Lemma 4.3.2). The tractor connection defines an injection $\mathcal{T}^* \to J^2\nu$,

$$(z_0, z_i, z_\infty) \mapsto (z_0, z_i, z_{ij}) \quad \text{where } z_{ij} = \Pi^k_{ij}z_k + \mathfrak{P}_{ij}z_0 - \mathsf{g}_{ij}z_\infty,$$

fibred over the identity on $J^1\nu$. That is to say, via the conformal tractor connection we may identify $\mathcal{T}^*$ with a vector sub-bundle of $J^2\nu$. Under this identification, $\widetilde{\mathcal{W}}^\circ M$ corresponds to the line bundle $L \subset J^2\nu$ which is generated by the section of $S^2T^*M \otimes \mathcal{D}M$ determined by g_{ij}, that is, by the conformal structure. We may write the equation for z_{ij} formally as

$$(\nabla_i\nabla_j - \mathfrak{P}_{ij})z_0 \propto \mathsf{g}_{ij}\,,$$

or equivalently

$$\text{trace-free part of } (\nabla_i\nabla_j - \mathfrak{P}_{ij})z_0 = 0\,,$$

which is the formulation of Bailey et al. [2] apart from the sign, which is due to differing conventions.

Chapter 11
Developments and Geodesics

In this chapter we recall and extend the concept of the development of a curve from Chap. 3. This will allow us to explain how the base manifold M of a Cartan geometry can be rolled, along a curve in it, on a fibre of E without slipping or twisting; the fibre of E can of course be considered as representative of the standard homogeneous-space fibre F. In suitable circumstances the same notion, of development, can be employed to define a geodesic in a Cartan geometry.

We shall also, as announced earlier, show that a vertical infinitesimal symmetry of an infinitesimal Cartan geometry must be trivial.

11.1 Developments

In Sect. 3.4 we introduced the idea of the development of a curve c, given in the base manifold of a fibre bundle $\pi : E \to M$, into the fibre $E_{c(0)}$; the construction made use of an attachment section $\varsigma : M \to E$ and a path connection Γ on the Lie groupoid of fibre morphisms of the bundle. The developed curve c_Γ was given by $c_\Gamma(t) = \big(c^\Gamma(t)\big)^{-1}\varsigma(c(t))$. Of course we have these ingredients available in a finite Cartan geometry.

In a Cartan geometry the manifold E is soldered to M along the attachment section $\varsigma : M \to E$, which is to say that there is a map $\mathfrak{s} : TM \to TE$, the soldering map, such that if $v \in T_x M$ then $\mathfrak{s}(v) \in T_{\varsigma(x)}E$, and that the restriction $\mathfrak{s}|_{T_x M}$ is an isomorphism $T_x M \to T_{\varsigma(x)} E_x$. The existence of the soldering map is a consequence of the nondegeneracy condition on the infinitesimal connection γ (see Sect. 6.5)

To see how soldering relates to the concept of the development of a curve we give an alternative illustration of the significance of the nondegeneracy condition.

Proposition 11.1.1 *With the soldering map $\mathfrak{s}$ defined as above, for each curve c in M the tangent vector $\dot{c}_\Gamma(0) \in T_{\varsigma(c(0))}E_{c(0)}$ to the developed curve c_Γ satisfies*

M. Crampin and D. Saunders, *Cartan Geometries and their Symmetries*,
Atlantis Studies in Variational Geometry 4, DOI 10.2991/978-94-6239-192-5_11

$$\dot{c}_\Gamma(0) = \mathfrak{s}(\dot{c}(0)),$$

where Γ *is the path connection corresponding to the infinitesimal connection* γ.

We say that the path connection Γ *respects the soldering* $\mathfrak{s}$ if, for each curve c, $\dot{c}_\Gamma(0) = \mathfrak{s}(\dot{c}(0))$. Thus the path connection corresponding to a nondegenerate infinitesimal connection respects the soldering the infinitesimal connection defines.

Proof If $\dot{c}(0) = 0$ then the result is immediate, so suppose $\dot{c}(0) \neq 0$. Then c is injective in a neighbourhood of zero, so by reparametrization we may suppose that it is injective on $[0, 1]$.

Set $c(0) = x$. Let $\boldsymbol{a} \in A_x\mathcal{G}_o$ be such that $\boldsymbol{a}_{\varsigma(x)} = \varsigma_*(\dot{c}(0))$: then

$$\mathfrak{s}(\dot{c}(0)) = \omega_x(\boldsymbol{a})_{\varsigma(x)} = \big(\boldsymbol{a} - \gamma(\dot{c}(0))\big)_{\varsigma(x)} = \varsigma_*(\dot{c}(0)) - \gamma(\dot{c}(0))_{\varsigma(x)}.$$

Now in terms of a local trivialization of $\pi : E \to M$ and the consequential identification of c^*E with $[0, 1] \times F$ that we used in the discussion leading up to Lemma 3.4.2 we have

$$\varsigma_*(\dot{c}(0)) \cong \left.\frac{\partial}{\partial t}\right|_{t=0}, \quad \gamma(\dot{c}(0))_{\varsigma(x)} \cong C(0, o),$$

and therefore

$$\mathfrak{s}(\dot{c}(0)) \cong \bar{C}(0, o).$$

On the other hand, $c_\Gamma(t) \cong (0, \bar{c}_\Gamma(t))$ where $\bar{c}_\Gamma(t) = \Psi(t)o$; and

$$\dot{\bar{c}}_\Gamma(t) = \frac{d}{dt}\big(\Psi(t)o\big)_{t=0} = \bar{C}(0, o),$$

and the result follows. □

Corollary 11.1.2 *If the infinitesimal connection is nondegenerate then the development of a regular curve is regular in a neighbourhood of zero.* □

The development c_Γ of a curve c in M is by definition a curve in $E_{c(0)}$. But this clearly involves an arbitrary choice of which fibre to develop into; and it is of interest to compare the results of developing into different fibres—one would expect that developments into different fibres would be rather simply related.

Let $\tilde{c}$ be the curve in M given by $\tilde{c}(s) = c(s + t)$ for some t, which will be fixed throughout the argument. Then by the reparametrization formula for path connections

$$c^\Gamma(s + t) = \tilde{c}^\Gamma(s) \circ c^\Gamma(t)$$

(where $\tilde{c}^\Gamma(s)$ is a fibre morphism $E_{c(t)} \to E_{c(s+t)}$ and $c^\Gamma(t)$ is a fibre morphism $E_{c(0)} \to E_{c(t)}$). It follows that the developments are related by

$$\tilde{c}_\Gamma(s) = c^\Gamma(t)\big(c_\Gamma(s + t)\big)$$

($\tilde{c}_\Gamma$ is a curve in $E_{c(t)}$, c_Γ a curve in $E_{c(0)}$, and $c^\Gamma(t)$ maps the latter fibre to the former one).

Now $c_\Gamma(0) = \varsigma\big(c(0)\big)$; and similarly $\tilde{c}(0) = \varsigma\big(\tilde{c}(0)\big) = \varsigma\big(c(t)\big)$. Furthermore,

$$\dot{c}_\Gamma(t) = \big(c^\Gamma(t)^{-1}\big)_* \dot{\tilde{c}}_\Gamma(0).$$

Since E is soldered to M along the image of ς and the path connection respects the soldering, for each t the map

$$C_t = \big(c^\Gamma(t)^{-1}\big)_* \circ \mathfrak{s} : T_{c(t)}M \to T_{c_\Gamma(t)}E_{c(0)}$$

is a linear isomorphism, and satisfies

$$C_t\big(\dot{c}(t)\big) = C_t\big(\dot{\tilde{c}}(0)\big) = \big(c^\Gamma(t)^{-1}\big)_* \mathfrak{s}\big(\dot{\tilde{c}}(0)\big) = \big(c^\Gamma(t)^{-1}\big)_* \dot{\tilde{c}}_\Gamma(0) = \dot{c}_\Gamma(t).$$

That is to say, to any curve $c : [0, 1] \to M$ there corresponds a curve $c_\Gamma : [0, 1] \to E_{c(0)}$ (its development), and for each $t \in [0, 1]$ we can canonically identify $T_{c(t)}M$ with $T_{c_\Gamma(t)}E_{c(0)}$ (by C_t) in such a way that $\dot{c}(t)$ is identified with $\dot{c}_\Gamma(t)$. We can imagine this correspondence as an active process, which we describe as *rolling M without slipping or twisting* along the curve c on the fibre $E_{c(0)}$ (see, for example, [36]). We may further identify $E_{c(0)}$ with the standard fibre F, and thereby roll M along c on F, though in general the identification of $E_{c(0)}$ with F will not be canonical but determined only up to the overall action of an element of G_o.

11.2 Vertical Infinitesimal Symmetries

Suppose that we have, not a finite Cartan geometry, but an infinitesimal geometry with Lie algebroids $\mathcal{A}$ and $\mathcal{A}_o$, attachment section ς and infinitesimal connection γ. Although we need not have a Lie groupoid of fibre morphisms, we can nevertheless construct diffeomorphisms between neighbourhoods of the attachment section in different fibres by using a version of parallel translation.

Let $\pi : E \to M$ be the bundle of the infinitesimal geometry, with standard fibre F where $\dim F = \dim M$. Let the horizontal distribution on E corresponding to γ be $v \mapsto v^{\mathsf{h}}$, $v \in T_xM$, $v^{\mathsf{h}} \in T_yE$, $x = \pi(y)$. The nondegeneracy condition is that $(T_xM)^{\mathsf{h}} \cap \varsigma_*(T_xM) = \{0\}$, both being subspaces of $T_{\varsigma(x)}E$.

Let I be an interval containing zero, and let c be a curve $I \to M$ which we assume is injective and has nowhere vanishing tangent. Then for any $y \in E_x$, $x = c(0)$, there is a unique curve $c^{\mathsf{h}}_y : I \to E$, the horizontal lift of c through y, such that $\pi \circ c^{\mathsf{h}}_y = c$, $c^{\mathsf{h}}_y(0) = y$, and $\dot{c}^{\mathsf{h}}_y(t) = \dot{c}(t)^{\mathsf{h}}$. We could, for example, construct c^{h}_y by realising c as an integral curve of some locally defined vector field X on M, and taking for c^{h}_y the integral curve of X^{h} through y. We may define a map $\tau_c(t) : E_x \to E_{c(t)}$

by $\tau_c(t)(y) = c^{\mathsf{h}}_y(t)$. Then for t sufficiently small $\tau_c(t)$ is a diffeomorphism of a neighbourhood U_y of any given $y \in E_x$ with a neighbourhood of $c^{\mathsf{h}}_y(t) \in E_{c(t)}$: this would be obvious from the vector field construction.

Take $y = \varsigma(x)$. For all t sufficiently small, $\tau_c(t)(U_{\varsigma(x)})$ contains $\varsigma(c(t))$. We may define a curve c_γ in E_x, the *development* of c, by $c_\gamma(t) = \tau_c(t)^{-1}(\varsigma(c(t)))$. It is evident that if γ is derived from a path connection Γ on a Lie groupoid of fibre morphisms then $c_\gamma(t) = c_\Gamma(t)$ for all t where both are defined.

We are now in a position to prove our advertised result about vertical infinitesimal symmetries.

Theorem 11.2.1 *Let ξ be a section of the algebroid $\mathcal{A}_o$ regarded as a vector field on E, and suppose that $\mathcal{L}_\xi X^{\mathsf{h}} = 0$ for every vector field X on M: that is to say, that ξ is an infinitesimal symmetry of the infinitesimal Cartan geometry. Suppose further that ξ is vertical. If the standard fibre F is connected then $\xi = 0$.*

Proof Let ψ_s be the flow of ξ. Then for any $y \in E$, ψ_s maps the horizontal subspace at $y \in E_{\pi(y)}$ to the horizontal subspace at $\psi_s(y) \in E_{\pi(y)}$, and therefore for the horizontal lift of any curve c,

$$c^{\mathsf{h}}_{\psi_s(y)} = \psi_s \circ c^{\mathsf{h}}_y.$$

It follows that

$$\tau_c(t) \circ \psi_s = \psi_s \circ \tau_c(t).$$

Now ξ is tangent to the image of ς (because it is a section of $\mathcal{A}_o$) and therefore vanishes on the image of ς (because it is vertical). Thus ψ_s is the identity on the image of ς, so for the development c_γ,

$$\psi_s(c_\gamma(t)) = \psi_s\big(\tau_c(t)^{-1}\big(\varsigma c(t)\big)\big) = \tau_c(t)^{-1}\big(\psi_s \varsigma c(t)\big) = \tau_c(t)^{-1}\big(\varsigma c(t)\big) = c_\gamma(t).$$

Thus the flow of a vertical infinitesimal symmetry ξ leaves developments fixed.

Next, we need to see that there is a neighbourhood of $\varsigma(M)$ such that each point of the neighbourhood lies on the development of some curve in M. But the argument in the proof of Proposition 6.7.2 for finite symmetries, using the exponential map of an arbitrary symmetric linear connection on M, applies equally to developments using an infinitesimal connection, so that for each $x \in M$ there is a neighbourhood $N_x \subset E_x$ of $\varsigma(x)$ such that every point $y \in N_x$ lies on the development of a curve in M. Then ψ_s is the identity on N_x, and ξ vanishes on N_x.

Now by definition the Lie algebra action of an infinitesimal Cartan geometry must be transitive and effective; and Proposition 6.3.2 asserts that, when F is connected, an action which is transitive and effective must be locally effective. Thus as ξ vanishes on an open subset of E_x it must vanish everywhere on E_x, and this for every x. Thus ξ vanishes everywhere on E. □

11.3 Geodesics in Cartan Affine and Projective Geometry

If the typical fibre F admits the concept of a 'straight line' then we say that a curve c is a *geodesic* of Γ if its development in $E_{c(0)}$ is a straight line. There are two obvious cases where this applies: affine and projective geometry (for this purpose we can treat Riemannian geometry as a special case of affine geometry).

It follows from the reparametrization property that being a geodesic is a local property, so that if $I \subset \mathbb{R}$ is an arbitrary nonempty interval then $c : I \to M$ is a geodesic if, and only if, for each $t \in I$ there is a neighbourhood $N_t \subset I$ such that the restriction $c|_{N_t}$ is a geodesic.

Proposition 11.3.1 *The geodesics of a Cartan affine geometry are those of the associated linear connection.*

Proof We have to consider the development of a curve in M into $T_x M$ where x is some point on it. We do this by using the method which culminates in Lemma 3.4.2, adapted to the present context.

Take a curve $c : I \to M$, with $0 \in I$. Choose a coordinate neighbourhood of $c(0)$, and write $c(t) = (x^i(t))$ for t sufficiently close to 0. Consider c^*TM, the pull-back of TM to I defined by c. We define a vector field C on this manifold by

$$C(t, y) = \gamma(\dot{c}(t)) = \frac{\partial}{\partial t} - \left(\dot{x}^i + \Gamma^i_{jk}\dot{x}^j y^k\right)\frac{\partial}{\partial y^i}, \quad y \in T_{c(t)}M.$$

According to Lemma 3.4.2 we have to solve the equation

$$l_{\Psi(t)^{-1}*} \left.\frac{d\Psi}{dt}\right|_t = \bar{C}(t)$$

where $t \mapsto \Psi(t)$ is a curve in the affine group and

$$\bar{C}^i_0 = \dot{x}^i, \quad \bar{C}^i_j = \Gamma^i_{kj}\dot{x}^k.$$

Thus the components of Ψ must satisfy

$$\frac{d\Psi^i_0}{dt} = \Psi^i_j \dot{x}^j$$
$$\frac{d\Psi^i_j}{dt} = \Psi^i_k \Gamma^k_{lj}\dot{x}^l.$$

The development is the curve in $T_{c(0)}M \cong \{0\} \times \mathbb{R}^n$ given by $t \mapsto \Psi(t)o$ (where o is the origin of $\mathbb{R}^n$), that is, $y^i(t) = \Psi^i_0(t)$. But

$$\frac{d^2\Psi^i_0}{dt^2} = \Psi^i_j\left(\ddot{x}^j + \Gamma^j_{kl}\dot{x}^k\dot{x}^l\right),$$

from which the result follows. □

We turn now to projective geometry. Here the geodesics do not come with a preferred parametrization in the first instance, so must be considered as paths.

Proposition 11.3.2 *The geodesic paths of a Cartan projective geometry are those of the associated projective equivalence class of linear connections.*

Proof Take a curve $c : I \to M$, with $0 \in I$; we shall construct the development of c into $\mathrm{P}\mathcal{W}_{c(0)}M$. We shall work in coordinates (x^i, y^i) as defined in Sect. 9.1, such that the image of the attachment section $[e]$ is the origin of the fibre coordinates.

Consider $c^*\mathrm{P}\mathcal{W}M$, the pull-back of $\mathrm{P}\mathcal{W}M$ to I defined by c. With $c(t) = (x^i(t))$, we define a vector field C on this manifold by

$$C(t, y) = \gamma(\dot{c}(t)) = \frac{\partial}{\partial t} - \left(\dot{x}^i + \Pi^i_{jk}\dot{x}^j y^k + (\Lambda_{jk}\dot{x}^j y^k)y^i\right)\frac{\partial}{\partial y^i},$$

where the Π^i_{jk} are the coefficients of the fundamental descriptive invariant of a projective equivalence class of linear connections. The flow φ of C takes the form

$$\varphi_s(t, y) = (s + t, \Phi(s, t)y)$$

where $\Phi(s, t)$ is a projective transformation $\mathrm{P}\mathcal{W}_{c(t)}M \to \mathrm{P}\mathcal{W}_{c(s+t)}M$, whose representation as a fractional-linear transformation is

$$(\Phi(s, t)y)^i = \frac{\Phi^i_j(s, t)y^j + \Phi^i_0(s, t)}{\Phi^0_k(s, t)y^k + 1}$$

say, where we have taken advantage of the fact that without loss of generality we may take $\Phi^0_0 = 1$. The tangent vector to the curve $s \mapsto \Phi(s, t)y$ at $s = 0$ is

$$\left(A^i_0 + A^i_j y^j - (A^0_j y^j)y^i\right)\frac{\partial}{\partial y^i} \quad \text{where} \quad A^a_b = A^a_b(t) = \left.\frac{\partial \Phi^a_b}{\partial s}\right|_{(0,t)}.$$

That is,

$$\left.\frac{\partial \Phi^i_0}{\partial s}\right|_{(0,t)} = -\dot{x}^i(t), \qquad \left.\frac{\partial \Phi^i_j}{\partial s}\right|_{(0,t)} = -\Pi^i_{kj}(x(t))\dot{x}^k(t), \qquad \left.\frac{\partial \Phi^0_i}{\partial s}\right|_{(0,t)} = \Lambda_{ji}(x(t))\dot{x}^j(t).$$

The one-parameter group property formula is not quite matrix multiplication, because one has to ensure that $\Phi^0_0 = 1$ always. In fact

$$\Phi_b^a(s_1+s_2,t) = \frac{\Phi_c^a(s_1,t+s_2)\Phi_b^c(s_2,t)}{\Phi_c^0(s_1,t+s_2)\Phi_0^c(s_2,t)}$$
$$= \frac{\Phi_k^a(s_1,t+s_2)\Phi_b^k(s_2,t)+\Phi_0^a(s_1,t+s_2)\Phi_b^0(s_2,t)}{\Phi_k^0(s_1,t+s_2)\Phi_0^k(s_2,t)+1}.$$

Moreover $\Phi_b^a(0,t) = \delta_b^a$, and $\varphi_s(t,y)$ is defined for (s,y) in a neighbourhood of $(0,[e](c(0)))$ in $I \times P\mathcal{W}_{c(t)}M$.

Set $\Psi(t) = \Phi(-t,t)$. It follows from the formula above, via Lemma 3.4.2, that

$$\frac{d\Psi_0^i}{dt} = \Psi_j^i\dot{x}^j - \left(\Psi_k^0\dot{x}^k\right)\Psi_0^i$$
$$\frac{d\Psi_j^i}{dt} = \Psi_k^i\Pi_{lj}^k\dot{x}^l - \left(\Psi_k^0\dot{x}^k\right)\Psi_j^i - \left(\Lambda_{kj}\dot{x}^k\right)\Psi_0^i$$
$$\frac{d\Psi_i^0}{dt} = \Psi_k^0\Pi_{ji}^k\dot{x}^j - \left(\Psi_k^0\dot{x}^k\right)\Psi_i^0 - \Lambda_{ji}\dot{x}^j,$$

where $\Psi_b^a(t) = \Phi_b^a(-t,t)$; we have $\Psi_b^a(0) = \delta_b^a$. All terms in these equations are functions of t, either directly or through $x^i(t)$. This is a system of simultaneous first-order ordinary differential equations which, together with the conditions $\Psi_j^i(0) = \delta_j^i$, $\Psi_0^i(0) = \Psi_i^0(0) = 0$, in principle determine the Ψ_b^a uniquely. In particular the final equation contains only terms in Ψ_i^0, and so in principle may be solved independently of the others. Notice that $d\Psi_b^a/dt$ depends linearly on dx^i/dt: this means that if one carries out a reparametrization of c, by changing x^i to $x^i \circ \chi$ say, then Ψ_b^a changes to $\Psi_b^a \circ \chi$ likewise.

The development is the curve $t \mapsto \Psi(t)[e](c(0))$ in $P\mathcal{W}_{c(0)}M$, where $[e]$ is the attachment section, so that $[e](c(0))$ is the origin of fibre coordinates. So the development is just $y^i(t) = \Psi_0^i(t)$. We have $y^i(0) = 0$, $\dot{y}^i(0) = \dot{x}^i(0)$. It follows from the remark above about reparametrization that development respects reparametrization.

Let $\sigma(t) = \int_0^t (\Psi_k^0\dot{x}^k)dt$: then

$$\frac{d\left(e^\sigma\Psi_0^i\right)}{dt} = e^\sigma\Psi_j^i\dot{x}^j$$
$$\frac{d\left(e^\sigma\Psi_j^i\right)}{dt} = e^\sigma\Psi_k^i\Pi_{lj}^k\dot{x}^l - \left(\Lambda_{kj}\dot{x}^k\right)e^\sigma\Psi_0^i.$$

Note that $\sigma(0) = 0$, $\dot{\sigma}(0) = 0$. With $\psi^i = e^\sigma\Psi_0^i$, $\psi_j^i = e^\sigma\Psi_j^i$, these equations may be written more succinctly as

$$\dot{\psi}^i = \psi_j^i\dot{x}^j$$
$$\dot{\psi}_j^i = \psi_k^i\Pi_{lj}^k\dot{x}^l - \left(\Lambda_{kj}\dot{x}^k\right)\psi^i,$$

whence

$$\ddot{\psi}^i + \left(\Lambda_{jk}\dot{x}^j\dot{x}^k\right)\psi^i = \psi^i_j\left(\ddot{x}^j + \Pi^j_{kl}\dot{x}^k\dot{x}^l\right);$$

moreover $\psi^i(0) = 0$, $\dot{\psi}^i(0) = \dot{x}^i(0)$.

Suppose that $t \mapsto (x^i(t))$ is a geodesic of the projective equivalence class of linear connections whose fundamental descriptive invariant is Π^k_{ij}. In view of the fact that development respects reparametrization, without loss of generality we may choose the parametrization so that $\ddot{x}^j + \Pi^j_{kl}\dot{x}^k\dot{x}^l = 0$. Let $\psi(t)$ be the unique solution of the differential equation $\ddot{\psi} + (\Lambda_{jk}\dot{x}^j\dot{x}^k)\psi = 0$ such that $\psi(0) = 0$, $\dot{\psi}(0) = 1$. Then the unique solution of the system $\ddot{\psi}^i + (\Lambda_{jk}\dot{x}^j\dot{x}^k)\psi^i = 0$ such that $\psi^i(0) = 0$, $\dot{\psi}^i(0) = \dot{x}^i(0)$ is $\psi^i(t) = \psi(t)\dot{x}^i(0)$. The development is then $y^i(t) = e^{-\sigma(t)}\psi(t)\dot{x}^i(0)$, and is a straight line.

Suppose, on the other hand, that the development is a straight line: again, without loss of generality we can choose the parametrization, so take $\psi^i(t) = \psi(t)\dot{x}^i(0)$ with ψ defined as before. Then $\ddot{\psi}^i + (\Lambda_{jk}\dot{x}^j\dot{x}^k)\psi^i = 0$, and therefore $\ddot{x}^j + \Pi^j_{kl}\dot{x}^k\dot{x}^l = 0$ and so the path determined by $t \mapsto (x^i(t))$ is geodesic. □

We can now see that in fact geodesics of a projective class of linear connections do have a preferred form of parametrization, namely that for which the development is a projective line, that is, for which $y^i(s) = \mu(s)v^i$, for some fixed (v^i), where μ is a Möbius function:

$$\mu(s) = \frac{as + b}{cs + d}$$

for some constants a, b, c and d with $ad - bc \neq 0$. The necessary and sufficient condition for μ to take this form is that its Schwarzian derivative $\mathrm{S}(\mu)$ vanishes:

$$\mathrm{S}(\mu) = \frac{d}{ds}\left(\frac{\mu''}{\mu'}\right) - \frac{1}{2}\left(\frac{\mu''}{\mu'}\right)^2,$$

where the prime denotes derivative with respect to s. We call such a parameter s a *projective parameter*.

Let us set $\chi(t) = e^{-\sigma(t)}\psi(t)$, the coefficient which occurs in the equation of the straight line in the proof above when the geodesic is parametrized so that $\ddot{x}^i + \Pi^i_{jk}\dot{x}^j\dot{x}^k = 0$ (and the dot denotes derivative with respect to t). We seek a reparametrization from t to a projective parameter s. We have $\mu(s(t)) = \chi(t)$ and we require that $\mathrm{S}(\mu) = 0$. Now

$$\frac{\mu''}{\mu'} = \frac{1}{\dot{s}}\left(\frac{\ddot{\chi}}{\dot{\chi}} - \frac{\ddot{s}}{\dot{s}}\right).$$

From the equation for $d\Psi^0_i/dt$ we obtain

$$\frac{d\left(\Psi_i^0 \dot{x}^i\right)}{dt} = \Psi_k^0\left(\ddot{x}^k + \Pi_{ij}^k \dot{x}^i \dot{x}^j\right) - \left(\Psi_k^0 \dot{x}^k\right)^2 - \Lambda_{ij}\dot{x}^i \dot{x}^j,$$

so that when $\ddot{x}^i + \Pi^i_{jk}\dot{x}^j\dot{x}^k = 0$, since $\dot{\sigma} = \Psi_i^0 \dot{x}^i$,

$$\ddot{\sigma} + \dot{\sigma}^2 = -\Lambda_{ij}\dot{x}^i\dot{x}^j.$$

It follows that

$$\begin{aligned}
\dot{\chi} &= e^{-\sigma}(\dot{\psi} - \dot{\sigma}\psi) \\
\ddot{\chi} &= e^{-\sigma}(\ddot{\psi} - 2\dot{\sigma}\dot{\psi} - (\ddot{\sigma} - \dot{\sigma}^2)\psi) \\
&= e^{-\sigma}\left(-\Lambda_{ij}\dot{x}^i\dot{x}^j\psi - 2\dot{\sigma}\dot{\psi} + (2\dot{\sigma}^2 + \Lambda_{ij}\dot{x}^i\dot{x}^j)\psi\right) \\
&= -2\dot{\sigma}e^{-\sigma}(\dot{\psi} - \dot{\sigma}\psi),
\end{aligned}$$

so that

$$\frac{\ddot{\chi}}{\dot{\chi}} = -2\dot{\sigma},$$

and therefore

$$\frac{\mu''}{\mu'} = -\frac{1}{\dot{s}}\left(2\dot{\sigma} + \frac{\ddot{s}}{\dot{s}}\right).$$

It follows that

$$\mathrm{S}(\mu) = -\frac{1}{\dot{s}^2}\left(2(\ddot{\sigma} + \dot{\sigma}^2) + \mathrm{S}(s)\right).$$

Thus a transformation from the parameter t to a projective parameter s is obtained by solving the equation

$$\mathrm{S}(s) = 2\Lambda_{ij}\dot{x}^i\dot{x}^j.$$

It is well known that if f is a Möbius function then $\mathrm{S}(f \circ s) = \mathrm{S}(s)$; thus if s is a reparametrization to a projective parameter then so is $f \circ s$ for any Möbius function f, which is as it should be.

The geodesic equations with respect to a projective parameter are

$$\ddot{x}^i + \Pi^i_{jk}\dot{x}^j\dot{x}^k = \lambda\dot{x}^i$$

where λ satisfies the Riccati equation

$$\dot{\lambda} - \tfrac{1}{2}\lambda^2 = -2\Lambda_{ij}\dot{x}^i\dot{x}^j.$$

(the dot now indicates derivative with respect to the projective parameter of course).

The projective parameter was introduced by Berwald in [3].

11.4 Generalized Geodesics

Geodesics are defined only for a special class of Cartan geometries, those for which it makes sense to talk about straight lines in the standard fibre F. But in the case of Cartan affine geometry a straight line in F can be defined in terms of the affine group G: a straight line is just an orbit of a one-parameter group of translations. This observation suggests a natural generalization of the concept of a geodesic: a curve c is a *generalized geodesic* if its development in $E_{c(0)}$ is (contained in) the orbit of $\varsigma(c(0))$ by a one-parameter subgroup of the vertex group $\mathcal{G}_{c(0)}$ acting on $E_{c(0)}$; or equivalently if its development is (contained in) the integral curve through $\varsigma(c(0))$ of a fixed element k of $K_{c(0)}\mathcal{G}$. (Sharpe [36, Definition 5.4.16] calls such a curve a generalized circle, while Blaom [5] calls it a geodesic (unqualified): we have adopted an intermediate name.)

Once again, the definition appears to depend on an arbitrary choice of fibre of $E \to M$ to develop c into. But recall that $\tilde{c}_\Gamma$, the development of c into $E_{c(t)}$, is related to c_Γ, its development into $E_{c(0)}$, by

$$\tilde{c}_\Gamma(s) = c^\Gamma(t)\big(c_\Gamma(s+t)\big).$$

Now $c^\Gamma(t) : E_{c(0)} \to E_{c(t)}$ is a fibre morphism, and therefore intertwines the action of the vertex group $\mathcal{G}_{c(0)}$ on $E_{c(0)}$ and the action of the vertex group $\mathcal{G}_{c(t)}$ on $E_{c(t)}$. It follows that $c^\Gamma(t)_* : TE_{c(0)} \to TE_{c(t)}$ induces an isomorphism $K_{c(0)}\mathcal{G} \to K_{c(t)}\mathcal{G}$; and if the development of c into $E_{c(0)}$ is the integral curve of $k \in K_{c(0)}\mathcal{G}$ through $\varsigma(c(0))$ then the development of c into $E_{c(t)}$ is the integral curve of $c^\Gamma(t)_*k \in K_{c(t)}\mathcal{G}$ through $\varsigma(c(t))$.

In this manner we associate with a generalized geodesic c a section κ of $K\mathcal{G}$ along c by $\kappa(t) = c^\Gamma(t)_*k$, where $k = \kappa(0) \in K_{c(0)}\mathcal{G}$. The fact that κ has that particular form may be conveniently specified in terms of ∇^γ.

Proposition 11.4.1 *Let $c : I \to M$ be a curve and κ any section of $K\mathcal{G}$ along c. Then $\kappa(t) = c^\Gamma(t)_*k$ if and only if $\nabla^\gamma_{\dot{c}}\kappa = 0$.*

We have defined ∇^γ in terms of vector fields, sections of $K\mathcal{G}$, and the Lie algebroid bracket: but we have demonstrated that it is a covariant derivative, so we know that $\nabla^\gamma_{\dot{c}}\kappa$ makes sense for any curve c in M and any section of $K\mathcal{G}$ along it. Our first task in proving this proposition will be to give some more practicable interpretation of ∇^γ.

Proof As before, we consider the pullback bundle $c^*E \to I$; and we think of κ as a vertical vector field on c^*E. Let $\tau = \gamma(\partial/\partial t)$, also considered as a vector field on c^*E. Then

$$\nabla^\gamma_{\dot{c}}\kappa = [\![\gamma(\partial/\partial t), \kappa]\!] = [\tau, \kappa],$$

a vertical vector field on c^*E (the Lie algebroid bracket is the ordinary bracket of vector fields on c^*E). In particular, if $u = \dot{c}(0)$ then

$$\nabla^{\gamma}_{u}\kappa = [\tau, \kappa]_{c(0)}.$$

Now let ϕ^{τ} be the flow of τ. Then the right-hand side is

$$\frac{d}{dt}\left(\phi^{\tau}_{-t*}\kappa(t)\right)_{t=0},$$

identifying bracket with Lie derivative. But $c^{\Gamma}(t) : E_{c(0)} \to E_{c(t)}$ is just ϕ^{τ}_{t} acting on $E_{c(0)}$. So we can write

$$\nabla^{\gamma}_{u}\kappa = \frac{d}{dt}\left(c^{\Gamma}(t)^{-1}_{*}\kappa(t)\right)_{t=0}.$$

In particular, if $\kappa(t) = c^{\Gamma}(t)_{*}k$ for some fixed $k \in \mathcal{K}_{c(0)}$ then $\nabla^{\gamma}_{\dot{c}}\kappa = 0$; and conversely. □

That is to say that $\kappa(t) = c^{\Gamma}(t)_{*}k$ precisely when κ is parallel along c. So a generalized geodesic has a section of $K\mathcal{G}$ defined along it which is parallel with respect to ∇^{γ}. But this property, while necessary, is by no means sufficient to pick out a special class of curves in M—indeed, for any curve c and any $k \in K_{c(0)}$ there is evidently a section κ of $K\mathcal{G}$ along c which is parallel and satisfies $\kappa(0) = k$. Regarding $K\mathcal{G}$ as an abstract Lie algebra bundle over M, we may think of κ simply as a curve in $K\mathcal{G}$ and c as its projection into M—its footprint, to use an evocative term of Blaom's.

Consider now instead a curve β in $A\mathcal{G}_{o}$ whose footprint is c. Then $\omega \circ \beta = \kappa$ is a curve in $K\mathcal{G}$ with footprint c, and

$$\nabla^{\gamma}_{\dot{c}}(\omega \circ \beta) = \omega \circ \nabla^{\mathrm{T}}_{\dot{c}}\beta.$$

In particular, β is parallel along c with respect to ∇^{T} if and only if $\omega \circ \beta$ is parallel along c with respect to ∇^{γ}.

Proposition 11.4.2 *If β is a curve in $A\mathcal{G}_{o}$ with footprint c such that $\rho \circ \beta = \dot{c}$ (where $\rho : A\mathcal{G}_{o} \to TM$ is the anchor) and $\nabla^{\mathrm{T}}_{\dot{c}}\beta = 0$ then c is a generalized geodesic; and every generalized geodesic is of this form.*

Proof We must consider the development c_{Γ} of c into $E_{c(0)}$. As we showed at the end of Sect. 11.1, for any curve c in M

$$\begin{aligned}
\dot{c}_{\Gamma}(t) &= C_{t}(\dot{c}(t)) \\
&= (c^{\Gamma}(t)^{-1})_{*}\mathfrak{s}(\dot{c}(t)) \\
&= (c^{\Gamma}(t)^{-1})_{*}\omega(\varsigma_{*}(\dot{c}(t))),
\end{aligned}$$

assuming of course that the soldering is given by the kernel projection.

Suppose that $\beta(t)$ is a vector field along $E_{c(t)}$ such that $\rho(\beta(t)) = \dot{c}(t)$. Since $\beta(t) \in A_{c(t)}\mathcal{G}_{o}$ it is tangent to the image of ς. It follows that $\varsigma_{*}(\dot{c}(t)) = \beta(t)_{\varsigma(c(t))}$, whence

$$\dot{c}_\Gamma(t) = (c^\Gamma(t)^{-1})_* \omega(\beta(t))_{\varsigma(c(t))};$$

this holds indeed for any β with footprint c such that $\rho \circ \beta = \dot{c}$. If in addition $\nabla^{\mathrm{T}}_{\dot{c}}\beta = 0$ then $\omega(\beta(t)) = c^\Gamma(t)_* k$ for some $k \in K_{c(0)}\mathcal{G}$, so

$$\dot{c}_\Gamma(t) = (c^\Gamma(t)^{-1})_* \big((c^\Gamma(t)_* k)_{\varsigma(c(t))}\big) = k_{c^\Gamma(t)^{-1}\varsigma(c(t))} = k_{c_\Gamma(t)},$$

which is to say that the development of c is the integral curve of k through $\varsigma(c(0))$.

If on the other hand c is a generalized geodesic then $\dot{c}_\Gamma(t) = k_{c_\Gamma(t)}$ for some $k \in K_{c(0)}\mathcal{G}$. There is then a unique curve β in $A\mathcal{G}_o$ such $\omega(\beta(t)) = c^\Gamma(t)_* k$; its footprint is c, and it satisfies $\nabla^{\mathrm{T}}_{\dot{c}}\beta = 0$. We show that $\rho \circ \beta = \dot{c}$. We have

$$\begin{aligned}\omega(\beta(t))_{\varsigma(c(t))} &= (c^\Gamma(t)_* k)_{\varsigma(c(t))} = c^\Gamma(t)_*(k_{c_\Gamma(t)}) = c^\Gamma(t)_*(\dot{c}_\Gamma(t)) \\ &= \omega(\varsigma_*(\dot{c}(t))),\end{aligned}$$

so that

$$\beta(t)_{\varsigma(c(t))} = \varsigma_*(\dot{c}(t)).$$

It follows that $\rho(\beta(t)) = \dot{c}(t)$, as required. □

By Proposition 7.4.2,

$$\omega \circ \big(\nabla^{\mathrm{T}}_{\dot{c}}\beta - \nabla^{\mathrm{B}}_{\dot{c}}\beta\big) = R^\gamma(\dot{c}, \rho \circ \beta).$$

Thus if $\rho \circ \beta = \dot{c}$ then $\nabla^{\mathrm{T}}_{\dot{c}}\beta = \nabla^{\mathrm{B}}_{\dot{c}}\beta$, and so in the proposition above it is immaterial whether we use ∇^{T} or ∇^{B}. It will emerge shortly that there are advantages, to do with symmetries, in using ∇^{B} rather than ∇^{T}. We therefore restate the result in terms of ∇^{B}.

Theorem 11.4.3 *If β is a curve in $A\mathcal{G}_o$ with footprint c such that $\rho \circ \beta = \dot{c}$ and $\nabla^{\mathrm{B}}_{\dot{c}}\beta = 0$ then c is a generalized geodesic; and every generalized geodesic is of this form.* □

11.5 Geodesic Sprays and Their Generalizations

A geodesic of a linear connection may be defined as a curve whose tangent vector is parallel along it with respect to the connection; but it may also be defined as a base integral curve (or footprint of an integral curve) of a certain vector field on the tangent bundle, the *geodesic spray* of the connection. Such a vector field S on TM satisfies the conditions $\Sigma \circ S = \Delta$ and $[\Delta, S] = S$ where Σ is the vertical endomorphism introduced in Sect. 5.5 and Δ is the Liouville field; if the connection has horizontal vector fields H_i with coefficients Γ^i_{jk} then the geodesic spray of the connection is

$$S = u^i H_i = u^i \frac{\partial}{\partial x^i} - \Gamma^i_{jk} u^j u^k \frac{\partial}{\partial u^i}.$$

For example, the TW-connection introduced in Sect. 9.4 is such a connection (although on TVM rather than on TM) and so the coordinate expression for its spray, the *TW-spray*, using the coefficients Γ^a_{bc} given there, is

$$u^0 \frac{\partial}{\partial x^0} + u^i \frac{\partial}{\partial x^i} + x^0 \Lambda_{ij} u^i u^j \frac{\partial}{\partial u^0} - \left(2\frac{u^0 u^i}{x^0} + \Pi^i_{jk} u^j u^k \right) \frac{\partial}{\partial u^i}.$$

We shall discuss sprays in more detail in Sect. 12.3.

In the previous section we saw that generalized geodesics in a Cartan geometry may be defined using parallel transport, in a generalization of the case for linear connections. We now show that there is also a natural generalization of the geodesic spray.

For our initial purposes all that is required is an anchored vector bundle: a vector bundle $\pi : V \to M$ with an anchor map, that is, a morphism $\rho : V \to TM$ of vector bundles over the identity on M.

Lemma 11.5.1 *Let $\pi : V \to M$ be an anchored vector bundle with anchor ρ, and let S be a vector field on V such that for all $v \in V$, $\pi_* S_v = \rho(v)$. Then if c^S is an integral curve of S and $c = \pi \circ c^S$ is its footprint,*

$$\dot{c} = \rho \circ c^S.$$

Proof

$$\dot{c}(t) = \pi_* \dot{c}^S(t) = \pi_* \big(S_{c^S(t)} \big) = \rho(c^S(t)).$$

□

Now suppose that in addition $\pi : V \to M$ is equipped with a linear connection, that is, with a covariant derivative operator ∇ on sections of π. From ∇ we can construct a horizontal distribution on V, that is, for each $v \in V$ a subspace H_v of $T_v V$ such that $\pi_* : H_v \to T_{\pi(v)} M$ is an isomorphism. There is a corresponding horizontal lift operation $T_x M \to H_v$ for each v with $\pi(v) = x$, denoted by $u \mapsto u^{\mathsf{h}}$; and $\pi_* u^{\mathsf{h}} = u$. There is an induced horizontal lift of curves $c \mapsto c^{\mathsf{h}}$—and the feature of the construction which really defines H is that c^{h}, considered as a V-vector field along c, is parallel with respect to ∇. We call a curve c in M a *generalized geodesic* of ∇ if there is a horizontal curve $t \mapsto v(t)$ in V such that c is the footprint of v, or in other words if there is a V-vector field v along c which parallel and satisfies $\rho \circ v = \dot{c}$. We shall see shortly how this relates to the definition given in the previous section.

We define a vector field S on V as follows:

$$S_v = \rho(v)^{\mathsf{h}}$$

(the horizontal lift is to v of course). Evidently $\pi_* S_v = \rho(v)$.

Proposition 11.5.2 *The generalized geodesics of* ∇ *are the footprints of integral curves of* S.

Proof Let c^S be an integral curve of S. Then since c^S is a horizontal curve, considered as a V-vector field along c it is parallel. Moreover, by Lemma 11.5.1, $\dot{c} = \rho \circ c^S$. So c is a generalized geodesic of ∇. □

Corollary 11.5.3 *There is a unique generalized geodesic for each* $v \in V$. □

We conclude that any generalized geodesic of the previous section—a curve c in M which develops into the integral curve through $\varsigma(c(0)) \in E_{c(0)}$ of a fixed element k of $K_{c(0)}\mathcal{G}$—is a base integral curve of the vector field S^{B} on $A\mathcal{G}_o$ determined by the covariant derivative ∇^{B}. In fact we don't need the full force of the definition of a Cartan geometry here: it will be enough to have a vector bundle $\mathcal{A}_o$ with the properties of an affine tractor bundle, for such a bundle is anchored—even transitively—and it carries a covariant derivative ∇^{B}. We call the vector field S^{B} on $\mathcal{A}_o$ defined by ∇^{B} the *generalized geodesic spray*, and any of the base integral curves of S^{B} a *generalized geodesic* of ∇^{B}.

The advantage of this approach is that it suggests a straightforward definition of completeness for an infinitesimal Cartan geometry: an infinitesimal Cartan geometry is *complete* if the generalized geodesic spray S^{B} of ∇^{B} on $\mathcal{A}_o$ is complete (as a vector field—so that its integral curves are all defined on the whole of $\mathbb{R}$). This definition is due to Blaom [5].

Symmetries of infinitesimal Cartan geometries fit neatly into this framework. A symmetry is a section ξ of $\mathcal{A}_o$ such that $\nabla^{\mathrm{B}}\xi = 0$. Then $\rho \circ \xi$ is a vector field on M: let c be any of its integral curves, and $\hat{c} = \xi \circ c$ the induced curve in $\mathcal{A}_o$. Then c is the footprint of $\hat{c}$, and $\rho \circ \hat{c} = \dot{c}$ because c is an integral curve of $\rho \circ \xi$. Moreover, $\hat{c}$ is horizontal with respect to the horizontal lift determined by ∇^{B}, simply because $\nabla^{\mathrm{B}}\xi = 0$ (in fact the image of the section $\xi : M \to \mathcal{A}_o$ is horizontal). Thus c is a generalized geodesic. In fact if ξ is a symmetry, S^{B} is tangent to the image of ξ.

Proposition 11.5.4 *In a complete infinitesimal Cartan geometry the action of the symmetry algebra* $\mathfrak{a}$ *on* M *is complete, that is, the vector field* $\rho \circ \xi$ *is complete for all symmetries* ξ. □

Palais showed [33] that if a Lie algebra $\mathfrak{g}$ acts on a manifold M, and the action is complete, then there is a connected Lie group G with Lie algebra $\mathfrak{g}$ which acts to the right on M such that the action of $\mathfrak{g}$ is that induced by G.

Theorem 11.5.5 *If* M *is a connected simply connected manifold over which is defined a complete locally symmetric infinitesimal Cartan geometry then* M *is a homogeneous space of a Lie group* G *whose Lie algebra is the Lie algebra of vector fields on* M *induced by the algebra of infinitesimal symmetries of the Cartan geometry.* □

For discussion of what happens when M is not simply connected we refer the reader to Blaom's article [5].

It's pretty easy to work out S^{B} for Cartan affine geometry, since we effectively know $\nabla^{\mathrm{B}}\xi$ from the discussion of symmetries in Sect. 8.1. We write there

$$\xi = X^i H_i + X^i_j y^j V_i$$

where H_i is horizontal for the *linear* connection. We may therefore take (X^i, X^i_j) as fibre coordinates on $\mathcal{A}_o$. Moreover,

$$\omega \circ \xi = (X^i + X^i_j y^j)\, V_i.$$

Then

$$\omega(\nabla^{\mathrm{B}}_{\partial/\partial x^i}\xi) = -(\mathcal{L}_\xi \gamma)\left(\frac{\partial}{\partial x^i}\right) = \left(\left(X^j_{|i} - X^j_i\right) + \left(X^j_{k|i} + R^j_{kli} X^l\right) y^k\right) V_j,$$

so that

$$\nabla^{\mathrm{B}}_{\partial/\partial x^i}\xi = \left(X^j_{|i} - X^j_i\right) H_j + \left(X^j_{k|i} + R^j_{kli} X^l\right) y^k V_j.$$

It follows that the horizontal lift to $\mathcal{A}_o$ determined by ∇^{B} is given by

$$\left(\frac{\partial}{\partial x^i}\right)^{\mathrm{h}} = \frac{\partial}{\partial x^i} + \left(X^j_i - \Gamma^j_{ki} X^k\right) \frac{\partial}{\partial X^j} + \left(\Gamma^l_{ki} X^j_l - \Gamma^j_{li} X^l_k - R^j_{kli} X^l\right) \frac{\partial}{\partial X^j_k},$$

and so

$$S^{\mathrm{B}} = X^i \frac{\partial}{\partial x^i} + X^i \left(X^j_i - \Gamma^j_{ki} X^k\right) \frac{\partial}{\partial X^j} + X^i \left(\Gamma^l_{ki} X^j_l - \Gamma^j_{li} X^l_k\right) \frac{\partial}{\partial X^j_k},$$

(notice the disappearance of the curvature, due to skew-symmetry in its last two indices). An integral curve of this vector field is a solution of the equations

$$\begin{aligned} \dot{x}^i &= X^i \\ \dot{X}^i + \Gamma^i_{jk} X^j X^k &= X^j X^i_j \\ \dot{X}^i_j + \Gamma^i_{lk} X^l_j X^k - \Gamma^l_{jk} X^i_l X^k &= 0, \end{aligned}$$

or in other terms

$$\begin{aligned} \nabla_{\dot{x}} \dot{x}^i &= \dot{x}^j X^i_j \\ \nabla_{\dot{x}} X^i_j &= 0. \end{aligned}$$

If X^i_j vanishes anywhere on a solution curve it vanishes everywhere; the footprint curve is then a geodesic of the linear connection. In fact $\mathcal{A}_o$ has TM as a vector sub-bundle: X^i and X^i_j are components of tensors, so it makes sense to set $X^i_j = 0$ and $X^i = y^i$, the canonical fibre coordinate on TM. Of course TM corresponds to the translation algebra: to be more precise, its image in $\mathcal{K}$ by ω is the translation algebra. From the formula above we see that S^{B} is tangent to $TM \subset \mathcal{A}_o$, and its restriction to TM can be written

$$y^i \frac{\partial}{\partial x^i} - \Gamma^i_{jk} y^j y^k \frac{\partial}{\partial y^i},$$

which is the geodesic spray of the linear connection.

We can apply these considerations to the TW-spray on $T\mathcal{V}M$ introduced at the beginning of this section. The base integral curves of the TW-spray (on $\mathcal{V}M$) are solutions of

$$\ddot{x}^0 - \left(\Lambda_{ij}\dot{x}^i\dot{x}^j\right) x^0 = 0, \quad \ddot{x}^i + \Pi^i_{jk}\dot{x}^j\dot{x}^k = -2\left(\frac{\dot{x}^0}{x^0}\right)\dot{x}^i.$$

These project to curves on M which are geodesics of the projective class given by Π^i_{jk}. The x^0-component determines the parametrization: it is in fact projective parametrization, as we shall see shortly.

Next, consider $\mathcal{W}M$. It is equipped with an anchor map (it is, as we mentioned in Sect. 4.3, the Atiyah algebroid of $\mathcal{V}M$), which is just

$$\rho(w^a) = w^i \frac{\partial}{\partial x^i}.$$

So there is a generalized spray on $\mathcal{W}M$,

$$w^i \tilde{H}_i = w^i \frac{\partial}{\partial x^i} + \Lambda_{ij} w^i w^j \frac{\partial}{\partial w^0} - \left(w^0 w^i + \Pi^i_{jk} w^j w^k\right) \frac{\partial}{\partial w^i}.$$

But we must be cautious, because this isn't the projection of the TW-spray: we have to make allowance for the term $u^0\partial/\partial x^0$ in the latter. The projection of the TW-spray is in fact

$$w^i \frac{\partial}{\partial x^i} + \left(\Lambda_{ij} w^i w^j - (w^0)^2\right) \frac{\partial}{\partial w^0} - \left(2w^0 w^i + \Pi^i_{jk} w^j w^k\right) \frac{\partial}{\partial w^i};$$

it differs from the generalized spray given above by a multiple of $\tilde{\Delta}$ (and after all $\Upsilon^{\mathrm{h}} \mapsto -\tilde{\Delta}$). The base integral curves of this vector field satisfy

$$\dot{w}^0 + (w^0)^2 = \Lambda_{ij}\dot{x}^i\dot{x}^j, \quad \ddot{x}^i + \Pi^i_{jk}\dot{x}^j\dot{x}^k = -2w^0\dot{x}^i.$$

The equations of a geodesic with respect to a projective parameter given earlier are

$$\ddot{x}^i + \Pi^i_{jk}\dot{x}^j\dot{x}^k = \lambda\dot{x}^i$$

where λ satisfies the Riccati equation

$$\dot{\lambda} - \tfrac{1}{2}\lambda^2 = -2\Lambda_{ij}\dot{x}^i\dot{x}^j.$$

The equations obtained from the generalized spray on $\mathcal{W}M$ are these with $\lambda = -2w^0$. The equations for the TW-spray have $\dot{x}^0/x^0 = w^0 = -\frac{1}{2}\lambda$: it is in fact a standard technique in the study of Riccati equations to make a substitution of this form to convert the nonlinear first-order equation to a linear second-order one.

Chapter 12
Cartan Theory of Second-Order Differential Equations

In the final chapter of this volume we investigate how our approach to Cartan geometries can be modified to encompass the geometry of second-order ordinary differential equations. Cartan analysed a single second-order ordinary differential equation using the methods of his connection theory, whereas we shall develop a theory of systems of such equations. It probably won't come as a surprise that we shall have to modify our approach somewhat to take on board the fact that we shall be dealing with *second*-order equations. In fact we shall have to work over base manifolds of the form $\tau_M^\circ : T^\circ M \to M$, where $T^\circ M$ denotes the slit tangent bundle of M (TM with the zero section deleted).

12.1 Affine Geometry over $T^\circ M$

Recall that one can construct Lie groupoids and algebroids by pullback. We consider the Lie groupoid $\mathcal{G}$ of affine fibre morphisms of TM, and pull it back over $\tau_M^\circ : T^\circ M \to M$. We denote by $\tau_M^{\circ *}(TM) \to T^\circ M$ the pullback bundle, with coordinates (x^i, y^i) in the base and z^i in the fibre. We denote the pullback Lie groupoid by $\tau_M^{\circ **}(\mathcal{G})$, and its Lie algebroid by $\tau_M^{\circ **}(A\mathcal{G})$ (Proposition 1.5.5). Let $\mathcal{A}^T = \Lambda(\tau_M^{\circ **}(A\mathcal{G}))$ (Theorem 3.2.1), so that an element of $\mathcal{A}^T$ is a vector field along a fibre of $\tau_M^{\circ *}(TM)$ of the form

$$X^i \frac{\partial}{\partial x^i} + Y^i \frac{\partial}{\partial y^i} + \left(Z^i + Z^i_j z^j\right) \frac{\partial}{\partial z^i}$$

(with constant coefficients). The anchor is projection onto the first two terms: $\mathcal{A}^T$ is of course transitive. We denote by $\mathcal{K}^T$ the kernel of $\mathcal{A}^T$.

Until now for the definition of a Cartan geometry we have required a fibre bundle $E \to M$ with fibre of the same dimension as M, with a global section along which E is soldered to M. These requirements clearly cannot be met for the pullback bundle; but modifications can be formulated in a natural manner, as we now show.

M. Crampin and D. Saunders, *Cartan Geometries and their Symmetries*,
Atlantis Studies in Variational Geometry 4, DOI 10.2991/978-94-6239-192-5_12

The zero section of $\tau^{\circ *}_M(TM) \to T^\circ M$ is well defined. On the zero section, for each $(x, y) \in T^\circ M$ the vertical lift defines a linear map $T_{(x,y)}(T^\circ M) \to T_0(T_x M)$, which is surjective and has the vertical subspace as its kernel. It defines a natural generalization of a soldering.

There is a second natural section of $\tau^{\circ *}_M(TM) \to T^\circ M$; to construct it we use the fact that the fibre of $\tau^{\circ *}_M(TM)$ over $(x, y) \in TM$ is just $T_x M$: the map $(x, y) \mapsto (x, y, y)$ is a section, given in coordinates by $z^i = y^i$. As a vector field along the projection τ°_M this is the total derivative, so we call it the *total derivative section*.

We denote by S_0 the image of the zero section and by S_1 the image of the total derivative section. A vector field along a fibre of $\tau^{\circ *}_M(TM) \to T^\circ M$ belonging to $\mathcal{A}^T$ is tangent to S_0 if and only if $Z^i = 0$. Such vector fields form a subalgebroid $\mathcal{A}^T_0$ of $\mathcal{A}^T$. A vector field along the fibre of $\tau^{\circ *}_M(TM) \to T^\circ M$ over (x, y) belonging to $\mathcal{A}^T_{(x,y)}$ is tangent to S_1 if and only if

$$Z^i + Z^i_j y^j = Y^i.$$

Such vector fields form a subalgebroid $\mathcal{A}^T_1$ of $\mathcal{A}^T$. Then $\mathcal{A}^T_0 \cap \mathcal{A}^T_1 = \mathcal{A}^T_o$ is a subalgebroid of $\mathcal{A}^T$, whose typical element is a vector field along the fibre of $\tau^{\circ *}_M(TM) \to T^\circ M$ over (x, y) of the form

$$X^i \frac{\partial}{\partial x^i} + Y^i \frac{\partial}{\partial y^i} + Z^i_j z^j \frac{\partial}{\partial z^i} \quad \text{where } Z^i_j y^j = Y^i.$$

Note that $\mathcal{A}^T_o$ is transitive (over $T^\circ M$). The requirement that $Z^i_j y^j = Y^i$ should be considered as a condition on Z^i_j with y^i and Y^i given: note that the vector (y^i) is nonzero, since the base is $T^\circ M$. The kernel $\mathcal{K}^T_{o,(x,y)}$ of $\mathcal{A}^T_{o,(x,y)}$ consists of vector fields along the fibre of $\tau^{\circ *}_M(TM) \to T^\circ M$ over (x, y) of the form

$$Z^i_j z^j \frac{\partial}{\partial z^i} \quad \text{where } Z^i_j y^j = 0.$$

For each $(x, y) \in T^\circ M$, $\mathcal{K}^T_{o,(x,y)}$ is a Lie subalgebra of the Lie algebra $\mathcal{K}^T_{(x,y)}$ of codimension $n^2 + n - (n^2 - n) = 2n$. So we have a transitive algebroid $\mathcal{A}^T_o$ over a $2n$-dimensional manifold $T^\circ M$ whose kernel $\mathcal{K}^T_o$ has fibre codimension $2n$ in $\mathcal{K}^T$, though we started with a fibre bundle $\tau^{\circ *}_M(TM)$ whose fibre has dimension just n.

Let us consider the Lie groups and algebras involved in this construction. Denote by G the affine group $\mathsf{A}(n)$ considered as the set of pairs (v, P) with $v \in \mathbb{R}^n$ and $P \in \mathsf{GL}(n)$, acting on $\mathbb{R}^n$ by $(v, P)x = v + Px$. The group G also acts on $T\mathbb{R}^n \cong \mathbb{R}^n \oplus \mathbb{R}^n$ via the derivative, explicitly by $(v, P)(x, y) = (v + Px, Py)$. This action is transitive on $T^\circ\mathbb{R}^n$. For any $y \in T_0\mathbb{R}^n$, $y \neq 0$, the isotropy group $G_{(0,y)}$ of $(0, y)$ consists of those $(v, P) \in G$ such that $v = 0$, $Py = y$, in other words, it is the subgroup of $\mathsf{GL}(n)$ consisting of elements which fix y. We may identify $G/G_{(0,y)}$ with $T^\circ\mathbb{R}^n$. Denote by $\mathfrak{g}$ the Lie algebra of G and $\mathfrak{g}_{(0,y)}$ the Lie algebra of $G_{(0,y)}$. We may identify $\mathfrak{g}_{(0,y)}$ with the subalgebra of $\mathfrak{gl}(n)$ consisting of matrices a such that

$ay = 0$. Let α be the linear map $\mathfrak{g} \to \mathbb{R}^n \oplus \mathbb{R}^n$ given by $\alpha(v, a) = (v, ay)$. Then we have a short exact sequence

$$0 \to \mathfrak{g}_{(0,y)} \to \mathfrak{g} \xrightarrow{\alpha} \mathbb{R}^n \oplus \mathbb{R}^n \to 0,$$

which identifies $\mathfrak{g}/\mathfrak{g}_{(0,y)}$ with $\mathbb{R}^n \oplus \mathbb{R}^n$ (as linear spaces). In the context of the algebroid structure the map $\mathcal{K}^T_{(x,y)} \to \mathcal{K}^T_{(x,y)}/\mathcal{K}^T_{o,(x,y)}$ corresponding to α is

$$\left(Z^i + Z^i_j z^j\right) \frac{\partial}{\partial z^i} \mapsto \left(Z^i + Z^i_j y^j\right) \frac{\partial}{\partial z^i}.$$

12.2 Cartan Affine Geometry over $T°M$

We now consider infinitesimal connections on such Lie algebroids.

Those elements of $\mathcal{A}^T$ which project onto vertical vectors on $T°M \to M$ (i.e. those with $X^i = 0$) form a subalgebroid, say $\mathcal{A}^{T\mathsf{v}}$. Now $\mathcal{A}^{T\mathsf{v}}$, considered as an algebroid over $\tau^{\circ *}_M(TM)$, is not transitive: however, we may consider its restriction to any fibre $T^\circ_x M$ of $T°M \to M$, say $\mathcal{A}^{T\mathsf{v}}_x$, and its restriction is transitive. Indeed, $\mathcal{A}^{T\mathsf{v}}_x$ is just the trivial Lie algebroid over the punctured vector space $T^\circ_x M$ with Lie algebra the affine algebra of $T_x M$.

Suppose given an infinitesimal connection γ on $\mathcal{A}^T$. If $v \in T(T°M)$ is vertical then $\gamma(v) \in \mathcal{A}^{T\mathsf{v}}$. Then for each $x \in M$, γ defines by restriction to $T^\circ_x M$ an infinitesimal connection γ_x on $\mathcal{A}^{T\mathsf{v}}_x$. Since $\mathcal{A}^{T\mathsf{v}}_x$ is trivial as an algebroid, we may require that γ_x is the trivial infinitesimal connection. If γ is such that γ_x is trivial for all $x \in M$ we say that γ is *vertically trivial*. If γ is vertically trivial it takes the coordinate form

$$\gamma\left(\frac{\partial}{\partial x^i}\right) = \frac{\partial}{\partial x^i} - \left(A^j_i + \Gamma^j_{ik} z^k\right) \frac{\partial}{\partial z^j}$$
$$\gamma\left(\frac{\partial}{\partial y^i}\right) = \frac{\partial}{\partial y^i}.$$

The first of these expressions is formally the same as the one for the infinitesimal connection for ordinary infinitesimal Cartan affine geometry, and indeed the coefficients A^i_j and Γ^i_{jk} transform as the components of a tensor and a connection: but they now may depend on y. For the kernel projection ω of a vertically trivial infinitesimal connection we have

$$\omega\left(X^i \frac{\partial}{\partial x^i} + Y^i \frac{\partial}{\partial y^i} + \left(Z^i + Z^i_j z^j\right) \frac{\partial}{\partial z^i}\right)$$
$$= \left(\left(Z^i + X^j A^i_j\right) + \left(Z^i_j + X^k \Gamma^i_{kj}\right) z^j\right) \frac{\partial}{\partial z^i}.$$

On the zero section ω defines, for each $(x, y) \in T^\circ M$, a linear map $T_{(x,y)}(T^\circ M) \to T_0(T_x M)$ by

$$X^i \frac{\partial}{\partial x^i} + Y^i \frac{\partial}{\partial y^i} \mapsto X^j A^i_j \frac{\partial}{\partial z^i}.$$

Provided that (A^i_j) is nonsingular this is surjective and has the vertical subspace as its kernel. We shall require that it coincides with the generalized soldering defined earlier, or in other words with the vertical lift; as in the ordinary affine case this means that $A^i_j = \delta^i_j$.

The restriction of the kernel projection ω to $\mathcal{A}^T_o$, which is given by

$$\omega\left(X^i \frac{\partial}{\partial x^i} + Y^i \frac{\partial}{\partial y^i} + Z^i_j z^j \frac{\partial}{\partial z^i}\right) = \left(X^i + \left(Z^i_j + X^k \Gamma^i_{kj}\right) z^j\right) \frac{\partial}{\partial z^i} \quad \text{where} \quad Z^i_j y^j = Y^i,$$

is a linear isomorphism $\mathcal{A}^T_o \to \mathcal{K}^T$: for both domain and image have fibre dimension $n^2 + n$, and if the right-hand side is 0 (as a vector field along a fibre) then $X^i = 0$, $Z^i_j = 0$, and therefore $Y^i = 0$.

We call such a Lie algebroid and infinitesimal connection an *infinitesimal Cartan affine geometry over $T^\circ M$*.

We show how to define the geodesics of an infinitesimal Cartan affine geometry over $T^\circ M$. The argument is very reminiscent of the corresponding one for the geodesics in ordinary Cartan affine geometry.

We can define the development of a curve in $T^\circ M$ in our set up—it is a curve in the fibre of $\tau^{\circ *}_M(T^\circ M)$ over a point of $T^\circ M$, so a curve in a linear space. We consider only the case in which the curve in $T^\circ M$ is the natural lift of a curve in M (with nowhere vanishing tangent vector).

Take a curve $c : I \to T^\circ M, c(t) = (x^i(t), \dot{x}^i(t))$, and consider $c^*(\tau^{\circ *}_M(TM))$, the pull-back of $\tau^{\circ *}_M(TM)$ to I defined by c. We define a vector field C on this manifold by

$$C(t, z) = \gamma(\dot{c}(t)) = \dot{x}^i(t)\left(\frac{\partial}{\partial x^i} - \left(\delta^j_i + \Gamma^j_{ik}(x(t), \dot{x}(t))z^k\right) \frac{\partial}{\partial z^j}\right) + \ddot{x}^i(t)\frac{\partial}{\partial y^i}.$$

The flow of C maps fibres to fibres, and consists of affine fibre-morphisms; $c^\Gamma(t)$ is the flow of C acting on $T_{\tau(c(0))}M$. We can write

$$C = \frac{\partial}{\partial t} - \left(\dot{x}^i + \Gamma^i_{jk}\dot{x}^j z^k\right) \frac{\partial}{\partial z^i}.$$

The integral curves of C are given by $t \mapsto (t, z^i(t))$ where the functions z^i are the solutions of the first-order linear inhomogeneous differential equations in z^i

$$\dot{z}^i + \Gamma^i_{jk}(x, \dot{x})\dot{x}^j z^k = -\dot{x}^i.$$

The solution of this equation with $z^i(0) = z_0^i$ takes the form $z^i(t) = z_p^i(t) + z_0^i(t)$ where $z_p^i(t)$ is the particular solution with $z_p^i(0) = 0$ and $z_0^i(t)$ is the solution of the equation $\dot{z}^i + \Gamma^i_{jk}(x, \dot{x})\dot{x}^j z^k = 0$ with $z_0^i(0) = z_0^i$. Since the equation is linear in z^i the latter depends linearly on the initial condition z_0^i. This confirms that the flow of C consists of affine fibre morphisms.

For $s \in I$ there is a unique solution such that $z^i(s) = 0$: denote this by $t \mapsto z^i(s, t)$ (so that $t \mapsto z^i(s, t)$, for fixed s, is a solution of the equation and satisfies $z^i(s, s) = 0$). Then $s \mapsto z^i(s, 0)$ (a curve in $T_{\tau(c(0))}M$) is the development of c.

Suppose that the development of c is a straight line. By differentiating the defining equation for z^i twice with respect to s we obtain

$$\frac{\partial}{\partial t}\left(\frac{\partial^2 z^i}{\partial s^2}\right) + \Gamma^i_{jk}\dot{x}^j \frac{\partial^2 z^k}{\partial s^2} = 0,$$

so that for each fixed s the functions

$$t \mapsto \left(\frac{\partial^2 z^i}{\partial s^2}(s, t)\right)$$

satisfy a linear system. But since the development is a straight line

$$\frac{\partial^2 z^i}{\partial s^2}(s, 0) = 0,$$

and therefore

$$\frac{\partial^2 z^i}{\partial s^2}(s, t) = 0.$$

So $z^i(s, t) = u^i(t)s + v^i(t)$, where $\dot{u}^i + \Gamma^i_{jk}\dot{x}^j u^k = 0$ and $\dot{v}^i + \Gamma^i_{jk}\dot{x}^j v^k = -\dot{x}^i$. But $z^i(t, t) = 0$, so $v^i(t) = -u^i(t)t$, whence $\dot{v}^i + \Gamma^i_{jk}\dot{x}^j v^k = -u^i = -\dot{x}^i$, and so

$$\ddot{x}^i + \Gamma^i_{jk}\dot{x}^j\dot{x}^k = 0.$$

So if the development of $c(t) = (X(t), \dot{x}^i(t))$ is a straight line then $x^i(t)$ satisfies the second-order system above. This looks formally the same as a set of geodesic equations: but the coefficients Γ^i_{jk} will in general be local functions on $T^\circ M$, not M.

12.3 The Cartan–Berwald Geometry of a Spray

We now specialize to Cartan affine geometries over $T^\circ M$ for which the coefficients Γ^i_{jk} have a special homogeneity property, namely that they satisfy

$$\Gamma^i_{jk}(x, \lambda y) = \Gamma^i_{jk}(x, y)$$

for every positive real number λ. The geodesics of the corresponding infinitesimal Cartan affine geometry on $T°M$ then have the property that if $t \mapsto c(t)$ is the geodesic with initial conditions $c(0) = x_0$, $\dot{c}(0) = v_0 \in T_{x_0}M$, and $\lambda > 0$, then the geodesic with initial conditions $c(0) = x_0$, $\dot{c}(0) = \lambda v_0$ is $t \mapsto c(\lambda t)$. This is of course a familiar property of geodesics of linear connections. Note however that we don't assume that another familiar property of linear connections, namely that the geodesic with initial conditions $c(0) = x_0$, $\dot{c}(0) = -v_0$ is $t \mapsto c(-t)$, holds. We don't assume that geodesics are reversible.

The geodesics are the base integral curves of a vector field on $T°M$, namely

$$y^i \frac{\partial}{\partial x^i} - \Gamma^i_{jk} y^j y^k \frac{\partial}{\partial y^i};$$

it is an example of a kind of vector field called for obvious reasons a second-order differential equation field.

A vector field S on $\tau^\circ_M : T°M \to M$ is a *second-order differential equation field* if it satisfies, for all $x \in M$ and $y \in T_xM$, $y \neq 0$,

$$\tau^\circ_{M*} S(x, y) = y;$$

equivalently $\Sigma \circ S = \Delta$ where Σ and Δ are the restrictions of the vertical endomorphism and the Liouville field to $T°M$. A second-order differential equation field is a *spray* S if

$$[\Delta, S] = S.$$

A second-order differential equation field takes the form

$$S = y^i \frac{\partial}{\partial y^i} - 2\Gamma^i \frac{\partial}{\partial y^i}, \quad \Gamma^i = \Gamma^i(x, y);$$

its integral curves are the natural lifts to $T°M$ of solutions of the system of second-order differential equations $\ddot{x}^i + 2\Gamma^i(x, \dot{x}) = 0$. A spray is a second-order differential equation field whose coefficients Γ^i are positively homogeneous of degree 2 in the y^i; if they are quadratic in the y^i (so that S is the geodesic field of a symmetric linear connection) then the spray is said to be *quadratic*. (We mentioned the geodesic spray of a linear connection in Sect. 11.5, and indeed generalized the concept there, but in a very different manner from what we propose here.) In fact a spray which extends smoothly from $T°M$ to TM is necessarily quadratic, which explains why we insist on restricting attention to $T°M$ here. If S is a spray then by Euler's theorem on homogeneous functions its coefficients satisfy

$$\Gamma^i = \tfrac{1}{2} y^j \frac{\partial \Gamma^i}{\partial y^j} = \tfrac{1}{2} y^j y^k \frac{\partial^2 \Gamma^i}{\partial y^j \partial y^k};$$

and if we set

$$\Gamma^i_j = \frac{\partial \Gamma^i}{\partial y^j}, \quad \Gamma^i_{jk} = \frac{\partial^2 \Gamma^i}{\partial y^j \partial y^k},$$

then Γ^i_j and Γ^i_{jk} are positively homogeneous of degrees 1 and 0 respectively.

Homogeneity occurs frequently in what follows and is always with respect to the y^i, so we will just say, for example, that some function is of degree 1. Moreover, the distinction between being positively homogeneous and being homogeneous without qualification won't be important here, so we won't repeat the qualifier 'positively'.

Suppose given a spray

$$S = y^i \frac{\partial}{\partial y^i} - 2\Gamma^i \frac{\partial}{\partial y^i}.$$

The infinitesimal Cartan affine geometry over $T^\circ M$ with

$$\Gamma^i_{jk} = \frac{\partial^2 \Gamma^i}{\partial y^j \partial y^k}$$

is called the *infinitesimal Cartan–Berwald geometry* of S. Note that $\Gamma^i_{kj} = \Gamma^i_{jk}$.

We shall now derive some of the properties of a Cartan–Berwald geometry. The reader who is already familiar with the differential geometry of sprays as expounded for example in Shen's book [37] or the more recent one by Szilasi et al. [38] will immediately recognise our results, though we obtain them by somewhat unconventional means.

We first make a remark about the construction of the Cartan–Berwald connection of a spray. We may see how this arises from S by taking the canonical vertical endomorphism Σ on $T^\circ M$ and considering the Lie derivative $\mathcal{L}_S \Sigma$. The operator $P_H = \frac{1}{2}(I - \mathcal{L}_S \Sigma)$ is a projection whose kernel is the vertical bundle $V\tau^\circ_M$, and so defines a nonlinear connection $X \mapsto X^{\mathsf{h}}$ for vector fields on $T^\circ M$; we may construct another nonlinear connection by subtracting a vertical lift, giving $X \mapsto X^{\mathsf{h}} - X^{\mathsf{v}}$. In coordinates we would have

$$\frac{\partial}{\partial x^i} \mapsto \frac{\partial}{\partial x^i} - \Gamma^j_i \frac{\partial}{\partial y^j} - \frac{\partial}{\partial y^i} = \frac{\partial}{\partial x^i} - \left(\delta^j_i + \Gamma^j_i\right) \frac{\partial}{\partial y^j}, \qquad \Gamma^j_i = \frac{\partial \Gamma^j}{\partial y^i}.$$

We now apply an affine version of the construction given in Sect. 2.7, writing this nonlinear connection as a section κ of $J^1\tau^\circ_M \to T^\circ M$ and then taking an affine fibre derivative at each point of $T^\circ M$ to give a map $\tau^{\circ *}_M(TM) \to J^1\tau^\circ_M$, $(y, z) \mapsto \kappa(y) + \kappa'_{\tau^\circ_M(y)}(z)$. In coordinates we have $\kappa^j_i = -(\delta^j_i + \Gamma^j_i)$ so that

$$\begin{aligned} y^k_i \left(\kappa(y) + \kappa'_{\tau^\circ_M(y)}(z)\right) &= -\left(\delta^j_i + \Gamma^j_i\right) - \frac{\partial \Gamma^j_i}{\partial y^k}(z^k - y^k) \\ &= -\left(\delta^j_i + \Gamma^j_{ik} z^k\right) \end{aligned}$$

using the homogeneity properties of Γ^j_i. It follows that the horizontal vector fields of the resulting affine connection on $\tau_M^{\circ *}(\tau_M^\circ) : \tau_M^{\circ *}(TM) \to T^\circ M$ are just the projectable vector fields defining the infinitesimal connection γ.

As we have just pointed out, a spray induces a horizontal distribution or nonlinear connection on $T^\circ M$, which is spanned locally by the vector fields

$$H_i = \frac{\partial}{\partial x^i} - \Gamma^j_i \frac{\partial}{\partial y^j}.$$

It is usual to work in terms of the horizontal vector fields H_i on $T^\circ M$ rather than the coordinate fields $\partial/\partial x^i$ since this ensures tensoriality. The spray itself is horizontal: $S = y^i H_i$. It will often be convenient to denote the vertical vector field $\partial/\partial y^i$ by V_i.

The covariant derivative operator ∇^γ is determined by its action on sections of $\mathcal{K}^T$. These take the form $X^i \partial/\partial z^i$, and we have

$$\nabla^\gamma_{H_i} \frac{\partial}{\partial z^j} = \Gamma^k_{ij} \frac{\partial}{\partial z^k}, \quad \nabla^\gamma_{V_i} \frac{\partial}{\partial z^j} = 0.$$

These arguments correspond to sections of the pullback bundle $\tau_M^{\circ *}(TM) \to T^\circ M$, or in other words vector fields along the projection $\tau_M^\circ : T^\circ M \to TM$. The *Berwald connection* (see for example [15]) associated with a spray S is a connection on $\tau_M^{\circ *}(TM) \to T^\circ M$. We shall write sections of $\tau_M^{\circ *}(TM)$ as $X^i \partial/\partial x^i$ where the coefficients X^i are local functions on $T^\circ M$ (rather than on M, as the notation might suggest). The Berwald connection can be specified by giving its covariant differentiation operator ∇ operating on $\partial/\partial x^i$, regarded again as a local section of $\tau_M^{\circ *}(TM)$, together with the usual rules of covariant differentiation: in fact

$$\nabla_{H_i} \frac{\partial}{\partial x^j} = \Gamma^k_{ij} \frac{\partial}{\partial x^k}, \quad \nabla_{V_i} \frac{\partial}{\partial x^j} = 0.$$

So the Berwald covariant derivative ∇ is just ∇^γ in a different guise. It is of course the close relation between the infinitesimal Cartan affine connection for a spray and the Berwald connection that suggested the name Cartan–Berwald connection.

Note that covariant differentiation with respect to the vertical vector field V_i of any tensor field along τ_M° amounts simply to partial differentiation of the components of the field with respect to y^i; and that therefore if one takes a tensor field along τ_M° and partially differentiates its components with respect to the y^i one obtains another tensor field, with one more covariant index. If T is a type $(1, 1)$ tensor along τ_M° and X a vector field on $T^\circ M$, $(\nabla_X T)^i_j$ is just written $\nabla_X T^i_j$. Moreover, where convenient we shall denote by (for example) $T^i_{j,k}$ the tensor component $\nabla_{V_k} T^i_j$, and $T^i_{j|k}$ the tensor component $\nabla_{H_k} T^i_j$.

The total derivative $\mathbf{T}$ is the vector field along τ_M° whose coordinate representation is $y^i \partial/\partial x^i$, and is just the section of $\tau_M^{\circ *}(TM) \to T^\circ M$ used to define the Lie subalgebroid $\mathcal{A}_1^T$. The covariant derivative of $\mathbf{T}$ in any horizontal direction vanishes, which is to say that $y^i_{|j} = 0$.

We next consider the curvature R^γ of γ. First, we have

$$R^\gamma(H_i, H_j) = R^k_{lij} z^l \frac{\partial}{\partial z^k}$$

where

$$R^k_{lij} = H_i(\Gamma^k_{jl}) - H_j(\Gamma^k_{il}) + \Gamma^k_{im}\Gamma^m_{jl} - \Gamma^k_{jm}\Gamma^m_{il}.$$

This is superficially similar to the usual formula for the Riemann curvature, and is therefore called the *Riemann curvature* of the spray. It has the usual symmetries, but one must remember that it may depend on y^i: in fact it is homogeneous of degree 0, and reduces to the ordinary curvature tensor when the spray is quadratic. Secondly,

$$R^\gamma(V_i, H_j) = B^k_{lij} z^l \frac{\partial}{\partial z^k}$$

where

$$B^k_{lij} = V_i(\Gamma^k_{jl}) = \frac{\partial^3 \Gamma^k}{\partial y^i \partial y^j \partial y^l}.$$

This component of the curvature evidently vanishes in the quadratic case: in fact its vanishing is the necessary and sufficient condition for the spray to be quadratic. It is completely symmetric in its lower indices, is homogeneous of degree -1, and satisfies $B^k_{lij} y^l = 0$ by Euler's identity. It is called the *Berwald curvature* of the spray. Finally, evidently

$$R^\gamma(V_i, V_j) = 0.$$

Since Γ^i_{jk} is symmetric there is no torsion in the usual sense, that is, R^γ depends linearly on z^i. We may also consider the torsion relative to $\mathcal{A}^T_o$, that is, the component of R^γ in $\mathcal{K}^T/\mathcal{K}^T_o$. Since $B^k_{lij} y^l = 0$, $R^\gamma(V_i, H_j) \in \sec(\mathcal{K}^T_o)$, and (perhaps surprisingly) this torsion is essentially the Riemann curvature, in the form $R^k_{ij} = R^k_{lij} y^l$. For a Cartan–Berwald geometry to be flat both components of its curvature must vanish, of course. If just $R^k_{ij} = 0$ the spray is said to be *R-flat*: thus the geometry is torsion free if and only if the spray is R-flat.

The Bianchi identities for R^γ can be obtained by applying Lemma 2.4.3. One of them is the usual first Bianchi identity,

$$R^l_{ijk} + R^l_{jki} + R^l_{kij} = 0.$$

Another relates the two curvatures:

$$R^l_{kij,m} = B^l_{jkm|i} - B^l_{ikm|j}.$$

It follows from this formula, together with the facts that $y^k_{|i} = 0$ and $B^l_{jkm} y^k = 0$, that

$$R^l_{kij,m} y^k = 0.$$

We shall also be concerned with the associated tensor $R^i_{jk} = R^i_{ljk} y^l$ that we met above. This type $(1, 2)$ tensor field may equally well be defined as

$$R^i_{jk} \frac{\partial}{\partial y^i} = -[H_j, H_k].$$

It contains the same information as the Riemann curvature, which can be recovered from it via the formula

$$R^i_{ljk} = R^i_{jk,l}.$$

Thus a spray is R-flat if and only if the full Riemann curvature R^l_{ijk} vanishes.

In fact the type $(1, 1)$ tensor $R^i_j = R^i_{jk} y^k$ also contains all the Riemann curvature information, since

$$R^i_{j,k} - R^i_{k,j} = R^i_{jk} + R^i_{kjl} y^l - R^i_{kj} - R^i_{jkl} y^l = 2R^i_{jk} + R^i_{ljk} y^l = 3R^i_{jk}.$$

This tensor governs geodesic deviation. We may think of the geodesic deviation equation as follows. Take an integral curve c of S and a vector field δ along it: then δ is a geodesic deviation vector if $\mathcal{L}_S \delta = 0$. We may write $\delta = X^{\mathsf{h}} + Y^{\mathsf{v}}$, where X, Y are vector fields on M along the geodesic obtained by projecting c to M. Now

$$\mathcal{L}_S(X^{\mathsf{h}}) = \mathcal{L}_S(X^i H_i) = \dot{X}^i H_i + X^i [y^j H_j, H_i] = (\nabla_S X^i) H_i + R^i_j X^j V_i,$$

while

$$\mathcal{L}_S(Y^{\mathsf{v}}) = \mathcal{L}_S(Y^i V_i) = \dot{Y}^i V_i + Y^i [y^j H_j, V_i] = -Y^i H_i + (\nabla_S Y^i) V_i.$$

So we require that $Y = \nabla_S X$ and $\nabla_S Y^i + R^i_j X^j = 0$, or

$$\nabla_S^2 X^i + R^i_j X^j = 0,$$

which is the geodesic deviation equation, also known as Jacobi's equation; the type $(1, 1)$ tensor field R^i_j is often called the *Jacobi endomorphism* of the spray.

Not every infinitesimal connection of the form

$$\gamma\left(\frac{\partial}{\partial x^i}\right) = \frac{\partial}{\partial x^i} - \left(\delta^j_i + \Gamma^j_{ik} z^k\right) \frac{\partial}{\partial z^j}$$
$$\gamma\left(\frac{\partial}{\partial y^i}\right) = \frac{\partial}{\partial y^i}$$

is the Cartan–Berwald connection of a spray. We can specify the requirements on γ that the connection coefficients Γ^i_{jk} come from a spray as follows.

First, homogeneity. It is clear that

$$\Delta = y^i \frac{\partial}{\partial y^i}$$

is a well-defined vector field on $\tau_M^{\circ *}(T^\circ M)$, which of course projects onto the usual vector field with the same name on $T^\circ M$. The first requirement is that $\mathcal{L}_\Delta \gamma = 0$. When this holds we say that γ is homogeneous.

Secondly, we want

$$\Gamma^i_{jk} = \frac{\partial^2 \Gamma^i}{\partial y^j \partial y^k}$$

for some functions Γ^i, homogeneous of degree 2. Now

$$R^\gamma\left(\frac{\partial}{\partial y^i}, \frac{\partial}{\partial x^j}\right) = \frac{\partial \Gamma^l_{jk}}{\partial y^i} z^k \frac{\partial}{\partial z^l},$$

so a necessary condition is that

$$R^\gamma\left(\frac{\partial}{\partial y^i}, \frac{\partial}{\partial x^j}\right) = R^\gamma\left(\frac{\partial}{\partial y^j}, \frac{\partial}{\partial x^i}\right).$$

When this holds the infinitesimal connection is said to be of *Berwald type*.

The translation part (z-independent part) of

$$R^\gamma\left(\frac{\partial}{\partial x^i}, \frac{\partial}{\partial x^j}\right)$$

is the torsion $\Gamma^k_{ij} - \Gamma^k_{ji}$. It is also necessary that the infinitesimal connection have zero torsion in this narrow sense.

Proposition 12.3.1 *The necessary and sufficient conditions for a vertically trivial infinitesimal Cartan affine connection on $T^\circ M$, of standard form (i.e. with $A^i_j = \delta^i_j$), to be the Cartan–Berwald connection of a spray are that it be*

- *homogeneous;*
- *torsion-free in the narrow sense;*
- *of Berwald type.*

Proof We have already shown necessity; it remains to show sufficiency.

Since the infinitesimal connection is vertcally trivial of standard form it is completely determined by the coefficients $\Gamma^i_{jk}(x, y)$, and these are positively homogeneous of degree zero in y, symmetric in j and k, and satisfy

$$\frac{\partial \Gamma^i_{jk}}{\partial y^l} = \frac{\partial \Gamma^i_{jl}}{\partial y^k}.$$

Define

$$\Gamma^i = \tfrac{1}{2}\Gamma^i_{jk} y^j y^k.$$

Then Γ^i is positively homogeneous of degree 2. Moreover

$$\frac{\partial \Gamma^i}{\partial y^j} = \tfrac{1}{2}\frac{\partial \Gamma^i_{kl}}{\partial y^j} y^k y^l + \Gamma^i_{jk} y^k = \tfrac{1}{2}\frac{\partial \Gamma^i_{kj}}{\partial y^l} y^k y^l + \Gamma^i_{jk} y^k = \Gamma^i_{jk} y^k,$$

and similarly

$$\frac{\partial^2 \Gamma^i}{\partial y^j \partial y^k} = \frac{\partial \Gamma^i_{jl}}{\partial y^k} y^l + \Gamma^i_{jk} = \Gamma^i_{jk}$$

as required. □

12.4 The Generalized Geodesic Spray of a Spray

We shall now compute the generalised geodesic spray of a spray

$$S = y^i \frac{\partial}{\partial y^i} - 2\Gamma^i \frac{\partial}{\partial y^i},$$

on $T^\circ M$.

The generalized geodesic spray of the covariant derivative ∇^{B} is a vector field on the vector bundle $\mathcal{A}^T_o$. We first choose a convenient coordinatization (which is really half the battle), The idea, as in the affine case, is to work relative to the horizontal distribution. But now there are two. There is the horizontal distribution associated with the spray, which is spanned by the vector fields

$$H_i = \frac{\partial}{\partial x^i} - \Gamma^j_i \frac{\partial}{\partial y^j}, \quad \Gamma^j_i = \frac{\partial \Gamma^j}{\partial y^i},$$

on $T^\circ M$. But there is also the horizontal distribution on $\tau^{\circ *}_M(TM)$ corresponding to the linear part of γ. On combining the two one obtains horizontal vector fields

$$\tilde{H}_i = \frac{\partial}{\partial x^i} - \Gamma^j_i \frac{\partial}{\partial y^j} - \Gamma^j_{ik} z^k \frac{\partial}{\partial z^j}.$$

We shall express the general element of $\mathcal{A}^T_o$ as

$$\xi = X^i \tilde{H}_i + Y^i V_i + Z^i_j z^j \frac{\partial}{\partial z^i};$$

it turns out that the condition relating the coefficients is still $Y^i = Z^i_j y^j$. We may take (X^i, Y^i, Z^i_j) as fibre coordinates on $\mathcal{A}^T$, and then $\mathcal{A}^T_o$ is the sub-bundle for which $Y^i = Z^i_j y^j$. With this set-up we find that

$$\nabla^{\mathrm{B}}_{H_i}\xi = \left(X^j_{|i} - Z^j_i\right) H_j + \left(Y^j_i + R^j_{ik}X^k\right) V_j + \left(Z^j_{k|i} + R^j_{kli}X^l + B^j_{kli}Y^l\right) z^k \frac{\partial}{\partial z^j},$$

$$\nabla^{\mathrm{B}}_{V_i}\xi = X^j_{,i} H_j + \left(Y^j_{,i} - Z^i_j\right) V_j + \left(Z^j_{k,i} - B^j_{kli}X^l\right) z^k \frac{\partial}{\partial z^j};$$

the results have been expressed in a form designed to facilitate comparison with the affine case. It follows that

$$\begin{aligned} S^{\mathrm{B}} &= X^i H_i + Y^i V_i \\ &\quad + \left(Z^i_j X^j - \Gamma^i_{jk} X^j X^k\right) \frac{\partial}{\partial X^i} + \left(Z^i_j Y^j - \Gamma^i_{jk} Y^j X^k\right) \frac{\partial}{\partial Y^i} \\ &\quad + \left(\Gamma^l_{kj} X^k Z^i_l - \Gamma^i_{lk} X^k Z^l_j\right) \frac{\partial}{\partial Z^i_j}; \end{aligned}$$

the curvature terms cancel as before.

Evidently S^{B} is tangent to the submanifold of $\mathcal{A}^T_o$ on which $Z^i_j = 0$, when necessarily $Y^i = 0$. It is furthermore tangent to the 'diagonal' $X^i = y^i \neq 0$, which can be identified with $T^\circ M$; and its restriction to $T^\circ M$ can be written

$$y^i \frac{\partial}{\partial x^i} - \Gamma^i_{jk} y^j y^k \frac{\partial}{\partial y^i},$$

which, since $\Gamma^i_{jk} y^j y^k = 2\Gamma^i$ by homogeneity, is the spray with which we started.

12.5 Symmetries of a Cartan–Berwald Geometry

We continue with a spray

$$S = y^i \frac{\partial}{\partial y^i} - 2\Gamma^i \frac{\partial}{\partial y^i}.$$

An infinitesimal symmetry of the Cartan–Berwald geometry of S is a section ζ of $\mathcal{A}^T_o \to T^\circ M$ such that $\mathcal{L}_\zeta \gamma = 0$.

Proposition 12.5.1 *The necessary and sufficient conditions for*

$$\xi = X^i \frac{\partial}{\partial x^i} + Y^i_j y^j \frac{\partial}{\partial y^i} + Y^i_j z^j \frac{\partial}{\partial z^i}$$

to be an infinitesimal symmetry of the Cartan–Berwald geometry of S are that

- X^i *and* Y^i_j *are independent of* y^i;
- $Y^i_j = \partial X^i/\partial x^j$;
- *the* X^i *satisfy*

$$\frac{\partial^2 X^i}{\partial x^j \partial x^k} - \frac{\partial X^i}{\partial x^l}\Gamma^l_{jk} + \frac{\partial X^l}{\partial x^j}\Gamma^i_{lk} + \frac{\partial X^l}{\partial x^k}\Gamma^i_{lj} + \left(X^l\frac{\partial}{\partial x^l} + \frac{\partial X^l}{\partial x^m}y^m\frac{\partial}{\partial y^l}\right)(\Gamma^i_{jk}) = 0.$$

The first two of these requirements may be condensed into one, namely that

$$X^i\frac{\partial}{\partial x^i} + Y^i_j y^j\frac{\partial}{\partial y^i} = X^{\mathrm{c}} \quad \text{where } X = X^i\frac{\partial}{\partial x^i} \text{ is a vector field on } M.$$

Then the third condition may be written

$$\frac{\partial^2 X^i}{\partial x^j \partial x^k} - \frac{\partial X^i}{\partial x^l}\Gamma^l_{jk} + \frac{\partial X^l}{\partial x^j}\Gamma^i_{lk} + \frac{\partial X^l}{\partial x^k}\Gamma^i_{lj} + X^{\mathrm{c}}(\Gamma^i_{jk}) = 0.$$

This is an obvious generalization of a formulation of the condition for a vector field to be an infinitesimal affine transformation of a linear connection which we quoted in Sect. 5.5.

Proof The condition

$$\mathcal{L}_{\xi\gamma}\left(\frac{\partial}{\partial y^i}\right) = 0$$

leads to the independence of X^i and Y^i_j from y^i. Setting to zero the terms independent of z^i in the condition

$$\mathcal{L}_{\xi\gamma}\left(\frac{\partial}{\partial x^i}\right) = 0$$

then gives $Y^i_j = \partial X^i/\partial x^j$, while those linear in z^i produce the final condition. □

Corollary 12.5.2 *The section ξ is an infinitesimal symmetry of the Cartan–Berwald geometry of the spray S if and only if the vector field X on M defined above satisfies* $[X^{\mathrm{c}}, S] = 0$.

Proof The condition $[X^{\mathrm{c}}, S] = 0$, when written in coordinates, is

$$\frac{\partial^2 X^i}{\partial x^j \partial x^k}y^j y^k + 2\frac{\partial X^l}{\partial x^j}\Gamma^j - 2X^{\mathrm{c}}(\Gamma^i) = 0.$$

This is easily seen to be equivalent to the condition given in Proposition 12.5.1, when account is taken of the homogeneity of Γ^i. □

We must now make what will appear to be a digresion, on the subject of sprays and their base integral curves. Let $t \mapsto c_{(x,y)}(t)$ be the base integral curve of the spray S with initial conditions $(x, y) \in T^{\circ}M$, that is, the projection of the integral curve of S through (x, y) from $T^{\circ}M$ to M. Then by the homogeneity property of S, for any $\lambda > 0$,

$$c_{(x,\lambda y)}(t) = c_{(x,y)}(\lambda t).$$

But we do not assume that the spray is reversible: that is, it need not be the case that

$$c_{(x,-y)}(t) = c_{(x,y)}(-t).$$

We concentrate therefore on the forward segment of $c_{(x,y)}$: that is, we assume that $t \in [0, \varepsilon)$ for some $\varepsilon > 0$, depending on x and, in particular, y. However, by restricting y, say, to lie on a coordinate Euclidean sphere B_r in T_xM, centred at the origin, of suitable radius $r > 0$, we may assume that $\varepsilon > 1$, so that in particular $c_{(x,y)}(1)$ is well defined. Then for any y in the interior B_r° of B_r, by homogeneity $c_{(x,y)}(t)$ is defined for $t = 1$ also, and we may validly set

$$\exp_x : B_r^{\circ} \subset T_xM \to M : \quad \exp_x(y) = c_{(x,y)}(1).$$

It is known [37] that $\exp_x$, so defined, is C^1 on its domain, and that its differential at $y = 0$ is the identity. It follows that there is an open neighbourhood of $x \in M$, every point of which lies on some forward base integral curve of S starting at x.

Proposition 12.5.3 *The maximum dimension of the algebra of infinitesimal symmetries of the Cartan–Berwald geometry of a spray over a connected manifold M is $n(n+1)$. If this dimension is attained the spray is quadratic, and defines a flat affine geometry.*

Proof Take any point x of M, and consider the map which associates with an infinitesimal symmetry the pair

$$\left(X^i(x), \left.\frac{\partial X^i}{\partial x^j}\right|_x \right) \in \mathbb{R}^n \oplus \mathfrak{gl}(n).$$

It is a linear map into a vector space of dimension $n(n+1)$. We show that it is injective. Assume that

$$X^i(x) = 0, \quad \left.\frac{\partial X^i}{\partial x^j}\right|_x = 0,$$

which is to say that $X^{\mathrm{c}} = 0$ on T_xM. Let ϕ_s^{c} be the flow of X^{c}, ψ_t the flow of S. By assumption, $\phi_s^{\mathrm{c}}(x, y) = (x, y)$ for all $y \in T_xM$. Since $[X^{\mathrm{c}}, S] = 0$,

$$\phi_s^{\mathrm{c}} \circ \psi_t = \psi_t \circ \phi_s^{\mathrm{c}}$$

when both sides make sense. It follows, by projection, that

$$\phi_s\big(c_{(x,y)}(t)\big) = c_{\phi_s^{\mathrm{c}}(x,y)}(t) = c_{(x,y)}(t),$$

where of course ϕ_s is the flow of X on M. That is to say, there is a neighbourhood of x in M on which ϕ_s is the identity, and therefore X vanishes. It follows that the set of points $x \in M$ such that

$$X^i(x) = 0, \quad \left.\frac{\partial X^i}{\partial x^j}\right|_x = 0,$$

is both open and closed, and therefore if not empty the whole of M. It follows that the map which associates with an infinitesimal symmetry the pair

$$\left(X^i(x), \left.\frac{\partial X^i}{\partial x^j}\right|_x\right)$$

is injective, as claimed.

Now by differentiating the differential condition for an infinitesimal symmetry with respect to y^l we obtain the following equation involving the Berwald curvature:

$$-\frac{\partial X^i}{\partial x^m}B^m_{jkl} + \frac{\partial X^m}{\partial x^j}B^i_{klm} + \frac{\partial X^m}{\partial x^k}B^i_{jlm} + \frac{\partial X^m}{\partial x^l}B^i_{jkm} + X^{\mathrm{c}}(B^i_{jkl}) = 0.$$

If the maximum dimension is attained then at each $x \in M$ there is an infinitesimal symmetry for which

$$X^i(x) = 0, \quad \left.\frac{\partial X^i}{\partial x^j}\right|_x = \delta^i_j.$$

With those choices we have

$$2B^i_{jkl} + \Delta(B^i_{jkl}) = 0.$$

But B^i_{jkl} is homogeneous of degree -1, so $\Delta(B^i_{jkl}) = -B^i_{jkl}$, and therefore $B^i_{jkl} = 0$. This holds everywhere: the spray is therefore quadratic, and defines an affine geometry. Its symmetry algebra is of maximum dimension, so the geometry is flat by Proposition 8.1.1. □

12.6 Projective Equivalence of Sprays

The Cartan–Berwald geometry of a spray may be regarded as specifying the geometry of a certain system of second-order differential equations, namely those determining the geodesics of the spray. But systems of this form comprise a pretty limited class. It turns out that we can deal with systems of second-order differential equations in far greater generality if we broaden our scope to cover projectively equivalent sprays.

Two sprays S, $\hat{S}$ are *projectively equivalent* if

$$\hat{S} - S = -2\alpha\Delta, \quad \text{or } \hat{\Gamma}^i = \Gamma^i + \alpha y^i,$$

where the function α is positively homogeneous of degree 1 in the y^i.

The base integral curves of a spray—its geodesics—are the solutions of the system of second-order ordinary differential equations

$$\ddot{x}^i + 2\Gamma^i(x, \dot{x}) = 0.$$

All sprays in a projective equivalence class have the same base integral curves up to a change of parameter which preserves sense. Thus a projective equivalence class of sprays determines a *path geometry*, that is, a collection of paths (unparametrized but oriented curves) in M with the property that there is a unique path of the collection through each point in each direction. In fact the notions of path geometry and projective equivalence class of sprays are interchangeable: it may be shown that any path geometry determines a projective equivalence class of sprays (see [18]).

A choice of spray in a projective equivalence class amounts to a choice of parametrization of the corresponding paths. In a local coordinate system we can choose to parametrize suitable paths of a projective class of sprays with one of the coordinates, say x^1; with such a parametrization $\dot{x}^1 = 1$, $\ddot{x}^1 = 0$, and the differential equations take the form[1]

$$\frac{d^2x^a}{d(x^1)^2} = f^a\left(x^1, x^b, \frac{dx^b}{dx^1}\right), \qquad a, b = 2, 3, \ldots, n.$$

In other words, there is always locally a member of the projective class for which $\Gamma^1 = 0$; then $f^a(x^i, u^b) = -2\Gamma^a(x^i, 1, u^b)$. Conversely, given a system of $n-1$ second-order differential equations in the variables x^a, with parameter x^1, we can locally recover a spray by setting

$$\Gamma^1 = 0, \quad \Gamma^a(x^i, y^i) = -\tfrac{1}{2}(y^1)^2 f^a(x^i, y^b/y^1).$$

Notice, however, that such a spray is reversible. In general sprays are required to be only positively homogeneous. A spray whose coefficients Γ^i are homogeneous of degree 2 without qualification, that is, satisfy $\Gamma^i(x, \lambda y) = \lambda^2\Gamma^i(x, y)$ for all non-zero λ, is said to be *reversible*. The geodesics of a reversible spray are reversible: that is, the corresponding path geometry has the property that given a point $x \in M$ and a line in T_xM there is a unique path (now an unparametrized and unoriented curve) through x whose tangent line at x is the given line.

Thus the study of systems of second-order ordinary differential equations of the form

[1] Our standard convention of using indices $a, b = 0, 1, 2, \ldots, n$ is suspended for this section.

$$\frac{d^2x^a}{dt^2} = f^a\left(t, x^b, \frac{dx^b}{dt}\right)$$

under point transformations—coordinate transformations in which no distinction is made between the dependent and independent variables—is subsumed by the projective geometry of sprays. (The equivalence problem for such systems under point transformations was solved by Fels in [24]. The relation between his work and ours may be inferred from [19].)

We shall embark on the construction of a Cartan projective geometry of sprays in the next section. But we first make a remark about the fundamental projective invariant.

It follows from the basic projective transformation rule that

$$\hat{\Gamma}^k_{ij} = \Gamma^k_{ij} + \left(\alpha_i\delta^k_j + \alpha_j\delta^k_i + \alpha_{ij}y^k\right), \qquad \alpha_i = \frac{\partial\alpha}{\partial y^i} \quad \alpha_{ij} = \frac{\partial^2\alpha}{\partial y^i\partial y^j}.$$

By taking a trace in this equation and writing Γ_i for Γ^j_{ij}, taking account of the homogeneity of α we obtain $\hat{\Gamma}_i = \Gamma_i + (n+1)\alpha_i$, whence the quantity

$$\Pi^k_{ij} = \Gamma^k_{ij} - \frac{1}{n+1}\left(\Gamma_i\delta^k_j + \Gamma_j\delta^k_i + B_{ij}y^k\right), \quad B_{ij} = B^l_{lij} = \frac{\partial\Gamma_i}{\partial y^j},$$

is projectively invariant. It is this that Douglas [21] calls the fundamental descriptive invariant: we used the corresponding object for a projective class of linear connections (or of quadratic sprays) extensively in Chap. 9. Note that $\Pi^j_{ij} = \Pi^j_{ij} = 0$.

12.7 The Pullback Projective Bundle

The Cartan–Berwald geometry of a spray is an infinitesimal affine Cartan geometry over $T^\circ M$. We now show how to construct a projective version of this geometry: that is, a Cartan–Berwald style connection for the projective geometry of sprays. We saw in Chap. 9 that the full version of Cartan projective geometry over M used $\pi : \mathrm{P}\mathcal{W}M \to M$ as its underlying bundle; here we shall use $\pi^*\mathrm{P}\mathcal{W}M \to \mathrm{P}\mathcal{W}M$, the pullback of $\mathrm{P}\mathcal{W}M$ over $\mathrm{P}\mathcal{W}M$.

Let us first consider $\tau^*\mathcal{W}M \to \mathcal{W}M$, the pullback of $\mathcal{W}M$ over $\mathcal{W}M$, where $\tau : \mathcal{W}M \to M$, with coordinates (x^i, w^a, v^a) where $a = 0, 1, 2, \dots, n$.

Consider further the Lie algebroid of fibre-linear vector fields, which consists of vector fields along fibres of $\tau^*\mathcal{W}M \to \mathcal{W}M$ of the form

$$X^i\frac{\partial}{\partial x^i} + Y^a\frac{\partial}{\partial w^a} + Z^a_b v^b\frac{\partial}{\partial v^a},$$

where X^i, Y^a and Z^a_b are constants. Call this algebroid $\mathcal{A}^{\mathcal{W}}$.

We define a subalgebroid $\mathcal{A}_o^{\mathcal{W}}$ of $\mathcal{A}^{\mathcal{W}}$ on the model of the construction in the discussion of the Cartan–Berwald connection. First, there is a global section of $\tau^*\mathcal{W}M \to \mathcal{W}M$, given in coordinates by $v^0 = 1$, $v^i = 0$. The subalgebroid $\mathcal{A}_0^{\mathcal{W}}$ of $\mathcal{A}^{\mathcal{W}}$ consisting of vector fields tangent to the image of this section consists of those for which $Z_0^a = 0$. There is a second submanifold (which will define a global section of $\pi^*\mathrm{P}\mathcal{W}M \to \mathrm{P}\mathcal{W}M$) consisting of those pairs (w, v) which are linearly dependent. For $v^0 \neq 0$, $w^0 \neq 0$, this is given by $v^0 w^i = w^0 v^i$, and the vector fields along fibres which are tangent to it are those for which

$$\begin{aligned} Y^i v^0 - Y^0 v^i &= Z_a^i v^a w^0 - Z_a^0 v^a w^i \\ &= Z_0^i v^0 w^0 + Z_j^i v^j w^0 - Z_0^0 v^0 w^i - Z_j^0 v^j w^i \end{aligned}$$

where $v^0 w^i = w^0 v^i$, that is, for which

$$Y^i w^0 - Y^0 w^i = Z_0^i (w^0)^2 + Z_j^i w^j w^0 - Z_0^0 w^0 w^i - Z_j^0 w^j w^i.$$

Such vector fields form a subalgebroid $\mathcal{A}_1^{\mathcal{W}}$ of $\mathcal{A}^{\mathcal{W}}$. We take $\mathcal{A}_o^{\mathcal{W}} = \mathcal{A}_0^{\mathcal{W}} \cap \mathcal{A}_1^{\mathcal{W}}$. It is a subalgebroid of $\mathcal{A}^{\mathcal{W}}$, consisting of vector fields

$$X^i \frac{\partial}{\partial x^i} + Y^a \frac{\partial}{\partial w^a} + Z_b^a v^b \frac{\partial}{\partial v^a}$$

where

$$Z_0^0 = Z_0^i = 0, \quad Z_j^i w^j w^0 - Z_j^0 w^j w^i = Y^i w^0 - Y^0 w^i;$$

we suppose Y^a given and regard the equations above as conditions on Z_b^a.

Now let us projectivize, in both w^a and v^a, which means setting $w^i/w^0 = y^i$ and $v^i/v^0 = z^i$. We must now think of Y^a as being defined along a ray in a fibre of $\mathcal{W}M \to M$, and being homogeneous of degree 1 in the w^a and determined only up to the addition of a multiple of the Liouville field. Then

$$Y^a \frac{\partial}{\partial w^a} \mapsto \frac{1}{(w^0)^2}(Y^i w^0 - Y^0 w^i)\frac{\partial}{\partial y^i} = \frac{1}{w^0}(Y^i - Y^0 y^i)\frac{\partial}{\partial y^i} = \bar{Y}^i \frac{\partial}{\partial y^i}$$

say, which is a well-defined (local) vector field on $\mathrm{P}\mathcal{W}M$ in virtue of the assumptions about Y^a. Likewise, in order to projectivize we must regard (Z_b^a) as being determined only up to the addition of a multiple of the identity, and we may use this freedom to require that $Z_0^0 = 0$. Then

$$Z_b^a v^b \frac{\partial}{\partial v^a} \mapsto \left(Z_0^i + Z_j^i z^j - (Z_j^0 z^j)\, z^i\right) \frac{\partial}{\partial z^i}.$$

So $\mathcal{A}^{\mathcal{W}}$ projects to $\mathcal{A}$, say, which consists of vector fields along fibres of $\pi^*\mathrm{P}\mathcal{W}M \to \mathrm{P}\mathcal{W}M$ of the form

$$X^i \frac{\partial}{\partial x^i} + \bar{Y}^i \frac{\partial}{\partial y^i} + \left(Z^i_0 + Z^i_j z^j - \left(Z^0_j z^j\right) z^i\right) \frac{\partial}{\partial z^i}.$$

Now the conditions $Z^a_0 = 0$ and $Z^i_j w^j w^0 - Z^0_j w^j w^i = Y^i w^0 - Y^0 w^i$ are consistent with the requirements on Z^a_b and Y^a which allow us to projectivize; and the latter condition, when expressed in terms of y^i, becomes $Z^i_j y^j - (Z^0_j y^j) y^i = \bar{Y}^i$. Thus $\mathcal{A}^{\mathcal{W}}_o$ projects to $\mathcal{A}_o$ which consists of vector fields along fibres of $\pi^*\mathrm{P}\mathcal{W}M \to \mathrm{P}\mathcal{W}M$ of the form

$$X^i \frac{\partial}{\partial x^i} + \bar{Y}^i \frac{\partial}{\partial y^i} + \left(Z^i_j z^j - \left(Z^0_j z^j\right) z^i\right) \frac{\partial}{\partial z^i} \quad \text{where } Z^i_j y^j - \left(Z^0_j y^j\right) y^i = \bar{Y}^i.$$

Then $\mathcal{A}_o$ is transitive over $\mathrm{P}\mathcal{W}M$. Its kernel $\mathcal{K}_o$ consists of vector fields of the form

$$\left(Z^i_j z^j - \left(Z^0_j z^j\right) z^i\right) \frac{\partial}{\partial z^i} \quad \text{where } Z^i_j y^j = (Z^0_j y^j) y^i.$$

For each $(x, y) \in \mathrm{P}\mathcal{W}M$, $\mathcal{K}_{o,(x,y)}$ is a Lie subalgebra of the Lie algebra $\mathcal{K}_{(x,y)}$ whose codimension is

$$\dim \mathcal{K}_{(x,y)} - \dim \mathcal{K}_{o,(x,y)} = (n^2 + 2n) - (n^2 + n - n) = 2n.$$

So $\mathcal{A}_o$ is a transitive algebroid over a $2n$-dimensional manifold $\mathrm{P}\mathcal{W}M$, whose kernel $\mathcal{K}_o$ has fibre codimension $2n$ in $\mathcal{K}$, the kernel of $\mathcal{A}$: that is, such that (fibre dimensions)

$$\dim \mathcal{A}_o = \dim \mathrm{P}\mathcal{W}M + \dim \mathcal{K}_o = 2n + \dim \mathcal{K} - 2n = \dim \mathcal{K}.$$

12.8 Connection and Curvature

We continue with the Lie algebroid of $\mathcal{A}$ of infinitesimal projective transformations of the pullback bundle $\pi^*\mathrm{P}\mathcal{W}M \to \mathrm{P}\mathcal{W}M$, and its Lie subalgebroid $\mathcal{A}_o$. The general infinitesimal connection on this bundle is given by

$$\gamma\left(\frac{\partial}{\partial x^i}\right) = \frac{\partial}{\partial x^i} - \left(\delta^j_i + \Pi^{\,j}_{ik} z^k + (\Lambda_{ik} z^k) z^j\right) \frac{\partial}{\partial z^j}$$

$$\gamma\left(\frac{\partial}{\partial y^i}\right) = \frac{\partial}{\partial y^i}.$$

Here (x^i, y^i) are strictly coordinates on $\mathrm{P}\mathcal{W}M$, but we shall treat the y^i as though they were fibre coordinates on $T^\circ M$. Likewise, the z^i are strictly fibre coordinates on $\pi^*\mathrm{P}\mathcal{W}M \to \mathrm{P}\mathcal{W}M$, but we shall treat them as fibre coordinates on the pullback of TM over $T^\circ M$. The quantities $\Pi^{\,i}_{jk}$ only make sense as a global object (that is, as local representatives of a global object) on $\pi^*\mathrm{P}\mathcal{W}M$; but given a projective equivalence class of sprays and any coordinate neighbourhood in M we can always choose a member of the class, at least locally, whose Berwald connection coefficients in that coordinate neighbourhood are the $\Pi^{\ i}_{jk}$. The corresponding Riemann curvature is $\mathfrak{R}^i_{jkl}$. The *projective curvature* is given by

$$P^i_{jkl} = \mathfrak{R}^i_{jkl} - \frac{1}{n-1}\left(\mathfrak{R}_{jl}\delta^i_k - \mathfrak{R}_{jk}\delta^i_l\right), \quad \mathfrak{R}_{ij} = \mathfrak{R}^k_{ikj};$$

it is known that $\mathfrak{R}_{ij}$ is symmetric. We shall also need the projective version of the Berwald tensor, namely the *Douglas tensor*,

$$D^i_{jkl} = \frac{\partial \Pi^{\ i}_{jk}}{\partial y^l};$$

D^i_{jkl} is completely symmetric in its lower indices, and since $\Pi^{\ i}_{jk}$ is homogeneous of degree zero, $D^i_{jkl}y^l = 0$.

For the normal connection we take

$$\Lambda_{ij} = -\frac{1}{n-1}\mathfrak{R}_{ij}.$$

The curvature of the infinitesimal connection γ is given by

$$R^\gamma\left(\frac{\partial}{\partial x^i}, \frac{\partial}{\partial x^j}\right) = \left(P^k_{lij}z^l + (\Lambda_{li|j} - \Lambda_{lj|i})z^l z^k\right)\frac{\partial}{\partial z^k}$$

$$R^\gamma\left(\frac{\partial}{\partial y^i}, \frac{\partial}{\partial x^j}\right) = \left(D^k_{ijl}z^l + \Lambda_{lj,i}z^l z^k\right)\frac{\partial}{\partial z^k}$$

$$R^\gamma\left(\frac{\partial}{\partial y^i}, \frac{\partial}{\partial y^j}\right) = 0.$$

As in Sect. 12.3 the rule indicates the Berwald covariant derivative in a horizontal direction, the comma the Berwald covariant derivative in a vertical direction; the latter is simply partial differentiation with respect to y^i. (These formulae agree with those for the infinitesimal projective connection in the quadratic case, and with those for the Cartan–Berwald connection of a spray when the Λ terms are ignored.)

The torsion of γ is the component of R^γ in $\mathcal{K}/\mathcal{K}_o$, and γ is torsion-free if R^γ takes its values in $\mathcal{K}_o$. We consider the normal connection only, and assume that $n > 2$.

Lemma 12.8.1 *The curvature component*

$$R^\gamma\left(\frac{\partial}{\partial y^k}, \frac{\partial}{\partial x^l}\right)$$

takes its values in $\mathcal{K}_o$.

Proof As we pointed out above, $D^i_{jkl}y^j = 0$. Furthermore, we showed in Sect. 11.3 that the Riemann tensor of any spray satisfies

$$R^l_{kij,m}y^k = 0.$$

In the present context we have

$$\Lambda_{jl,k}y^j = -\frac{1}{n-1}\mathfrak{R}_{jl,k}y^j = -\frac{1}{n-1}\mathfrak{R}^m_{jml,k}y^j = 0.$$

That is, $\zeta^i_j y^j = 0$ and $\zeta^0_j y^j = 0$, which establishes the result. □

Proposition 12.8.2 *The infinitesimal connection γ is torsion-free if and only if $P^i_{jkl} = 0$.*

In the proof of this proposition certain results about isotropic sprays from [16] will be used. Recall first that for any spray, R^l_{kij}, $R^l_{ij} = R^l_{kij}y^k$ and $R^l_k = R^l_{ikj}y^i y^j$ all contain essentially the same information, since R^l_{kij} and R^l_{ij} are both completely determined by R^l_k, as follows:

$$R^l_{ij} = \tfrac{1}{3}\left(R^l_{i,j} - R^l_{j,i}\right), \qquad R^l_{kij} = R^l_{ij,k}.$$

The results about isotropic sprays we shall need are the following.

- A spray is *isotropic* if and only if there is a function ρ and 1-form τ_i such that

$$R^i_j = \rho\delta^i_j + \tau_j y^i.$$

  (This is the basic definition.)
- Equivalently, a spray is isotropic if and only if there is a 1-form θ_i and skew-symmetric tensor ϕ_{ij} such that

$$R^i_{jk} = \theta_k\delta^i_j - \theta_j\delta^i_k + \phi_{jk}y^i.$$

- A spray is isotropic if and only if $P^i_{jkl} = 0$. (This shows that being isotropic is a projective property: if it holds for one spray in a projective equivalence class it holds for all. This isn't obvious from the basic definition.)
- A spray is isotropic if and only if it is projectively R-flat, that is, if and only if it is projectively equivalent to a spray whose Riemann tensor vanishes.

Proof We have to show that

$$R^\gamma\left(\frac{\partial}{\partial x^k}, \frac{\partial}{\partial x^l}\right)$$

takes its values in $\mathcal{K}_o$ if and only $P^i_{jkl} = 0$.

If this component of R^γ takes its values in $\mathcal{K}_o$ then

$$P^i_{jkl}y^j = \frac{1}{n-1}(\mathfrak{R}_{jk|l} - \mathfrak{R}_{jl|k})y^j y^i = \phi_{kl}y^i$$

say. Now

$$P^i_{jkl}y^j = \mathfrak{R}^i_{kl} - \frac{1}{n-1}\left(\mathfrak{R}_{jl}y^j\delta^i_k - \mathfrak{R}_{jk}y^j\delta^i_l\right),$$

and so if $P^i_{jkl}y^j = \phi_{kl}y^i$ then

$$\mathfrak{R}^i_{kl} = \theta_l\delta^i_k - \theta_k\delta^i_l + \phi_{kl}y^i, \qquad \theta_k = \frac{1}{n-1}\mathfrak{R}_{jk}y^j,$$

and the spray is isotropic. But then $P^i_{jkl} = 0$.

Suppose, conversely, that $P^i_{jkl} = 0$. We show that then $\mathfrak{R}_{jk|l} - \mathfrak{R}_{jl|k} = 0$, whence R^γ takes its values in $\mathcal{K}_o$ and γ is torsion-free. From the definition of P^i_{jkl} we have

$$\mathfrak{R}^i_{jkl} = \frac{1}{n-1}\left(\mathfrak{R}_{jl}\delta^i_k - \mathfrak{R}_{jk}\delta^i_l\right).$$

Now $\mathfrak{R}^i_{jkl}$ satisfies the Bianchi identity $\mathfrak{R}^i_{jkl|m} + \mathfrak{R}^i_{jlm|k} + \mathfrak{R}^i_{jmk|l} = 0$. Thus

$$\mathfrak{R}_{jl|m}\delta^i_k - \mathfrak{R}_{jk|m}\delta^i_l + \mathfrak{R}_{jm|k}\delta^i_l - \mathfrak{R}_{jl|k}\delta^i_m + \mathfrak{R}_{jk|l}\delta^i_m - \mathfrak{R}_{jm|l}\delta^i_k = 0.$$

On taking the trace on i and m we obtain $(n-2)(\mathfrak{R}_{jk|l} - \mathfrak{R}_{jl|k}) = 0$. □

Corollary 12.8.3 *The following statements are equivalent.*

1. *The normal infinitesimal connection γ is torsion-free.*
2. *The sprays of the projective equivalence class are isotropic.*
3. $P^i_{jkl} = 0$.
4. *There is a spray in the projective equivalence class which is R-flat.*

□

The systems of second-order ordinary differential equations corresponding to torsion-free path geometries have many interesting properties. They have been investigated in depth by Grossman in [25], and again from the point of view of Cartan connection theory by Čap in [7].

12.9 BTW-Connections

Finally in this chapter we shall copy, as far as possible, the construction of the TW-connection from an infinitesimal Cartan projection geometry over P$\mathcal{W}M$ to the case of a Cartan–Berwald geometry over π^*P$\mathcal{W}M$. This will involve the construction of a Berwald connection on $\tau^{\circ *}_{\mathcal{V}M}T\mathcal{V}M$, the pullback of $T\mathcal{V}M$ over $T^\circ\mathcal{V}M$. The resulting connection will be called the *Berwald-Thomas-Whitehead connection* or *BTW-connection* corresponding to the Cartan–Berwald geometry. We have the following set-up.

$$\begin{array}{ccc}
\tau^{\circ *}_{\mathcal{V}M}T\mathcal{V}M & \xrightarrow{\tilde{\chi}} & \pi^*\mathrm{P}\mathcal{W}M \\
\downarrow & & \downarrow \\
T^\circ\mathcal{V}M & \xrightarrow{\bar{\chi}} & \mathrm{P}\mathcal{W}M \\
\tau^\circ_{\mathcal{V}M}\downarrow & & \downarrow\pi \\
\mathcal{V}M & \xrightarrow[\nu]{} & M
\end{array}$$

The map $\tilde{\chi}$ is the composition of $\bar{\chi}$ and projectivization of the fibre of $\tau^{\circ *}_{\mathcal{V}M}T\mathcal{V}M \to T^\circ\mathcal{V}M$.

We denote by (x^a, u^a, v^a), $a = 0, 1, 2, \ldots, n$, coordinates on $\tau^{\circ *}_{\mathcal{V}M}T\mathcal{V}M$.

Let Z be a vector field on $T^\circ\mathcal{V}M$ which is projectable to $\mathcal{V}M$. Then there is a natural lift of Z to a vector field $\tilde{Z}$ on $\tau^{\circ *}_{\mathcal{V}M}T\mathcal{V}M$, which is a form of complete lift, given in coordinates by

$$\tilde{Z} = X^a\frac{\partial}{\partial x^a} + Z^a\frac{\partial}{\partial u^a} + v^b\frac{\partial X^a}{\partial x^b}\frac{\partial}{\partial v^a},$$

where

$$Z = X^a\frac{\partial}{\partial x^a} + Z^a\frac{\partial}{\partial u^a}.$$

If Z is vertical with respect to $\tau^\circ_{\mathcal{V}M}$, so that $X^a = 0$, then $\tilde{Z}$ has formally the same coordinate representation as Z. Now Υ^{c} is of course projectable to $\mathcal{V}M$, so we may apply this construction to it. We denote the resulting vector field on $\tau^{\circ *}_{\mathcal{V}M}T\mathcal{V}M$ by $\tilde{\Upsilon}$ for simplicity of notation. Likewise $\tilde{\Delta}$ lifts to $\tau^{\circ *}_{\mathcal{V}M}T\mathcal{V}M$, and its lift is formally identical: we denote the lift by the same symbol, again for ease of writing. Finally we denote by D the Liouville field on the fibres of $\tau^{\circ *}_{\mathcal{V}M}T\mathcal{V}M \to T^\circ\mathcal{V}M$. Then the three vector fields $\tilde{\Upsilon}$, $\tilde{\Delta}$ and D span an integrable distribution on $\tau^{\circ *}_{\mathcal{V}M}T\mathcal{V}M$ whose leaves are the fibres of the map $\tilde{\chi} : \tau^{\circ *}_{\mathcal{V}M}T\mathcal{V}M \to \pi^*\mathrm{P}\mathcal{W}M$.

We wish to define a Berwald connection, conceived of as a horizontal lift operation from vectors on $T^\circ\mathcal{V}M$ to vectors on $\tau^{\circ *}_{\mathcal{V}M}T\mathcal{V}M$ linear in the fibre coordinates v^a, which maps to γ under $\tilde{\chi}$. As was the case for the definition of the TW-connection we shall need an additional condition to define this horizontal lift uniquely. The condition we imposed before was that $[X^{\mathrm{h}}, \Upsilon^{\mathrm{v}}] = X^{\mathrm{v}}$ for all vector fields X on $\mathcal{V}M$, which is equivalent to Υ being a radius vector field. We shall use a version of this same condition here: the concept of a radius vector field as originally formulated does not apply, but the version just given does.

We must first explain how to define a vertical lift. Now any vector on $\mathcal{V}M$ has a natural vertical lift to $\tau^{\circ *}_{\mathcal{V}M} T\mathcal{V}M$, given by

$$\xi^a \frac{\partial}{\partial x^a} \mapsto \xi^a \frac{\partial}{\partial v^a};$$

and this vertical lift construction extends to vectors on $T^\circ \mathcal{V}M$ by projecting first, that is, by mapping vector vertical with respect to $\tau^\circ_{\mathcal{V}M}$ to zero. So we may impose formally the same condition in this more general case.

Theorem 12.9.1 *Given an infinitesimal Cartan–Berwald projective geometry on M there is a unique Berwald connection on $T^\circ \mathcal{V}M$ such that*

1. *for every vector field X on $T^\circ \mathcal{V}M$ which is projectable to $\mathrm{P}\mathcal{W}M$, X^{h} is projectable to $\pi^*\mathrm{P}\mathcal{W}M$ and*

$$\tilde{\chi}_*(X^{\mathsf{h}}) = \gamma(\bar{\chi}_* X)$$

2. $[X^{\mathsf{h}}, \Upsilon^{\mathsf{v}}] = X^{\mathsf{v}}$,

where $X \mapsto X^{\mathsf{h}}$ is the horizontal lift operator of the Berwald connection on $T^\circ \mathcal{V}M$ and γ is the infinitesimal connection of the Cartan–Berwald projective geometry.

Proof As before we show first that if there is such a connection on $T^\circ \mathcal{V}M$ it is unique. Let $X \mapsto X^{\mathsf{h}}$ be the horizontal lift of the given Berwald connection. The horizontal lift of any other connection which projects to γ must take the form $X \mapsto X^{\mathsf{h}} + \theta(X)D$, where θ is a 1-form on $T^\circ \mathcal{V}M$. By assumption, $[X^{\mathsf{h}}, \Upsilon^{\mathsf{v}}] = X^{\mathsf{v}}$. We require that a similar property holds for the modified horizontal lift; that is to say, we require that $[\theta(X)D, \Upsilon^{\mathsf{v}}] = 0$. Now

$$[\theta(X)D, \Upsilon^{\mathsf{v}}] = -\Upsilon^{\mathsf{v}}(\theta(X))D - \theta(X)\Upsilon^{\mathsf{v}};$$

but D and Υ^{v} are almost everywhere linearly independent, so we must have $\theta = 0$, and the connection is unique.

We next show that there is such a connection over any coordinate neighbourhood in $\mathcal{V}M$ adapted to the bundle structure $\mathcal{V}M \to M$. Now $\tilde{\chi}$ is given in coordinates by

$$(x^a, u^a, v^a) \mapsto (x^i, y^i, z^i) \quad \text{where } y^i = \frac{x^0 u^i}{u^0}, \quad z^i = \frac{x^0 v^i}{v^0}.$$

The coefficients $\Pi^{\,i}_{jk}$ and Λ_{ij} of γ are functions of y^i (as well as x^i of course): if we pull them back to the coordinate neighbourhood in $T^\circ \mathcal{V}M$, by substituting $x^0 u^i/u^0$ for y^i, we obtain functions which are homogeneous of degree 0 in the u^a. With this understanding we define horizontal lifts of the coordinate vector fields as follows:

$$\left(\frac{\partial}{\partial x^0}\right)^{\mathsf{h}} = \frac{\partial}{\partial x^0} - \frac{v^i}{x^0}\frac{\partial}{\partial v^i}$$

$$\left(\frac{\partial}{\partial x^i}\right)^{\mathsf{h}} = \frac{\partial}{\partial x^i} + x^0 \Lambda_{ij} v^j \frac{\partial}{\partial v^0} - \left(\frac{v^0}{x^0}\delta_i^j + \Pi^j_{ik} v^k\right) \frac{\partial}{\partial v^j}$$
$$\left(\frac{\partial}{\partial u^a}\right)^{\mathsf{h}} = \frac{\partial}{\partial u^a}.$$

The horizontal lift of Υ^{c} must map to zero under $\tilde{\chi}$. But

$$\begin{aligned}(\Upsilon^{\mathsf{c}})^{\mathsf{h}} &= x^0 \frac{\partial}{\partial x^0} + u^0 \frac{\partial}{\partial u^0} - v^i \frac{\partial}{\partial v^i} \\ &= x^0 \frac{\partial}{\partial x^0} + u^0 \frac{\partial}{\partial u^0} + v^0 \frac{\partial}{\partial v^0} - v^a \frac{\partial}{\partial v^a} \\ &= \tilde{\Upsilon} - D,\end{aligned}$$

so that condition is satisfied. Furthermore,

$$\left(u^a \frac{\partial}{\partial u^a}\right)^{\mathsf{h}} = \tilde{\Delta}$$

which also maps to zero under $\tilde{\chi}$ as required.

From the coordinate expression for $\tilde{\chi}$ we find that

$$\begin{aligned}
\left.\frac{\partial}{\partial x^0}\right|_{(x^a,u^a,v^a)} &\mapsto \left(\frac{u^i}{u^0}\right) \left.\frac{\partial}{\partial y^i}\right|_{(x^i,y^i,z^i)} + \left(\frac{v^i}{v^0}\right) \left.\frac{\partial}{\partial z^i}\right|_{(x^i,y^i,z^i)} \\
\left.\frac{\partial}{\partial u^0}\right|_{(x^a,u^a,v^a)} &\mapsto -\left(\frac{x^0 u^i}{(u^0)^2}\right) \left.\frac{\partial}{\partial y^i}\right|_{(x^i,y^i,z^i)} \\
\left.\frac{\partial}{\partial u^i}\right|_{(x^a,u^a,v^a)} &\mapsto \left(\frac{x^0}{u^0}\right) \left.\frac{\partial}{\partial y^i}\right|_{(x^i,y^i,z^i)} \\
\left.\frac{\partial}{\partial v^0}\right|_{(x^a,u^a,v^a)} &\mapsto -\left(\frac{x^0 v^i}{(v^0)^2}\right) \left.\frac{\partial}{\partial z^i}\right|_{(x^i,y^i,z^i)} \\
\left.\frac{\partial}{\partial v^i}\right|_{(x^a,u^a,v^a)} &\mapsto \left(\frac{x^0}{v^0}\right) \left.\frac{\partial}{\partial z^i}\right|_{(x^i,y^i,z^i)}.
\end{aligned}$$

We note that

$$\begin{aligned}
\tilde{\Upsilon} &= x^0 \frac{\partial}{\partial x^0} + u^0 \frac{\partial}{\partial u^0} + v^0 \frac{\partial}{\partial v^0} \mapsto 0 \\
\tilde{\Delta} &= u^0 \frac{\partial}{\partial u^0} + u^i \frac{\partial}{\partial u^i} \mapsto 0 \\
D &= v^0 \frac{\partial}{\partial v^0} + v^i \frac{\partial}{\partial v^i} \mapsto 0
\end{aligned}$$

as expected. Now $\partial/\partial x^i$ is projectable to $\mathrm{P}\mathcal{W}M$ and

$$\left(\frac{\partial}{\partial x^i}\right)^{\mathsf{h}} \mapsto \frac{\partial}{\partial x^i} - \left(\delta^i_j + \Pi^{\,j}_{ik} y^k + (\Lambda_{ik} y^k) y^j\right)\frac{\partial}{\partial y^j} = \gamma\left(\frac{\partial}{\partial x^i}\right).$$

Moreover $(u^0/x^0)\partial/\partial u^i$ projects to $\mathrm{P}\mathcal{W}M$; indeed

$$\left(\left(\frac{u^0}{x^0}\right)\frac{\partial}{\partial u^i}\right)^{\mathsf{h}} = \left(\frac{u^0}{x^0}\right)\frac{\partial}{\partial u^i} \mapsto \frac{\partial}{\partial y^i} = \gamma\left(\frac{\partial}{\partial y^i}\right).$$

So this connection satisfies the first of the conditions in the statement of the theorem on the coordinate neighbourhood. For the second, we have

$$\Upsilon^{\mathsf{v}} = x^0 \frac{\partial}{\partial v^0},$$

and therefore

$$\begin{aligned}
\left[\left(\frac{\partial}{\partial x^0}\right)^{\mathsf{h}}, \Upsilon^{\mathsf{v}}\right] &= \frac{\partial}{\partial v^0} = \left(\frac{\partial}{\partial x^0}\right)^{\mathsf{v}} \\
\left[\left(\frac{\partial}{\partial x^i}\right)^{\mathsf{h}}, \Upsilon^{\mathsf{v}}\right] &= -x^0\left(-\frac{1}{x^0}\frac{\partial}{\partial v^i}\right) = \frac{\partial}{\partial v^i} = \left(\frac{\partial}{\partial x^i}\right)^{\mathsf{v}} \\
\left[\left(\frac{\partial}{\partial u^a}\right)^{\mathsf{h}}, \Upsilon^{\mathsf{v}}\right] &= 0 = \left(\frac{\partial}{\partial u^a}\right)^{\mathsf{v}},
\end{aligned}$$

so the second condition is satisfied also. But as before this means that there is a globally defined Berwald connection with the required properties, because the connections defined on coordinate neighbourhoods must agree on overlaps by uniqueness. □

The BTW-connection is the Berwald connection of a spray on $T^\circ\mathcal{V}M$, namely

$$u^a\frac{\partial}{\partial x^a} + x^0\Lambda\frac{\partial}{\partial u^0} - \left(2\left(\frac{u^0u^i}{x^0}\right) + \Pi^i\right)\frac{\partial}{\partial u^i},$$

where

$$\Lambda = \tfrac{1}{2}\Lambda_{ij}u^iu^j, \quad \Pi^i = \tfrac{1}{2}\Pi^{\,i}_{jk}u^ju^k.$$

The coefficients Λ and Π^i are homogeneous in the u^a of degree 2 and independent of u^0, so that

$$\frac{\partial^2\Lambda}{\partial u^i\partial u^j} = \Lambda_{ij}, \quad \frac{\partial^2\Pi^i}{\partial u^j\partial u^k} = \Pi^{\,i}_{jk}.$$

The base integral curves (on $\mathcal{V}M$) of this spray are the solutions of the second-order differential equations

$$\ddot{x}^i + \Pi^i = -2\left(\frac{\dot{x}^0}{x^0}\right)\dot{x}^i, \quad \ddot{x}^0 - \Lambda x^0 = 0\,.$$

We may write these equations equivalently as

$$\ddot{x}^0 - (\Lambda_{ij}\dot{x}^i\dot{x}^j)x^0 = 0, \quad \ddot{x}^i + \Pi^{\,i}_{jk}\dot{x}^j\dot{x}^k = -2\left(\frac{\dot{x}^0}{x^0}\right)\dot{x}^i,$$

which makes it apparent that they generalize the equations for the geodesics of a projective class of linear connections with projective parametrization.

Bibliography

1. D.V. Alekseevsky, P.W. Michor, Differential geometry of $\mathfrak{g}$-manifolds. Diff. Geom. Appl. **5**, 371–403 (1995)
2. T.N. Bailey, M.G. Eastwood, A.R. Gover, Thomas's structure bundle for conformal, projective and related structures. Rocky Mt. J. Math. **24**, 1191–1217 (1994)
3. L. Berwald, On the projective geometry of paths. Ann. Math. **37**, 879–898 (1936)
4. A.D. Blaom, Geometric structures as deformed infinitesimal symmetries. Trans. Am. Math. Soc. **358**, 3651–3671 (2006)
5. A.D. Blaom, The infinitesimalization and reconstruction of locally homogeneous manifolds. SIGMA **9**, 074 (2013)
6. R. Brown, From groups to groupoids: a brief survey. Bull. Lond. Math. Soc. **19**, 113–134 (1987)
7. A. Čap, Two constructions with parabolic geometries, in *Proceedings of the 25th Winter School on Geometry and Physics, Srni 2005*. Rendiconti del Circolo Matematico di Palermo Supplement Series II, **75**, 11–37 (2006)
8. A. Čap, Infinitesimal automorphisms and deformations of parabolic geometries. J. Eur. Math. Soc. **10**, 415–437 (2008)
9. A. Čap, A.R. Gover, Tractor bundles for irreducible parabolic geometries. Société Mathématique de France, Séminaires et Congrès **4**, 129–154 (2000)
10. A. Čap, A.R. Gover, Tractor calculi for parabolic geometries. Trans. Am. Math. Soc. **354**(4), 1511–1548 (2001)
11. É. Cartan, Les espaces à connexion conforme. Ann. Soc. Pol. Math. **2**, 171–221 (1923)
12. É. Cartan, Sur les variétés à connexion projective. Bull. Soc. Math. Fr. **52**, 205–241 (1924)
13. É. Cartan, Les récentes généralisations de la notion d'espace. Bull. Sci. Math. **2**(48), 205–241 (1924)
14. C. Chevalley, *Theory of Lie groups* (Princeton University Press, Princeton, 1946)
15. M. Crampin, Connections of Berwald type. Publ. Math. Debrecen **57**, 455–473 (2000)
16. M. Crampin, Isotropic and R-flat sprays. Houston J. Math. **33**, 451–459 (2007)
17. M. Crampin, Cartan connections and Lie algebroids. SIGMA **5**, 061 (2009)
18. M. Crampin, T. Mestdag, D.J. Saunders, Hilbert forms for a Finsler metrizable projective class of sprays. Diff. Geom. Appl. **31**, 63–79 (2013)
19. M. Crampin, D.J. Saunders, Projective connections. J. Geom. Phys. **57**, 691–727 (2007)
20. M. Crampin, D.J. Saunders, Holonomy of a class of bundles with fibre metrics. Publ. Math. Debrecen **81**, 199–234 (2012)
21. J. Douglas, The general geometry of paths. Ann. Math. **29**, 143–168 (1928)
22. C. Ehresmann, Les connexions infinitésimales dans un espace fibré différentiable. Séminaire N. Bourbaki **1**, 153–168 (1948–1951)

M. Crampin and D. Saunders, *Cartan Geometries and their Symmetries*,
Atlantis Studies in Variational Geometry 4, DOI 10.2991/978-94-6239-192-5

23. C. Ehresmann, Catégories topologiques et catégories différentiables. Coll. Géom. Diff. Glob. Bruxelles, 137–150 (1959)
24. M.E. Fels, The equivalence problem for systems of second-order ordinary differential equations. Proc. Lond. Math. Soc. **71**, 221–240 (1995)
25. D.A. Grossman, Torsion-free path geometries and integrable second-order ODE systems. Selecta Math. **6**, 399–442 (2000)
26. V.W. Guillemin, S. Sternberg, An algebraic model of transitive differential geometry. Bull. Am. Math. Soc. **70**(1), 16–47 (1964)
27. J. Hebda, C. Roberts, Examples of Thomas-Whitehead projective connections. Diff. Geom. Appl. **8**, 87–104 (1998)
28. S. Kobayashi, On connections of Cartan. Can. J. Math. **8**, 145–156 (1956)
29. S. Kobayashi, K. Nomizu, *Foundations of differential geometry*, vol. 1 (Wiley: Interscience, 1963)
30. K.C.H. Mackenzie, *General theory of Lie groupoids and Lie algebroids*, LMS Lecture Note Series vol. 213 (Cambridge University Press, Cambridge, 2005)
31. C.-M. Marle, The works of Charles Ehresmann on connections: from Cartan connections to connections on fibre bundles, in *Geometry and Topology of Manifolds*, ed. by Kubarski, Pradines, Rybicki, Wolak. Banach Center Publications vol. 76 (Polish Academy of Sciences, Warszawa, 2007), pp. 65–86
32. K. Ogiue, Theory of conformal connections. Kōdai Math. Sem. Rep. **18**, 193–234 (1967)
33. R.S. Palais, A global formulation of the Lie theory of transformation groups. Mem. Am. Math. Soc. **22** (1957)
34. J. Pradines, Théorie de Lie pour les groupoïdes différentiables. Calcul différentiel dans la catégorie des groupoïdes infinitésimaux. C. R. Acad. Sci. Paris **264A**, 245–248 (1967)
35. C. Roberts, The projective connections of T.Y. Thomas and J.H.C. Whitehead applied to invariant connections. Diff. Geom. Appl. **5**, 237–255 (1995)
36. R.W. Sharpe, *Differential Geometry: Cartan's generalization of Klein's Erlanger Program* (Springer, New York, 1997)
37. Z. Shen, *Differential Geometry of Spray and Finsler Spaces* (Kluwer Academic Publishers, Dordrecht, 2001)
38. J. Szilasi, R.L. Lovas, D.Cs. Kertész, *Connections, Sprays and Finsler Structures* (World Scientific, London, 2014)
39. N. Tanaka, Projective connections and projective transformations. Nagoya Math. J. **12**, 1–24 (1957)
40. N. Tanaka, Conformal connections and conformal transformations. Trans. Am. Math. Soc. **92**, 168–190 (1959)
41. T.Y. Thomas, On the projective and equi-projective geometries of paths. Proc. Natl. Acad. Sci. **11**, 199–203 (1925)
42. T.Y. Thomas, A projective theory of affinely connected manifolds. Math. Zeit. **25**, 723–733 (1926)
43. T.Y. Thomas, On conformal geometry. Proc. Natl. Acad. Sci. **12**, 352–359 (1926)
44. F.W. Warner, *Foundations of differentiable manifolds and Lie groups* (Springer, New York, 1983)
45. J.H.C. Whitehead, The representation of projective spaces. Ann. Math. **32**, 327–360 (1931)
46. K. Yano, *The theory of Lie derivatives and its applications*, Bibliotheca Mathematica, vol. III (North-Holland, Amsterdam, 1957)

Index

M. Crampin and D. Saunders, *Cartan Geometries and their Symmetries*,
Atlantis Studies in Variational Geometry 4, DOI 10.2991/978-94-6239-192-5

Printed in the United States
By Bookmasters